Advances in Horticulture

Advances in Horticulture

Editor: Thelma Bosso

www.callistoreference.com

Callisto Reference,
118-35 Queens Blvd., Suite 400,
Forest Hills, NY 11375, USA

Visit us on the World Wide Web at:
www.callistoreference.com

ISBN: 978-1-63239-787-4 (Hardback)

The publisher's policy is to use permanent paper from mills that operate a sustainable forestry policy. Furthermore, the publisher ensures that the text paper and cover boards used have met acceptable environmental accreditation standards.

Trademark Notice: Registered trademark of products or corporate names are used only for explanation and identification without intent to infringe.

Printed in the United States of America.

Cataloging-in-publication Data

Advances in horticulture / edited by Thelma Bosso.
       p. cm.
Includes bibliographical references and index.
ISBN 978-1-63239-787-4
1. Horticulture. 2. Plant products--Biotechnology. 3. Agricultural pests--Control. I. Bosso, Thelma.
SB319.5 .A38 2017
635--dc23

Table of Contents

Preface

Horticulture is the science of growing plants. Advancements in horticultural studies have branched into various sub-disciplines such as plant conservation, arboriculture and landscape design and restoration. The objective of this book is to give a general view of the different areas of horticulture and its applications. Horticulture opens the way for many types of agricultural practices and it has a big impact on agricultural sciences as well as soil sciences. This book contains some path-breaking studies in the field of horticulture. It elucidates new techniques and their applications in a multidisciplinary approach. This text provides significant information of this discipline to help develop a good understanding of horticulture and related fields. As this field is emerging at a fast pace, this book will help the readers to better understand the concepts of horticulture.

The researches compiled throughout the book are authentic and of high quality, combining several disciplines and from very diverse regions from around the world. Drawing on the contributions of many researchers from diverse countries, the book's objective is to provide the readers with the latest achievements in the area of research. This book will surely be a source of knowledge to all interested and researching the field.

In the end, I would like to express my deep sense of gratitude to all the authors for meeting the set deadlines in completing and submitting their research chapters. I would also like to thank the publisher for the support offered to us throughout the course of the book. Finally, I extend my sincere thanks to my family for being a constant source of inspiration and encouragement.

Editor

Influence of Vegetation Restoration on Topsoil Organic Carbon in a Small Catchment of the Loess Hilly Region, China

Yunbin Qin, Zhongbao Xin*, Xinxiao Yu, Yuling Xiao

Institute of Soil and Water Conservation, Beijing Forestry University, Beijing, China

Abstract

Understanding effects of land-use changes driven by the implementation of the "Grain for Green" project and the corresponding changes in soil organic carbon (SOC) storage is important in evaluating the environmental benefits of this ecological restoration project. The goals of this study were to quantify the current soil organic carbon density (SOCD) in different land-use types [cultivated land, abandoned land (cessation of farming), woodland, wild grassland and orchards] in a catchment of the loess hilly and gully region of China to evaluate the benefits of SOC sequestration achieved by vegetation restoration in the past 10 years as well as to discuss uncertain factors affecting future SOC sequestration. Based on soil surveys (N = 83) and laboratory analyses, the results show that the topsoil (0–20 cm) SOCD was 20.44 Mg/ha in this catchment. Using the SOCD in cultivated lands (19.08 Mg/ha) as a reference, the SOCD in woodlands and abandoned lands was significantly higher by 33.81% and 8.49%, respectively, whereas in orchards, it was lower by 10.80%. The correlation analysis showed that SOC and total nitrogen (TN) were strongly correlated ($R^2 = 0.98$) and that the average C:N (SOC:TN) ratio was 9.69. With increasing years since planting, the SOCD in woodlands showed a tendency to increase; however, no obvious difference was observed in orchards. A high positive correlation was found between SOCD and elevation ($R^2 = 0.395$), but a low positive correlation was found between slope and SOCD ($R^2 = 0.170$, $P = 0.127$). In the past 10 years of restoration, SOC storage did not increase significantly (2.74% or 3706.46 t) in the catchment where the conversion of cultivated land to orchards was the primary restoration pattern. However, the potential contribution of vegetation restoration to SOC sequestration in the next several decades would be massive if the woodland converted from the cropland is well managed and maintained.

Editor: Manuel Reigosa, University of Vigo, Spain

Funding: This study was supported by the Fundamental Research Funds for the Central Universities (NoTD2011-2), the Open Foundation of Key Laboratory of Soil and Water Loss Process and Control on the Loess Plateau of Ministry of Water Resources (201301), the National Natural Science Foundation of China (No. 41001362) and the College Student Scientific Research Training Project of Beijing Forest University (No. 201210022013). The funders had no role in study design, data collection and analysis, decision to publish, or preparation of the manuscript.

Competing Interests: The authors have declared that no competing interests exist.

* Email: xinzhongbao@126.com

Introduction

Afforestation and other vegetation restoration techniques have been considered effective practices for the sequestration of carbon (C) to mitigate carbon dioxide (CO_2) concentrations in the atmosphere [1–4]. Soil plays an important role in the global carbon cycle. The soil C pool to a one metre depth has been estimated to sequester approximately three times the amount of carbon sequestered by the atmospheric pool and about four times that in the biotic/vegetation pool [5]. Therefore, a relatively small change in the soil C pool can significantly mitigate or enhance CO_2 concentrations in the atmosphere [6–7].

Land-use change can significantly influence the accumulation and release of SOC. When an ecosystem is disturbed by land-use change, the original equilibrium of the soil carbon pool is broken and a new equilibrium is created. During this process, soil may act as either a source or a sink of carbon depending on the ratio between inflows and outflows [8]. Many studies have reviewed the effect of land-use change on SOC. For instance, deforestation for agricultural purposes is the primary reason for the SOC loss. It

was reported that approximately 25% of SOC was lost by conversion of primary forest into cropland [9]. Houghton (1999) estimated that 105 PgC was released into the atmosphere due to the conversion of forests to agricultural lands between 1850 and 1990 [10]. However, afforestation and reforestation on agricultural lands have been cited as effective methods for increasing SOC pool and reducing the atmospheric CO_2 concentration [3,11]. Morris et al. (2007) observed that placing agricultural soils in deciduous and conifer forests resulted in soil carbon accumulations of 0.35 and 0.26 $MgCha^{-1} yr^{-1}$, respectively. Based on a meta analysis of 33 recent publications, Laganière et al. (2010) reported that afforestation increased SOC stocks by 26% for croplands. Some studies found that croplands that were converted to abandoned lands or grasslands could also increase the SOC storage [9,13–14]. Therefore, understanding the influence of land-use changes on soil organic carbon is an important step in predicting climatic change and developing potential future CO_2 mitigation strategies.

To improve the carbon sink status of afforeseation or other vegetation restoration methods on agricultural land, it is necessary to understand the control mechanisms of SOC dynamics to allow more carbon storage in soils [1]. A variety of factors will affect the quantity and quality of SOC after land-use changes. For example, climate variations have a significant effect on SOC, including temperature, precipitation, and potential evapotranspiration changes [11,15]. Laganière et al. (2010) reported that SOC restoration after afforestation was found to vary with the climate zone, with the temperate maritime zone having a higher SOC increase than others by approximately 17%. In addition, landscape and elevation also have a pronounced effect on SOC change at the catchment scale. Slope is an important topographical factor that affects soil erosion and also has an important influence on the soil nutrient loss of the slope surface [16]. Wang et al. (2012) reported that the SOCD was higher in shady slopes than in sunny slopes, and gentle slopes would generally sequester more SOC than steep slopes. Elevation differences can cause climatic and biological changes in the soil-forming environment and can influence the vertical distribution of SOC [18]. In natural ecosystems, it has been extensively documented that carbon and nitrogen are closely related [4,19]. Increased nitrogen retention may increase the carbon sequestration potential [1]. In Panama, Batterman et al. (2013) found that symbiotic N_2 fixation has potentially important implications for the ability of tropical forests to sequester CO_2 [20]. N_2-fixing tree species accumulated carbon up to nine times faster per individual than neighbouring non-fixing trees. The soil C/N ratio also has a significant effect on the rate of decomposition of organic compounds by soil microorganisms [21].

Soil erosion, as the most widespread form of soil degradation, has a large impact on the global C cycle, causing a severe depletion of SOC pools in the soil [22–24]. The total amount of C released by soil erosion each year has been estimated at approximately 4.0–6.0 Pg/year [23]. China's Loess Plateau, covering approximately 6.2×10^5 km^2, has the world's most severe soil erosion because of its unusual geographic landscape, soil and climatic conditions, and long history (over 5000 years) of human activity. Over 60–80% of the land in the Loess Plateau has been affected by soil erosion, with an average annual soil erosion of 2000 to 20000 $t\,km^{-2}\,yr^{-1}$ [25–26]. Severe soil erosion has resulted in land degradation, which was manifested primarily in the thinning of the soil layer, nutrient loss and fertility reduction, which has directly caused decreases in the local farmers' income and has economically and socially hindered sustainable development [27–29]. Unreasonable human activities, such as deforestation and tillage on slopes, have further intensified soil erosion and land degradation in the Loess Plateau [30].

Since the 1950s, the Chinese government has launched many large-scale projects to attempt to control soil erosion and restore vegetation in the Loess Plateau, including large-scale afforestation in the 1970s and comprehensive control of soil erosion on the watershed scale in the 1980s and 1990s. Despite these efforts, there have not been significant increases in ecological benefits, which is largely due to the limitations and influences of bad natural conditions and unreasonable ecological restoration techniques. To control soil erosion and improve the quality of the local environment, "Grain for Green" was initiated by the government in the Loess Plateau in 1999. The government demanded that the agricultural lands with a slope of over 25 degrees be converted to forest, terrace orchards or grassland. To compensate famers for their economic loss, they will be given grain, cash and planting stocks by the government as subsidies and incentives for converting cultivated land back to forest, orchards or grassland. Currently, the Loess Plateau environment appears to be experiencing a recovery following more than 10 years of vegetation restoration [31–32].

Over the past decade, large-scale vegetation restoration efforts have brought obvious land-use changes to the Loess Plateau, and also have significantly influence on SOC sequestration. Therefore, understanding the effects of these dynamic changes and the corresponding changes in SOC storage caused by land-use change driven by the implementation of the Grain for Green project is important in evaluating the environmental benefits of this ecological restoration project. Recently, many studies have focused on SOC changes induced by the Grain for Green project in the Loess Plateau. These studies have mainly focused on the SOCD of different land-use types, the SOC pool, the rate of SOC change, the variation in SOC among different land-use conversions and factors which have influenced SOC sequestration in the Loess Plateau after vegetation restoration efforts [32–34]. However, for most of the researches, the study time period was shorter than 10 years as vegetation restoration was driven by the Grain for Green project. The study areas have been mostly located at the gully region of the Loess Plateau [2,17,35–36]. Little is known about the gully area of the Loess hilly region. To gain a complete understanding of the SOC change after restoration in this region, we selected the Luoyugou catchment of the Loess Plateau as the study area, which has the typical geomorphologic characteristics of the gully area of the Loess hilly region.

In this study, we hypothesised that the SOCD varied with land-use types in this catchment, and there was an increase in SOC storage of the total catchment after 10 years of restoration. Therefore, the objectives of this study were to (i) quantify the SOCD of different land-use types differ significantly in the study catchment, (ii) analyze those factors affecting SOCD, (iii) estimate the contribution of land-use conversions on SOC sequestration in the study area, in the past 10 years of restoration, and (iv) discuss the uncertainties in potential SOC sequestration during future ecological restoration efforts in the Loess Plateau.

Materials and Methods

Ethics statement

The administration of the Tianshui Experiment Station of Soil and Water Conservation, Yellow River Conservancy Commission and local farmers which are the owners of study lands gave permission for this research at each study site. We confirm that the field studies did not involve endangered or protected species.

Study area

Tianshui city (104°35′–106°44′E, 34°05′–35°10′N) of Gansu province, China, is located in the western side of the Qinling Mountains, which is the transitional zone between the Qinling Mountains and the Loess Plateau and also belongs to the third sub-region of the Loess hilly region in the middle portion of the Yellow River and the second class tributary of the Wei River. This region has the typical geomorphology of the gully area of the Loess hilly region, which includes earth-rocky mountainous areas, Loess ridges and hilly areas. In this area, the catchment landscape is an agroforestry landscape, and the climate has obvious transitional characteristics. Because of these special geomorphologic characteristics, Walter Lowermilk, deputy director of the US Department of Soil Conservation Service, chose this region in which to build an experimental station for soil and water conservation in 1941. Tianshui station, Suide station and Xifeng station now are three the best-known experimental stations for soil and water conservation in the Loess Plateau.

The Luoyugou catchment (105°30′–105°45′E, 34°34′–34°40′N), located in northern Tianshui, is a part of the observation area of the Tianshui soil and water conservation station. Its total area is 72.79 km^2, with a range from 1165–1895 m above sea level. The main topographic type is loess ridge landform, and the average slope is 19°. It has a typical continental monsoon climate with a mean annual precipitation of 548.9 mm (1986–2004). Approximately 78% of the rainfall is concentrated between July and September. The average annual temperature is 11.4°C (1986–2004), and the annual evaporation is 1293.3 mm. The main soil type in this catchment is mountain grey cinnamon soil, which, according to the Food and Agriculture Organization of the United Nations Educational, Scientific and Cultural Organization (FAO-UNESCO), belongs to the Cambisol soil group and accounts for 91.7% of the soil in the region [37–38] (Figure 1).

The main land-use types for the Luoyugou catchment include woodland, wild grassland, orchards, cultivated land, and abandoned land. In the woodland, the major tree species is black locust (*Robinia pseudoacacia L.*), most of which was artificially planted. Woodlands that are more than 30 years old have good water condition and low density because planting is located mainly beside the place beside water ditches and shady slope. They are the retained trees of the planting projects in the late 1950s to 1970s, without human management now. And most of woodland at the age of 10 to 30 years since planting is found in Fenghuang forest farm located in the gully head of this catchment, which has good management and protecting. Since 1999, the government of Tianshui city has widely implemented the "Grain for Green" project. A lot of cultivated land in the ridge top and slope top has been converted into woodland. Most of the woodland is less than 10 years since planting, and has a high density and bad management. Wild grassland is usually found on stone mountain of the northern catchment where the soil layer is thin and slope is steep. Human activities are restricted in there, only sometimes sheep may reach. The major species are dahurian bushclover (*Lespedeza dahurica (Laxm.) Schindl.*), russian wormwood (*Artemsia sacrorum Ledeb.*), digitate goldenbeard (*Bothriochloa ischaemun (L.) Keng*) and bunge needlegrass (*Stipa bungeana Trin.*), etc. The main fruit species under orchards are cherry (*Cerasus pseudocerasus (Lindl.) G. Don*), apple (*Malus pumila Mill.*) and apricot trees (*Armeniaca vulgaris Lam.*). The management system of orchards is the conventional tillage which is that the weeds beneath the trees are eliminated with herbicides and dead leaves, dried fruit and twigs are removed by manual blowers. Besides, there are not irrigation facilities and the tree water demand all comes from the rain. Fertiliser and pesticide are also applied in the orchards. This catchment is a rainfed agriculture region with a long cultivated history. The major species in the cultivated land are wheat (*Triticum aestivum L.*), maize (*Zea mays L.*) and edible rape (*Brassica campestris L.*), etc. Abandoned land is the cultivated land has been the ceased farming

activities, due to the demand of "Grain for Green" project, the bad environment conditions, the shortage of rural labour and the damage by wild animals, etc. After cultivated land abandoned, old field are spontaneously colonised by various plants, while a secondary succession process will gradually develop during different plant communities. Because of short time since abandoned, the main species in the abandoned land are annual and biennial herb and a bit of perennial herb, such as virgate wormwood (*Artemisia scoparia Waldst. et Kit.*) and green bristlegrass herb (*Utricularia australis R. Br.*).

Soil sampling and laboratory analysis

Based on the main land-use types and the percentage of each land-use-type area in the total catchment area (2008), we randomly selected 83 land-use blocks that included cultivated land (34 samples), abandoned cropland (9 samples), orchards (13 samples), wild grassland (8 samples) and woodland (19 samples) as investigation plots. Their latitudes and longitudes are shown in Table 1. Spatially separated plots were located at least 3 km from one another to help avoid pseudo-replication. In July 2012, the selected plots were surveyed and collected in the field using the Global Positioning System (GPS). The information recorded from the sampling plots in the field included geographic coordinates, slope, elevation, plant species, and water conservation measures. The years since planting or abandoned of the sampling plots were evaluated by recording tree diameters and heights, plant composition, the degree of decomposition of the topsoil crust and the crop's residual body, and by interviewing local farmers. In each sampling plot — woodland and orchards (10×10 m), cultivated land, abandoned land and wild grass land (5×5 m) — we randomly collected three soil samples using a stainless steel cutting ring 5.0 cm high and 5.0 cm in diameter to measure the soil bulk density (BD) of the topsoil layer (0–20 cm) and by taking five random soil samples between 0 and 20 cm depth with a 20 cm long soil auger. The five soil samples were manually homogenised to form a composite sample for each sampling plot, and a quarter of this sample was taken to the laboratory. These samples were air-dried and passed through a 2 mm sieve, while gravel and roots were removed from each soil sample. A quarter of each sub-sample was completely passed through a 0.25 mm sieve to determine SOC and TN. SOC was determined by the $K_2Cr_2O_7$-H_2SO_4 Walkey-Black oxidation method [39]. TN was measured using the micro-Kjeldahl procedure [40].

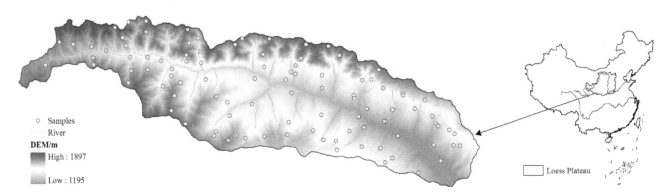

Figure 1. Location of the study area and distribution of the sampling points.

Table 1. Attributes of the studied sites.

Land-use	Sample sites	Years since planting (abandoned)/a	Elevation/m	Slope/(°)	Bulk density/(g/cm³)	Plants
Cultivated land	34	—	1536.94	4.85	1.29	Wheat, maize, rape
Abandoned cropland	9	8–10a	1559.40	4.22	1.35	*Artemisia scoparia, Leymus secalinus*
Orchards	13	<5a, 5–10a, >10a	1446.45	1.92	1.32	Cherry, apple, pear
Woodland	24	<10a, 10a–30a, >30a	1639.14	12.13	1.39	Robinia
Wild grassland	8	—	1526.55	13.00	1.44	*Artemisia sacrorum var. incana mattf, Stipa bungeana, Bothriochloa ischaemun*

Statistical analysis

The SOCD was calculated using the following equation:

$$SOCD = SOC \times D \times BD \tag{1}$$

where SOCD was the soil organic carbon density (Mg ha^{-1}), SOC was the soil organic carbon content (%), D was the soil layer depth (cm), BD was the soil bulk density (g cm^{-3}).

The soil organic carbon storage (SOCS) was calculated using the following equation:

$$SOCS = SOCD_i \times A_i \times 100 \tag{2}$$

where SOCS was the soil organic carbon storage (t), SOCD$_i$ was the soil organic carbon density in the i land-use (Mg ha^{-1}), and A$_i$ was the area of the i land-use (km^2).

The increase in SOCS caused by land-use conversion was calculated using the following equation:

$$\text{Increase of SOCD (t)} = SOCD_i \times (A_{i2} - A_{i1}) \times 100 \tag{3}$$

$$\text{Sequestration amount of SOCS (t)} = (SOCD_i - SOCD_c) \times (A_{i2} - A_{i1}) \times 100 \tag{4}$$

where SOCD$_i$ was the SOCD in the i land-use (Mg ha^{-1}), A$_{i1}$ and A$_{i2}$ were the i land-use area in 2002 and 2008, respectively (km^2), and SOCD$_c$ was the SOCD of the cultivated land (Mg ha^{-1}). All SOCD values of the different land-use types in 2002 and 2008 were the average SOCD values of the soil data collected in 2012.

All data were analysed using Excel 2007 and SPSS (Statistical Package for the Social Sciences) 18.0 software. Analysis of variance (ANOVA) was used to determine the significance of the mean difference. Fisher's LSD test was used to compare the mean values of soil variables when the results of ANOVA were significant at $p < 0.05$. Pearson's test was used to analyse the correlation between SOCD and the soil variables. For all analyses, a $p < 0.05$ was used to test statistical significance. Finally, SOCD, land-use type, elevation and slope values were used to establish the multiple regression model. SOCD was the dependent variable, and others were the independent variables.

Satellite data acquisition and processing

In our study, two types of SPOT5 multispectral images were acquired on August 25, 2002 and May 5, 2008, which were multi-spectral images (resolution: 10 m) and panchromatic images (resolution: 2.5 m). To validate the final classification result, we selected 17 points located in the Luoyugou catchment to use as the observed data based on the land-use map and some thematic maps. In July 2012, these monitoring points were collected in the field to validate the panchromatic map, while we recorded 83 points with information of land-use type and longitude and latitude to validate the accuracy of the classification results. Using ArcGIS and ERDAS software, we chose the common pre-classification, which is the supervised classification with a Maximum Likelihood Decision (MLD), to classify the prepro-cessed images and acquire the data of different land-use types in 2002 and 2008. The overall accuracy and Kappa coefficient of the initial classification images were 87.50% and 0.85 in 2002, respectively, and 90.10% and 0.87 in 2008, respectively. The accuracy of the 88 typical types of land-use information used to validate the classification results was 95.74%, which implied that this classification result could meet the analysis demands.

Results

Topsoil organic carbon density under different land-uses

The one-way ANOVA indicated that land-use type had a significant effect on the SOCD ($F = 8.34$, $P < 0.01$). However, only the woodland was significantly different from the other land-uses ($P < 0.05$), while no significant differences were found between other land-use types. The average SOCD of the topsoil layer (0–20 cm) in the Luoyugou catchment was 20.44 Mg/ha, and was highest in the woodland and lowest in the orchards (Table 2). Using the SOCD in the cultivated land (19.08 Mg/ha) as a reference, the SOCD in the woodland and abandoned cropland significantly increased by 33.81% and 8.49%, respectively, while the SOCD in the orchards decreased by 10.80%. The SOCD in the wild grassland was close to that of cultivated land. In this area, the coefficient of variation of the SOCD was 26.46%, and was highest (27.08%) in the woodland and smallest (16.75%) in the orchards.

Change of topsoil SOC storage in the catchment

Based on the interpretation of the remote sensing data (2000), the results showed that the total area of the four main land-use types including woodland, orchards, cultivated land and wild grassland was 67.5 km^2, accounted for 92.09% of the total region of the Luoyugou catchment. The difference of 7.02% included river, residential points, and roads that were not taken into account in the calculation of the topsoil SOC storage. From 2002 to 2008, the net decrease in the cultivated land area was

Table 2. Topsoil (0–20 cm) SOCD of different land-uses.

Land-use	Total	Woodland	Orchards	Abandoned cropland	Wild grassland	Cultivated land
Sample sites/N	83	19	13	9	8	34
SOCD Mg/ha	20.44	25.53 a	17.02 b	20.70 b	19.35 b	19.08 b
Standard deviation	5.41	6.91	2.85	4.59	3.99	3.66
Minimum	10.47	16.45	11.03	16.55	12.59	10.47
Maximum	40.71	40.71	20.72	31.14	26.02	28.86
Coefficient of variation (%)	26.46	27.08	16.75	22.17	20.62	19.16

* A different letter means a difference significant at 0.05 level.

15.11 km^2, where 7.95 km^2 of this land was converted into the woodland, and 6.51 km^2 was converted into the orchards. The increased areas of land were accounted for 52.65% and 43.12% of the total conversion area, respectively. In 2008, the SOC storage in this catchment was 1.39×10^5 t and was higher than the SOC storage in 2002 (1.35×10^5 t) by 3706.46 t. When the cultivated land was converted into the woodland, the SOC sequestration contribution was 4484.83 t; however, when the cultivated land was converted into the orchards, the contribution was −723.02 t (Table 3).

Relationship of soil organic carbon and total nitrogen content

In this catchment, soil organic carbon and total nitrogen revealed a significant positive correlation ($R^2 = 0.978$, $p<0.01$) that increased as total nitrogen content and soil organic carbon increased (Figure 2). A one-way ANOVA indicated that land-use had significant effect on the TN ($F = 3.07$. $P = 0.021$), and that had no significant effect on the soil C/N ratio ($F = 0.86$, $P = 0.48$). The average soil C/N was 9.69, with a range of 8.73–10.77. The soil C/N ratio in the woodland was the largest at 9.80 and that in the orchards was the smallest at 9.57. Other land-uses followed the order of abandoned cropland (9.76)>cultivated land (9.66)>wild grassland (9.59).

Relationship of years since planting and topsoil SOCD

For the woodland, years since planting had a significant effect on SOCD ($F = 13.00$, $P<0.001$). SOCD increased as years since planting increased. However, SOC was lost initially after afforestation in the cultivated land. The SOCD of the woodland that was older than 30 years was higher than that of the woodland under 10 years old (17.45 Mg/ha) and the cultivated land (19.08 Mg/ha), improving by 74.44% and 59.54% respectively (Figure 3a). However, years since planting had no significant effect on SOCD in the orchards ($F = 2.01$, $P = 0.146$), where the SOCD at >10a was 18.04 Mg/ha and was 10.13% higher at <10a. But, they all were below that of the cultivated land (Figure 3b).

Relationship of elevation and topsoil SOCD

The correlation analysis showed that there was a significant positive correlation between elevation and SOCD ($R^2 = 0.395$ $P< 0.01$), using the land-use type as the covariate. The SOCD was calculated as the average value with each 100 m used as an elevation gradient. The results showed that with increasing elevation, SOCD had an increasing trend (Figure 4). At an elevation gradient of ≥1700 m, the SOCD was 26.57 Mg/ha,

which was higher than that at the elevation gradient of <1300 m by approximately 58.10%.

Relationship of slope and topsoil SOCD

A one-way ANOVA indicated that land-use had a significant effect on slope ($F = 2.87$, $P = 0.028$) where the average slopes of the woodland and wild grassland all were more than 11.00°, and the average slopes of the cultivated land, abandoned cropland and orchards all were less than 5.00° (Table 4). The correlation analysis showed a low positive correlation between slope and SOCD ($R^2 = 0.170$, $P = 0.127$). A negative correlation was found between the slope and SOCD in the cultivated land, abandoned cropland and orchards ($R^2 = -0.210$, $F = 0.123$), and a positive correlation was found in the woodland and wild grassland ($R^2 = 0.250$, $F = 0.209$). For all correlations, the land-use type was used as the covariate.

Influence of multivariates on topsoil SOCD

Through normalisation processing of SOCD, elevation and slope and setting land-use dummy variables, the multiple regression model was established using SPSS software. The results were as follows:

$$y = 0.150 + 0.233x_1 + 0.054x_2 + 0.004s_1 - 0.025s_2 + 0.044s_3 + 0.168s_4 \left(R^2 = 0.329, \ p < 0.01 \right) \quad (5)$$

where y was the topsoil SOCD, x_1 was elevation, x_2 was slope, s_1 was wild grassland, s_2 was orchards, s_3 was abandoned land, and s_4 was woodland. The cultivated land was used as the reference when the value of s_{1-4} was 0.

According to the multiple model, elevation, slope and land-use accounted for 32.9% of the SOCD variation using the cultivated land as the reference (Eqn. 5).

Discussion

SOC sequestration in different ecological restoration types

The different types of conversion result in different trends in SOC [8]. In this study, the highest SOCD in the woodland implied that the conversion from the cultivated land to the woodland may lead to the accumulation of more SOC than when cultivated land is converted into the orchards or abandoned cropland in this area (Table 2). Some previous research has also found that the conversion from cultivated land to woodland can increase SOC levels [3,17]. This is mainly because the conversion from cultivated land to woodland can increase topsoil SOCD

Table 3. Change in SOC storage from 2002 to 2008 in the Luoyugou catchment.

Land-use	2002			2008			SOC	
	Area/km^2	SOC		Area/km^2	SOC			
		Storage/t	Percentage/%		Storage/t	Percentage/%	Increased/t	Sequestration amount/t
Immature woodland	8.16	19764.05	14.62	15.48	37480.99	26.98	17716.94	3759.91
Mature woodland	2.35	7145.31	5.28	2.99	9087.78	6.54	1942.48	724.92
Orchards	4.11	7377.46	5.46	10.62	19082.52	13.74	11705.06	−723.02
Wild grassland	0.71	1374.26	1.02	1.31	2536.62	1.83	1162.36	16.22
Cultivated land	52.18	99557.47	76.63	37.07	70737.09	50.92	−28820.38	–
Total	67.50	135218.55	100.00	67.46	138925.00	100.00	3706.46	3778.03

* Because the area of this catchment in 2008 was less than that in 2002, by 0.04 km^2, so that about 71.58 t of SOC wasn't taken into the amount of SOC storage. Immature woodland refers to trees under 30 years of age since planting (SOCD = 24.22 Mg/ha), and mature woodland refers to trees more than 30 years of age since planting (SOCD = 30.44 Mg/ha).

through increasing biomass inputs into the soil, and reducing soil erosion [9,12]. In addition, compared to frequent human activity such as tillage and grazing, in the cultivated land and abandoned cropland, less human disturbance in the woodland also enhanced SOC accumulation.

Our research found that the SOCD in the immature woodland (<10 a) was lower than that in the cultivated land. However, the SOCD of the mature woodland (>30 a) was higher than that of the cultivated land, increasing by 59.54% (Figure 3a). Therefore, in the initial conversion from the cultivated land to the woodland, SOCD may decrease, and from 10a after afforestation SOCD increase rapidly and significantly. Other studies have found similar trends [4,15]. Degryze et al. (2004) in Michigan found no difference in topsoil (0–25 cm) soil C during the first ten years after afforesting in a cropland; however, soil C of the native forest (48.6 t/ha) was significantly greater than the cropland (31.9 t/ha) [41]. Lu et al. (2013), studying afforestation in the Loess Plateau, found that the time of the SOC source to sink transition was 3 to 8 years after afforestation. The decrease of SOC in the first few years following afforestation has been attributed to low net primary productivity of plants, decreased litter inputs and increased decomposition rates [12,29].

In this study, the topsoil SOCD in the orchards was less than that in the cultivated land, which implied that a decrease in SOC may occur when the cultivated land is converted into the orchards (Figure 3b). This result is similar to the results of Yang et al. in the gully region of the Loess Plateau [42]. However, Xue et al. (2011) reported that when the slope farmland was converted into the orchards, the SOC content increased slowly with increasing years since planting and reached its peak between 20 to 30 years [43]. Thirty years later, the SOC content of the converted orchards was 4.96 g/kg and was higher than that in the slope farmland by 97%. One difference between our study and that by Xue et al. (2011) is the different reference used. The reference in their study was slope cropland, whereas we used the average value of the slope and terrace cropland as the reference. Moreover, management measures in their study were better than ours. For example, they interplanted crops into the fruit trees at the seedling stage and used more fertiliser.

Management measures are one of the main factors affecting the loss of SOC in orchards. In this catchment, fruit, trimmed branches, and litter are removed from orchards, which can cause less organic compounds to enter the soil. Clear tillage in the orchards also causes more soil erosion. Therefore, in some farmland regions that are returning to their historic uses and thus need to develop fruit trees for economic purposes, methods for reducing the loss of SOC and enhancing SOC sequestration have become important. To manage this problem, some researchers have found that orchards managed with conservation practices such as growing grass and leguminous cover crop, mulching the ground, and frequently using organic and inorganic fertilisers could reduce soil erosion and improve the SOC content [44–45]. In addition, establishing new orchard management models oriented to SOC sequestration is also very necessary.

Many studies have found that the land-use change from the cultivated land to the abandoned land can increase the SOC stored in soil [13–14]. In our study, our results also showed that the SOCD may improve when the cultivated land is converted into the abandoned land (Table 2). With the termination of human disturbance, vegetation in the abandoned land may begin the process of self-succession [14], gradually forming original vegetation communities of their region, which are secondary wild grassland or secondary forest communities. However, this process of succession is slow, as the process that turns the abandoned land into the top secondary bunge needlegrass (*Stipa bungeana Trin.*) takes 40 to 50 years [46]. Further studies are needed to understand the methods needed to accelerate community succession and better sequester SOC in the abandoned land.

The research of Li et al. (2007) in the northern region of the Loess Plateau found that the SOCD of abandoned land at approximately 10 years of age was slightly lower than that in the secondary wild land where planted grassland (*Medicago sativa*) changed in the same number of years. After 6 to 10 years of restoration, degraded artificial grassland may form secondary bunge needlegrass (*Stipa bungeana Trin.*) [47–48]. Therefore, planting artificial grass in the abandoned land can shorten the time of succession, which ensures ecological benefits and increase economic benefits at the same time. However, in this study, the SOCD of the abandoned land at 8–10 years of age was higher

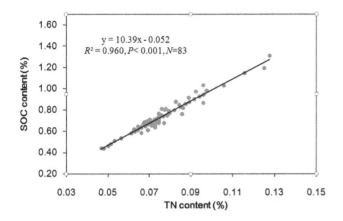

Figure 2. Relationship of topsoil soil organic carbon and total nitrogen content.

than in the wild grassland, which was different to the results of previous studies [47]. The main reason for this may be that these grasslands were distributed in the barren stony mountainous area. The average slope of these regions was $13°$, and soil erosion was relatively severe. These factors led to low SOCD.

Benefits of SOC sequestration in the entire region

The SOC storage is determined by the dynamic equilibrium of SOC input and output. After approximately 10 years of ecological restoration, the area of cultivated land greatly decreased, while the area of woodland and orchards increased. From 2002 to 2008, the SOC storage of the entire area only improved by 2.74% (3706.46 t) (Table 3). Therefore, the ecological restoration has an increase in SOC storage, but not significant in the short-term. The main reason for this is that an increase in the SOC resulting from the woodland accumulation was offset by a loss in the SOC resulting from the orchards and the initial conversion stage. So, the SOC storage of the total area is in a relatively stable state. If all of the woodland values are calculated using the SOCD value of the woodland more than 30 years old (30.44 Mg/ha), then considering the loss of SOC by the orchards, the SOC storage of the total area can improve by 7.12% (9625.59 t). Therefore, SOC storage has great potential to increase in the future if the woodland converted from the cropland is well manage and maintain.

Although SOC storage has not increased in the past 10 years of restoration, with increasing ecological restoration, soil erosion has been controlled and the quality of the ecological environment, such as the air, water, vegetable coverage has gradually improved in this catchment.

Influencing factors on the SOC

Results from this study showed a strong correlation between SOC and TN (Table 2), which was consistent with earlier findings [33,36]. The trend of C:N ratio varying with the land-use types was similar to that of SOC. The woodland had a higher C:N ratio than other land-use types due to increased above and below ground biomass. The most prevalent tree in the woodland is the black locust, which has N-fixing capabilities, increased TN can promote tree growth, which results in an increase in SOC; it was found that the SOC increase was greater than the TN increase [33]. In catchments, elevation is a governing variable because of its effect on various environmental factors, such as temperature and precipitation. Our study showed a significant positive correlation between elevation and the SOCD (Figure 4). With rising elevation, topsoil SOC increased. This phenomenon was similar to results of other studies [18,49], which can be explained by the lower temperatures and increased precipitation with higher elevations. Increased precipitation can promote vegetation growth, which increases the accumulation of humus, and lower temperatures can limit the decomposition and turnover of SOC, leading to enhanced SOC storage.

The slope is one of the main topographical features affecting the soil erosion intensity as well as SOC loss and enrichment [16]. In the present study, land-use types significantly affected the correlation between the slope and SOCD, where there was a positive correlation in woodland and wild grassland and a negative correlation in cultivated land, abandoned land and orchards (Table 4). These outcomes were due mainly to the influence of vegetation type and coverage. Vegetation restoration on slopes such as planting trees and grass can increase surface vegetation coverage and effectively control soil erosion, further reducing the loss of SOC. These results have been found in other regions [17,30]. Thus, the slope land should be converted to woodland and grassland, or changed to terrace to maintain and increase SOC levels. The multiple regression analysis showed that land-use, slope and elevation accounted for 32.9% of the SOCD variation. The unexplained variation may be caused by the soil particle size,

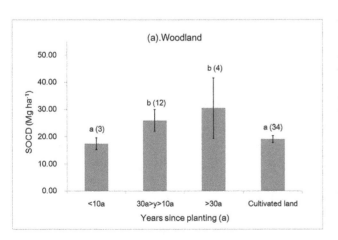

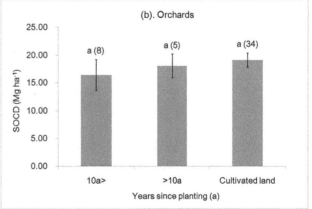

Figure 3. Relationship of years since planting and topsoil SOCD. * The error bars are the standard errors of the mean of SOCD and values above the bars is the number of observations (in parentheses). A different letter means a difference significant at $P<0.05$.

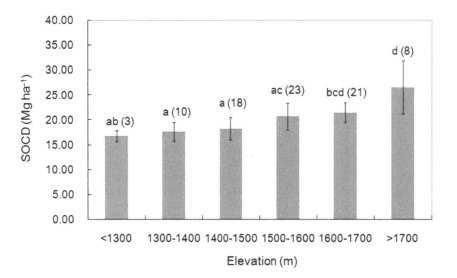

Figure 4. Relationship of elevation and topsoil SOCD. * The error bars are the standard errors of the mean of SOCD and values above the bars is the number of observations (in parentheses). A different letter means a difference significant at $P<0.05$.

soil aggregates, the aspect and the years of utilisation. Therefore, a benefits evaluation of SOC sequestration achieved through ecological restoration efforts should fully consider those influencing factors.

Uncertainties over SOC sequestration from future ecological restoration efforts

With ecological restoration efforts increasing, soil erosion in the Loess Plateau has significantly decreased, and ecological benefits such as vegetation coverage have gradually increased. However, in the long run, there are still uncertainties that will persist if we use the present SOC sequestration achievements from ecological restoration to evaluate possible future SOC sequestration benefits.

(I)   Influence of national policies: with the economy developing, the present subsidy standard for returning farmland to forests in the Loess Plateau has been increasingly unable to meet the basic living demands of local farmers. The competitiveness of this subsidy is also growing increasingly weak due to the influence of factors which are raising grain prices and creating policies which benefit farmers, including the abolition of the agricultural tax and the enhancement of subsidies for farming. Based on the results of 2000 farmers' surveys, Cao et al. (2009) found that approximately 37.2% of farmers planned to return to cultivating forested areas and grasslands once the project's subsidies end in 2018. The smaller the subsidy amount to farmers, the fewer the farmers who are willing to participate in them. If the government is unable to improve farmers' income, much of the vegetation restored during the "Grain for Green" project is at risk of being re-converted into farmland and rangeland at the end of this project [50]. Therefore, it is very important and urgent to determine how to effectively consolidate the achievements of the "Grain for Green" project, especially once the project ends, while continually completing the task of returning farmland to forest. However, the government does not do well in this respect. Therefore, uncertainties exist concerning the benefits from SOC sequestration by ecological restoration in the future.

(II)   Influence of the environment: Most regions converted from cultivated land to forest are experiencing drought and soil impoverishment. There are no rich water resources in the Loess Plateau, and large-scale vegetation construction will certainly consume massive amounts of water. Lü et al. (2012) found that over half of the Loess

Table 4. Slope of different land-uses.

Land-use	Total	Woodland	Orchards	Abandoned cropland	Wild grassland	Cultivated land
Sample sites/N	83	19	13	9	8	32
Slope/°	6.65	11.58 a	1.92 b	4.22 ab	13.00 a	4.85 b
Standard deviation	10.85	14.76	4.80	8.39	12.10	8.93
Minimum	0	0	0	0	0	0
Maximum	45.00	45.00	15.00	20.00	30.00	30.00
Coefficient of variation %	163.16	127.46	250.00	198.82	93.08	184.12
Correlation between slope and SOCD/R^2	0.170	0.269	−0.169	−0.459	0.480	−0.229

* A different letter means a difference significant at 0.05 level. The signification of correlation between slope and SOCD all were more than 0.05 ($P>0.05$).

Plateau experienced a decrease in runoff (2–37 mm/year) with an average of 10.3 mm/year after the implementation of the "Grain for Green" project [51]. Therefore, with water consumption increases resulting from vegetation, some regions are likely to face a deficit of water resources or even a possible depletion. Drought, decreased rainfall and the excessive depletion of deep soil water by planted vegetation causes the formation of the dried soil layer [52]. The dried soil layer can lead to land degradation and heavily influences vegetation growth, which forms large-scale, low-efficiency production forests [53–54]. Once the dried soil layer appears, it is very difficult for recovery to occur. In the process of implementing "Grain for Green" some improper phenomena have occurred, including the improper selection of vegetation restoration types and excessive close planting conducted in the blind pursuit of economic benefits and political achievement [55], and this has accelerated the formation of the dried soil layer. Therefore, in the long run, ecological restoration in the Loess Plateau is likely to face water storage restrictions. Thus, the benefits of SOC sequestration may be uncertain in the future.

(III)   Influence of private and local government interests: Many regions would prefer that cultivated land be converted into orchards when the cultivated land needs to be converted due to this conversion type may generate a higher land value. However, the SOC sequestration benefits resulting from this type of conversion are lower than in the conversion from cultivated land to woodland, and this conversion type may even cause SOC loss. Over time, farmers may remove old trees in orchards converted from cultivated land, or abandoned land. In addition, according to changes in market demand, farmers may replace other tree species before the original trees have matured. The non-ecological woodland converted from cultivated land also can be harvested under reasonable conditions. However, the SOC sequestration effects of these changes have not been studied. In the future, many woodlands and orchards converted from cultivated land will almost certainly undergo cutting and regeneration. The benefits of SOC sequestration are uncertain in the future unless we can understand the SOC variation under these changes.

(IV)   The influence of urbanization and economic interests: since the reforms and open policies were established, labour exports from rural areas have increased year after year. Lack of labour has caused the abandonment of large areas of rural cultivated lands [56]. Recent research has shown that the SOC content could increase when cultivated land is converted into abandoned land. However, given the increasing speed of abandonment, it is unclear how long this phenomenon can continue. In addition, these labours may choose to go home to engage in farming in the future, affecting by the economic crisis of the original working regions and other reasons such as older age and homesickness. This transform may bring new pressure on the local ecological environment, which may cause the loss of SOC. Therefore, the future benefits of SOC sequestration are uncertain.

(V)   Influence of climate change: Here, climate change mainly refers to precipitation decline, temperature increase and elevated carbon dioxide levels. According to the results of

Xin et al. (2011), the drying trend of the Loess Plateau was highly significant, and annual rainfall showed an obvious decreasing trend over the past five decades (1956–2008) by approximately −1.4 mm/a [57]. The reduced precipitation aggravated the water resource shortage situation in the Loess Plateau, which led to the poor tree growth. This outcome may affect SOC sequestration by vegetation restoration, because the temperature increased and elevated carbon dioxide levels have opposite effects on SOC sequestration. Increasing temperatures may promote SOC decomposition; however, elevated carbon dioxide levels may also improve the net primary productivity of plants and increase both the residual body of vegetation and SOC storage [58–59]. Over the last 50 years (1961–2010), the annual mean temperature has significantly increased by 1.91°C in the Loess Plateau [60]. The atmospheric concentration of CO_2 has increased from 280 p.p.m. to 385 p.p.m. between the pre-industrial era and 2008 [61–62]. There is no clear research concerning what functions of the increasing temperatures and elevated carbon dioxide levels for SOC storage are more pronounced. Therefore, some uncertainties exist regarding the possible benefits of SOC sequestration in the future.

Generally speaking, there are many factors that may affect SOC sequestration in the future and may increase or decrease SOC storage. Factors that may be detrimental for SOC storage include: (i) the low national subsidy standard for returning farmland into forest or grasslands and the low income of local farmers; (ii) the persistent, deteriorating environment situation and; (iii) land-use changes caused by private and local government interests. Because of these disadvantageous factors, it is very important to determine how to effectively increase SOC storage. Creating new agriculture products by the implementation of more modern agriculture techniques and offering more work opportunities in urban areas for farmers to increase their income could partially solve these problems. Additionally, the large-scale afforestation that may be exacerbating the soil water shortage in the Loess Plateau should be controlled, especially in vulnerable arid and semi-arid regions, and should fully consider the affordability of such environmental efforts before vegetation restoration is conducted.

Conclusions

In this study, land-use type had a significant influence on the topsoil SOCD. Compared with the average SOCD of cultivated land, the average SOCD in woodland was significantly larger by 33.81%, while the average SOCD in the orchards decreased by 10.80%. Therefore, the SOCD may improve when cultivated land is converted into woodland and may decline when cultivated land is converted into orchards. Over time, there was an increasing trend of SOCD in woodland, but only a small change of SOCD in the orchards. Based on the remote sensing data on land-use change, we found that the "Grain for Green" project did not significantly increase SOC storage in the Luoyugou catchment in the past 10 years of restoration. This is mainly because the increase of SOC storage by woodland was offset by a SOC storage loss in the orchards and decreases in SOC storage in the initial stage of land-use conversion. If all of the woodland area was calculated as 30.44 t/km² of SOCD, the potential contribution to SOC sequestration in this catchment would increase by 9625.59 t, improving the SOC storage ratio by 7.12%. Therefore, it has a huge potential for future environmental restoration efforts if we could manage and maintain the woodland converted from the cropland well. However, we cannot ignore other factors that affect SOC sequestration, including climate change and national policies.

Author Contributions

Conceived and designed the experiments: YBQ ZBX XXY YLX. Performed the experiments: YBQ ZBX. Analyzed the data: YBQ ZBX. Contributed reagents/materials/analysis tools: YBQ ZBX XXY YLX. Wrote the paper: YBQ ZBX XXY YLX.

References

1. Morris SJ, Bohm S, Haile-Mariam S, Paul EA (2007) Evaluation of carbon accrual in afforested agricultural soils. Global Change Biology, 13: 1145–1156.
2. Wang YF, Fu BJ, Lü YH, Chen LD (2011) Effects of vegetation restoration on soil organic carbon sequestration at multiple scales in semi-arid Loess Plateau, China. Catena, 85: 58–66.
3. Sauer TJ, James DE, Cambardella CA, Hernandez-Ramirez G (2012) Soil properties following reforestation or afforestation of marginal cropland. Plant Soil, 360: 375–390.
4. Li DJ, Niu SL, Luo YQ (2012) Global patterns of the dynamics of soil carbon and nitrogen stocks following afforestation: a meta-analysis. New Phytologist, 195: 172–181.
5. Lal R (2004) Agricultural activities and the global carbon cycle. Nutrient Cycling in Agroecosystems, 70: 103–116.
6. Powlson D (2005) Will soil amplify climate change? Nature, 433: 204–205.
7. Smith P, Martino D, Cai ZC, Gwary D, Janzen H, et al (2008) Greenhouse gas mitigation in agriculture. Philosophical Transactions of the Royal Society, Series B, 363: 789–813.
8. Guo LB, Gifford RM (2002) Soil carbon stocks and land-use change: a meta analysis. Global Change Biology, 8: 345–360.
9. Don A, Schumacher J, Freibauer A (2011) Impact of tropical land-use change on soil organic carbon stocks- a meta-analysis. Global Change Biology, 17: 1658–1670.
10. Houghton RA (1999) The annual net flux of carbon to the atmosphere from changes in land-use 1850–1990. Tellus, 51B: 298–313.
11. Lal R (2005) Forest soils and carbon sequestration. Forest Ecology and Management, 220: 242–258.
12. Laganière J, Angers AD, Paré D (2010) Carbon accumulation in agricultural soils after afforestation: a meta-analysis. Global Change Biology, 16: 439–453.
13. Raiesi F (2012) Soil properties and C dynamics in abandoned and cultivated farmlands in a semi-arid ecosystem. Plant Soil, 351: 161–175.
14. Novara A, Gristina L, Mantia TL, Rühl J (2013) Carbon dynamics of soil organic matter in bulk soil and aggregate fraction during secondary succession in a Mediterranean environment. Geoderma, 193–194: 213–221.
15. Paul KI, Ploglase PJ, Nyakuengama JG, Khanna PK (2002) Change in soil carbon following afforestation. Forest Ecology and Management, 168: 241–257.
16. Wang BQ, Liu GB (1999) Effects of relief on soil nutrient losses in sloping fields in hilly region of Loess Plateau. Soil Erosion and Soil and Water Conservation, 5: 18–22 (in Chinese).
17. Wang Z, Liu GB, Xu MX, Zhang J, Wang Y, et al (2012) Temporal and spatial variations in soil organic carbon sequestration following revegetation in the hilly Loess Plateau, China. Catena, 99: 26–33.
18. Zhou Y, Xu XG, Ruan HH, Wang JS, Fang YH, et al (2008) Mineralization rates of soil organic carbon along an elevation gradient in Wuyi Mountain of Southeast China. Ecology, 27: 1901–1907 (in Chinese).
19. Luo YQ, Hui DF, Zhang DQ (2006) Elevated CO$_2$ stimulates net accumulations of carbon and nitrogen in land ecosystems: a meta-analysis, Ecology, 87: 53–63.
20. Batterman SA, Hedin LO, Breugel Mv, Ransijn J, Craven DJ, et al (2013) Key role of symbiotic dinitrogen fixation in tropical forest secondary succession. Nature, 502: 224–227.
21. Huang CY (2000) Soil Science. Beijing: China Agriculture Press, 311p.
22. Gregorich EG, Greer KJ, Anderson DW, Liang BC (1998) Carbon distribution and losses: erosion and deposition effects. Soil & Tillage Research, 47: 291–302.
23. Lal R (2003) Soil erosion and the global carbon budget. Environment International, 29: 437–450.
24. Chartier MP, Rostagno CM, Videla LS (2013) Selective erosion of clay, organic carbon and total nitrogen in grazed semiarid rangelands of northeastern Patagonia, Argentina. Journal of Arid Environments, 88: 43–49.
25. Fu BJ (1989) Soil erosion and its control in the Loess Plateau of China. Soil Use and Management, 5: 76–81.
26. Shi H, Shao MA (2000) Soil and water loss from the Loess Plateau in China. Journal of Arid Environments, 45: 9–20.
27. Fu BJ, Chen LX, Qiu Y (2002) Land-use structure and ecological processes in the Loess Plateau. Beijing: The Commercial Press, pp: 1–12 (in Chinese).
28. Feng XM, Wang YF, Chen LD, Fu BJ, Bai GS (2010) Modeling soil erosion and its response to land-use change in hilly catchments of the Chinese Loess Plateau. Geomorphology, 118: 239–248.
29. Lu N, Liski J, Chang RY, Akujärvi A, Wu X, et al (2013) Soil organic carbon dynamics following afforestation in the Loess Plateau of China. Biogeosciences Discuss, 10: 11181–11211.
30. Zheng FL (2006) Effect of Vegetation Changes on soil erosion on the Loess Plateau. Pedosphere, 16: 420–427.
31. Gong J, Chen LD, Fu BJ, Huang Y, Huang Z, et al (2006) Effect of land-use on soil nutrients in the loess hilly area of the Loess Plateau, China. Land Degradation & Development, 17: 453–465.
32. Chen LD, Gong J, Fu BJ, Huang ZL, Huang YL, et al (2007) Effect of land-use conversion on soil organic carbon sequestration in the loess hilly area, loess plateau of China. Ecological Research, 22: 641–648.
33. Fu XL, Shao MA, Wei XR, Horton R (2010) Soil organic carbon and total nitrogen as affected by vegetation types in Northern Loess Plateau of China. Geoderma, 155: 31–35.
34. Chang RY, Fu BJ, Liu GH, Liu SG (2011) Soil carbon sequestration potential for "Grain for Green" project in Loess Plateau, China. Environment Management, 48: 1158–1172.
35. Wei J, Cheng JM, Li WJ, Liu WG (2012) Comparing the effect of naturally restored forest and grassland on carbon sequestration and its vertical distribution in the Chinese Loess Plateau. PLoS ONE, 7: e40123. doi:10.1371/journal.pone.0040123
36. Lei D, ShangGuan ZP, Sweeney S (2013) Changes in soil carbon and nitrogen following land abandonment of farmland on the Loess Plateau, China. PLoS ONE 8: e71923. doi:10.1371/journal.pone.0071923
37. Yu XX, Zhang XM, Niu LL, Yue YJ, Wu SH, et al (2009) Dynamic evolution and driving force analysis of land-use/cover change on loess plateau catchment. Transaction of the CSAE, 25: 219–225 (in Chinese).
38. Zhao Y, Yu XX (2013) Effects of climate variation and land-use change on runoff-sediment yield in typical watershed of loess hilly-gully region. Journal of Beijing Forestry University, 35: 39–45 (in Chinese).
39. Nelson DW, Sommers LE (1982) Total carbon, organic carbon, and organic matter. In: Page AL, Miller RH, Keeney DR (eds) Methods of soil analysis, Part 2, Chemical and microbial properties. Agronomy Society of America, Agronomy Monograph 9, Madison, Wisconsin, pp 539–552.
40. Institute of Soil Sciences, Chinese Academy of Sciences (ISSCAS) (1978) Physical and chemical analysis methods of soil. Shanghai: Shanghai Science Technology Press, pp 7–15.
41. Degryze S, Six J, Paustian K, Morris SJ, Paul EA, et al (2004) Soil organic carbon pool changes following land-use conversions. Global Change Biology, 10: 1120–1132.
42. Yang YL, Guo SL, Ma YH, Chen SG, Sun WY (2008) Changes of orchard soil carbon, nitrogen and phosphorus in gully region of Loess Plateau. Plant Nutrition and Fertilizer Science, 14: 685–691 (in Chinese).
43. Xue S, Liu GB, Zhang C, Zhang CS (2011) Analysis of effect of soil quality after orchard established in hilly Loess Plateau. Scientia Agricultura Sinica, 44: 3154–3161 (in Chinese).
44. Umali BP, Oliver DP, Forrester S, Chittleborough DJ, Hutson JL, et al (2012) The effect of terrain and management on the spatial variability of soil properties in an apple orchard. Catena, 93: 38–48.
45. Guimarães DV, Gonzaga MIS, Silva TOd, Silva TLd, Dias NdS, et al (2013) Soil organic matter pools and carbon fractions in soil under different land-uses. Soil & Tillage Research, 126: 177–182.
46. Zou HY, Cheng JM, Zhou L (1998) Natural recoverage succession and regulation of the prairie vegetation on the Loess Plateau. Research of Soil and Water Conservation, 5: 126–138 (in Chinese).
47. Li YY, Shao MA, Shang Guan ZP, Fan J, Wang LM (2006) Study on the degrading process and vegetation succession of Medicago sativa grassland in North Loess Plateau, China. Acta Prataculturae Sinica, 15: 85–92 (in Chinese).
48. Li YY, Shao MA, Zhang JY, Li QF (2007) Impact of grassland recovery and reconstruction on soil organic carbon in the northern Loess Plateau. Acta Ecologica Sinica, 27: 2279–2287 (in Chinese).
49. Leifeld J, Bassin S, Fuhrer J (2005) Carbon stocks in Swiss agricultural soils predicted by land-use, soil characteristics, and elevation. Agriculture, Ecosystems and Environment, 105: 255–266.
50. Cao SX, Xu CG, Chen L, Wang XQ (2009) Attitudes of farmers in China's northern Shaanxi Province towards the land-use changes required under the Grain for Green Project, and implications for the project's success. Land-use Policy, 26: 1182–1194.
51. Lü YH, Fu BJ, Feng XM, Zeng Y, Liu Y, et al (2012) A policy-driven large scale ecological restoration: quantifying ecosystem services changes in the Loess Plateau of China. PLoS ONE, 7: e31782, doi:10.1371/journal.pone.0031782
52. Pan ZB, Zhang L, Yang R, Li SB, Dong LG, et al (2012) Overview on research progress of soil drought in semiarid regions of the Loess Plateau. Research of Soil and Water Conservation, 19: 287–291, 298 (in Chinese).
53. Wang L, Shao MA (2004) Soil desiccation under the returning farms to forest on the Loess Plateau. World Forestry Research, 17: 57–60 (in Chinese).
54. Chen HS, Shao MA, Li YY (2008) Soil desiccation in the Loess Plateau of China. Geoderma, 143: 91–100.
55. Niu JJ, Zhao JB, Wang SY (2007) A study on plantation soil desiccation in the upper reaches of the Fenhe River basin based on deep soil experiments. Geographical Research, 26: 773–781 (in Chinese).

56. Duan FL, Lin Z, Xiong YQ (2007) Analysis on the phenomenon of farmland abandoned by the reason of rural laborers moving out for work. Rural Economy, 16–19 (in Chinese).

57. Xin ZB, Yu XX, Li QY, Lu XX (2011) Spatiotemporal variation in rainfall erosivity on the Chinese Loess Plateau during the period 1956–2008. Regional Environmental Change, 11: 149–159.

58. William HS (1999) Carbon and agriculture: carbon sequestration in soils. Science, 284: 2095.

59. Guo GF, Zhang CY, Xu Y (2006) Effects of climate change on soil organic carbon storage in terrestrial ecosystem. Ecology, 25: 435–442 (in Chinese).

60. Wang QX, Fan XH, Qin ZD, Wang MB (2012) Change trends of temperature and precipitation in the Loess Plateau Region of China, 1961–2010. Global and Planetary Change, 92–93: 138–147.

61. Intergovernmental Panel on Climate Change (2007) Climate Change 2007: The Science of Climate Change. Cambridge University Press, Cambridge.

62. Lal R (2009) Challenges and opportunities in soil organic matter research. European Journal of Soil Science, 60: 158–169.

2

Stability in and Correlation between Factors Influencing Genetic Quality of Seed Lots in Seed Orchard of *Pinus tabuliformis* Carr. over a 12-Year Span

Wei Li[1], Xiaoru Wang[2], Yue Li[1]*

1 National Engineering Laboratory for Forest Tree Breeding, Key Laboratory for Genetics and Breeding of Forest Trees and Ornamental Plants of Ministry of Education, Beijing Forestry University, Beijing, People's Republic of China, **2** State Key Laboratory of Systematic and Evolutionary Botany, Institute of Botany, Chinese Academy of Sciences, Beijing, People's Republic of China

Abstract

Coniferous seed orchards require a long period from initial seed harvest to stable seed production. Differential reproductive success and asynchrony are among the main factors for orchard crops year-to-year variation in terms of parental gametic contribution and ultimately the genetic gain. It is fundamental in both making predictions about the genetic composition of the seed crop and decisions about orchard roguing and improved seed orchard establishment. In this paper, a primary Chinese pine seed orchard with 49 clones is investigated for stability, variation and correlation analysis of factors which influence genetic quality of the seed lots from initial seed harvest to the stable seed production over a 12 years span. Results indicated that the reproductive synchrony index of pollen shedding has shown to be higher than that of the strobili receptivity, and both can be drastically influenced by the ambient climate factors. Reproductive synchrony index of the clones has certain relative stability and it could be used as an indication of the seed orchard status during maturity stage; clones in the studied orchard have shown extreme differences in terms of the gametic and genetic contribution to the seed crop at the orchard's early production phase specifically when they severe as either female or male parents. Those differences are closely related to clonal sex tendency at the time of orchard's initial reproduction. Clonal gamete contribution as male and female parent often has a negative correlation. Clone utilization as pollen, seed or both pollen and seed donors should consider the role it would play in the seed crop; due to numerous factors influencing on the mating system in seed orchards, clonal genetic contribution as male parent is uncertain, and it has major influence on the genetic composition in the seed orchard during the initial reproductive and seed production phase.

Editor: Daniel J. Kliebenstein, University of California, United States of America

Funding: This work was supported by grants from The Fundamental Research Funds for the Central Universities (No.JD2011-4) and Special Fund for Forestry Scientific Research in the Public Interest (No. 201104022). The funders had no role in study design, data collection and analysis, decision to publish, or preparation of the manuscript.

Competing Interests: The authors have declared that no competing interests exist.

* E-mail: liyue@bjfu.edu.cn

Introduction

Seed orchards are the most common means of making available a stable supply of genetically improved seed and it constitutes an important component in most coniferous species improvement programs in the world [1]. Genetic gain of seed orchards' crops depends on the section differential between the orchards' parental populations and that of unselected seed sources as well as orchards' parental population actual gamete contribution to the harvested seed crops [2]. The attainment of balanced gametic contribution from orchards' parents is hardly observed due to many factors including differential male and female reproductive success and asynchrony as well as the degree of selfing and pollen introgression from outside unselected sources [3] [4]. Moreover coniferous trees have a long lifespan, thus seed orchards require a long period from the initial establishment until reproductive maturity and the steady production of a stable seed yield. Differential reproductive success and asynchrony are among the main factors for orchard crops year-to-year variation in terms of parental gametic contribution and ultimately the genetic gain [5]. Therefore, the time span from

initial seed harvest to stable seed production shows substantial parental gametic contribution instability and is the subject of intense research [6] [7].

Chinese pine (*Pinus tabuliformis* Carr.) is an important native coniferous tree species to northern China and is naturally distributed across 14 northern provinces and autonomous regions, with a land cover of close to 3 million km^2 [8]. Because of the important ecological and economy status of Chinese pine as indigenous conifer tree species in northern China, a tree improvement program was initiated in the 1970 s that was followed by clonal seed orchard establishment [9]. Many of the basic aspects of species' genetic improvement such as grafting technology, pollen dynamics and parental gamete contribution have been comprehensively and systematically studied [10] [11]. Currently, it is inadequate in addressing the flowering synchronization between clones, annual relevant gamete contribution, stability of genetic gain of the seed lots, factors influencing the genetic structure of the seed lots and relationships between factors influencing the genetic quality of seed lots during the period of from initial seed harvest to the stable seed production in the primary Chinese pine seed orchard.

In this paper, we evaluate clonal reproductive success and synchrony among 49 Chinese pine clones growing in a seed located in Xingcheng City (Liaoning Province, China). In particular, we investigate the flowering phases, quantity of female and male flowers as well as the average number of full seeds in each cone of selected clones and their impact on seed crops breeding value over a period of 12 years. Results are expected to provide the theoretical foundation for the seed orchard management, genetic composition prediction and advanced seed orchard establishment.

Materials and Methods

Ethics statement

This research only involved in coniferous trees and general station of forest seedling management of Liaoning Province has agreed our observational and field studies. We have published several other papers with the same material.

Materials

This study was carried out in a primary clonal Chinese pine seed orchard which located in Xingcheng City Liaoning Province of China (40°44′N, 120°34′E, 100 m above sea level). The seed orchard is composed of 49 clones planted at 7 m×7 m fixed block design. Clones were grafted in 1974 using two-year rootstock and planted in 1975 [12]. Samples and investigation were carried out in the 10th, 13th, 18th, and 21st years after grafting spanning 12 years covering the time from initial flowering to stable seed production.

Methods

Four ramets were randomly selected to represent each clone. Within each ramet, 10 female strobili were marked at the central part of sunny side of the selected plants representing the female component. Additionally the male component was represented by 10 male strobili branches located in the short middle branches on the shady side where most male strobili were positioned. Efforts were made to reach a level of sampling consistency throughout the study period [13].

In the years of observation, the marked strobili, male and female, were monitored at fixed time (noon) throughout the reproductive stage of each clone, and based on this the number of female and male strobili in each sample plant was calculated. At the same time, the average number of filled seeds in a sample of 30 seed-cones was determined (note: this was done the year following pollination as pine cones require 2 years to reach maturity).

Flowering process was divided 4 stages which have been described in other conifer tree species. The female stages were described by Matziris [14] and Codesido *et al.* [15]: stage 1, the female bud is increasing in size, becomes cylindrical, but is still completely covered by the bud scales (0% female receptivity); stage 2, the apex of the enlarged cylindrical bud is opened and the first ovuliferous scales appear (20% female receptivity); stage 3, the scales of the female conelet are gradually separated and almost form right angles with the axis of the conelet (100%), and stage 4, the ovuliferous scales increase in size and thickness (0% female receptivity). The male stages were described as follows [15] [16]: stage 1, the round brown strobili are covered by the bud scales (0% pollen shedding); stage 2, the male strobili burst through the bud scales and elongate (0% pollen shedding); stage 3, the yellow strobili start shedding their pollen (100% pollen shedding) and stage 4, end of pollen shedding. The male strobili wither and fall down (0% pollen shedding).

Date analysis

Index of measuring phonological overlap (PO_{ij}) proposed by Askew and Blush [17] was used for quantifying the degree of reproductive synchronization. When mating pairs have exactly same flowering phases, PO_{ij} value will reach the maximum 1. Conversely, when the flowering phases do not overlap, PO_{ij} value equals to minimum 0. When the flowering phases partially overlapped, PO_{ij} value will range between 0 and 1. If $i = j$, PO_{ij} means selfing flowering synchronization within the same clone.

According to the method described by El-Kassaby and Askew [18] and Askew [19], the relative gametic reproduction of a specific clone as female or male was calculated by the ratio of the average quantity of female or male strobili in all sampled ramets to the total number overall strobili average of all clones in the seed orchard. Clonal breeding value was determined from progeny testing and gain values of tree height at age 20 were considered in the following analyses.

Parental gamete contribution to the seed lots of seed orchard was estimated with the following method:

$$PM_i = PO_i.M_i \bigg/ \sum_{i=1}^{n} PO_i.M_i \qquad (1)$$

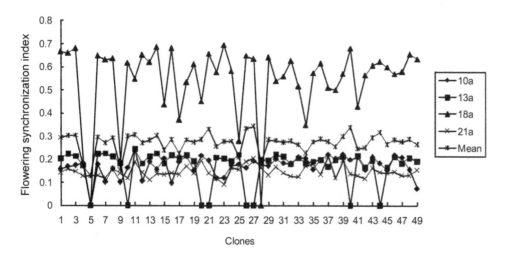

Figure 1. Clonal flowering synchronization as female in *Pinus tabuliformis* seed orchard.

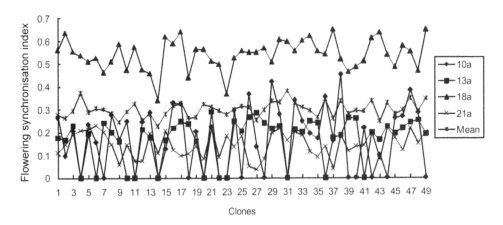

Figure 2. Clonal flowering synchronization as male in *Pinus tabuliformis* seed orchard.

$$PF_j = PO._jF_j \bigg/ \sum_{i=1}^{n} PO._jF_j \qquad (2)$$

$$GA_j = 1 \bigg/ 2\left[\left(\sum_{i=1}^{t} P_{ij}G_i\right) + G_j\right] \qquad (4)$$

$$PC_{ij} = PO_{ij}M_iF_j \bigg/ \sum_{i=1}^{n}\sum_{j=1}^{n} PO_{ij}M_iF_j \qquad (3)$$

$$PL_{ij} = PO_{ij} \times POL_i \qquad (5)$$

$$PL_{ij} = PO_{ij} \times POL_i \qquad (6)$$

in which, PM_i and PF_j are the gamete ratio of the clones i and j as male and female parent respectively in the seed orchard; PC_{ij} is the gamete ratio of the clone i and j as mating pair in seed orchard; PO_i. and PO_j are the average flowering synchronization indices of the clone i and j as male and female parent respectively in the seed orchard; M_i and F_j are the pollen loading and ovules number of the clone i and j in the seed orchard.

For a clonal seed orchard with t clones, the relative genetic contribution of the clone j as female parent (GA_j) can be formulated as following:

in which, G_i and G_j represent the relative breeding value of the clones i and j, respectively; P_{ij} is the seed ratio in the strobilus of clone j which pollinated by clone i; PL_{ij} is calculated with the relative male flowers quantity (POL_i) and flowering synchronization index (PO_{ij}).

Correlation analysis was estimated as Pearson correlation (r) and was used to determine the extent of relationship between all the indexes. Special analysis software developed by Huang and Chen [20] was used for the date processing.

Table 1. Within clone male and female flowering synchrony in *Pinus tabulaeformis* seed orchard.

	M10	F10	MF10	M13	F13	MF13	M18	F18	MF18	M21	F21
F10	0.01										
MF10	0.79**	0.15									
M13	0.42*	0.05	0.28								
F13	0.07	0.10	−0.02	0.10							
MF13	0.22	0.09	0.43*	0.55**	0.04						
M18	−0.50*	0.36	−0.34	−0.19	0.04	−0.23					
F18	0.18	−0.11	−0.06	0.07	0.34	0.04	−0.45*				
MF18	−0.12	0.02	−0.15	−0.04	0.01	0.12	0.11	0.63**			
M21	0.35*	−0.05	0.03	−0.08	−0.07	0.02	−0.10	0.08	0.34		
F21	−0.05	0.38*	0.21	−0.10	0.94**	0.21	0.34*	−0.39**	0.56	−0.05	
MF21	0.44**	0.13	0.32	−0.12	0.75**	0.24	0.14	0.54**	0.14	−0.27	0.56**

Note: M10 means clonal gametic contribution as male in the 10th year after clone grafting in the seed orchard; F10 means clonal gametic contribution as female in the 10th year after clone grafting in the seed orchard; MF10 means clonal gametic contribution as both male and female in the 10th year after clone grafting in the seed orchard; the same with others.
*represents significant at 0.05 level;
**represents significant at 0.01 level; similar comments apply to below tables.

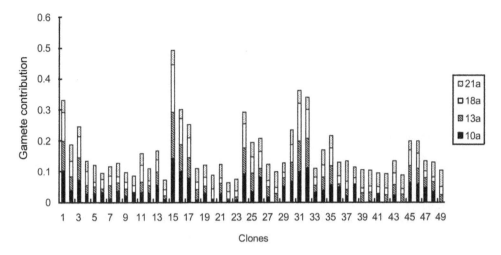

Figure 3. Gametic contribution of clones in *Pinus tabuliformis* seed orchard.

Results

Variation in clonal reproductive synchrony

Reprodductive synchrony index among clones as female parents ranged from 0.146 to 0.563 with average of 0.249. Across observation years, reproductive synchrony index as female parent was below 0.2; however, the 18th year produced an index of 0.563 and that was mainly caused by the damp cool weather that caused prolonged receptivity period (Fig. 1). Male reproductive synchrony index ranged from 0.145 to 0.297. Except for the abnormal data of the 21st year, all values were over 0.2 and it was higher than that of female parents (Fig. 2).

Correlation analysis shows that the flowering synchronization index of male parent in the 10th year has a significant positive correlation with that in the 13th and 21st years, significant negative correlation with that in the 18th year, and no significant correlation with the observations in other years. Conversely, the flowering synchronization index of female parents has a positive correlation between most of the observed years and some correlations reach to significance level. The flowering synchronization index of selfing in the 10th year has a significant correlation with that in the 13th year, and those in the other years have no significant correlation. The

flowering synchronism index of selfing has a positive correlation with that of as male parent, and reachs to significance level in the 10th and 13th years. The study showed that reproductive synchrony index of the clones has certain relative stability and thus could be used as an indication of the seed orchard status during maturity (Table 1).

Clonal gametic contribution variation

The average gamete contribution of each clone served as both male and female parents is about 0.04 across study years. At the initial stage of seed production, great variations occurred between the clones and the variation coefficient reached 55.20% among the annual average gamete contributions of all the clones. For the studied 49 clones, clone 15 had the largest gamete contribution with an annual average of about 7.8 times greater than that of clone 22 (the smallest contributer) (Fig. 3). But the gamete contribution is also very different when the clones served as single male and female parent (Fig. 4 and 5). Variation of the gamete contribution among the clones as female parent is much big than that of male parent across study years and the variation coefficient in the annual female gamete contribution is greater than 1, indicating that there is a substantial difference among clonal gametic contribution specifically at the initial seed production stage (Table 2).

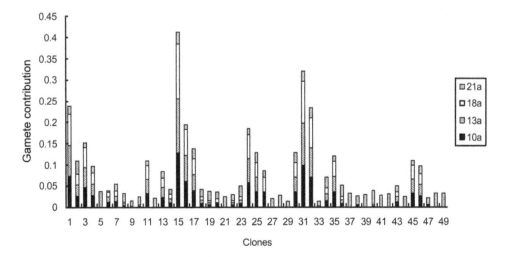

Figure 4. Clonal gametic contribution as female in *Pinus tabuliformis* seed orchard.

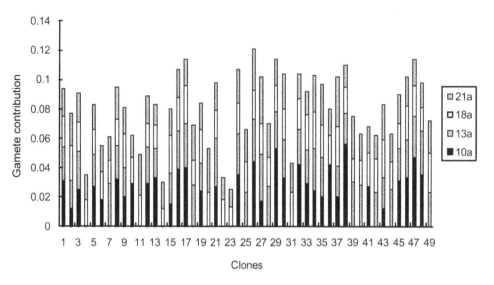

Figure 5. Clonal gametic contribution as male in *Pinus tabuliformis* **seed orchard.**

Table 2. Gametic contribution related values in *Pinus tabuliformis* seed orchard.

Clone	Breeding value	Flowering synchronization			Gamete contribution	Genetic contribution	Clone	Breeding value	Flowering synchronization			Gamete contribution	Genetic contribution
		Father	Mother	Both					Father	Mother	Both		
1	0.16	0.277	0.268	0.0414	0.0414	0.0132	26	0.13	0.306	0.332	0.0261	0.0261	0.0068
2	0.15	0.285	0.272	0.0233	0.0233	0.0070	27	0.12	0.260	0.300	0.0152	0.0152	0.0036
3	0.14	0.294	0.272	0.0306	0.0306	0.0086	28	0.14	0.246	0.163	0.0122	0.0122	0.0034
4	0.06	0.372	0.167	0.0166	0.0166	0.0020	29	0.08	0.335	0.252	0.0159	0.0158	0.0025
5	0.23	0.294	0.128	0.0151	0.0151	0.0069	30	0.10	0.328	0.259	0.0292	0.0292	0.0059
6	0.18	0.304	0.274	0.0119	0.0119	0.0043	31	0.15	0.380	0.250	0.0453	0.0453	0.0136
7	0.22	0.301	0.229	0.0144	0.0144	0.0064	32	0.10	0.327	0.251	0.0425	0.0425	0.0085
8	0.13	0.283	0.256	0.0160	0.0160	0.0042	33	0.14	0.312	0.243	0.0135	0.0135	0.0038
9	0.12	0.234	0.146	0.0121	0.0121	0.0029	34	0.17	0.294	0.208	0.0147	0.0147	0.0072
10	0.17	0.269	0.298	0.0108	0.0108	0.0037	35	0.09	0.280	0.238	0.0271	0.0271	0.0049
11	0.08	0.324	0.273	0.0197	0.0197	0.0032	36	0.21	0.343	0.254	0.0163	0.0163	0.0068
12	0.19	0.253	0.236	0.0137	0.0137	0.0052	37	0.14	0.274	0.252	0.0168	0.0168	0.0047
13	0.16	0.280	0.259	0.0208	0.0208	0.0067	38	0.13	0.336	0.232	0.0173	0.0173	0.0037
14	0.13	0.226	0.266	0.0089	0.0089	0.0023	39	0.18	0.281	0.281	0.0132	0.0132	0.0047
15	0.15	0.280	0.221	0.0616	0.0616	0.0185	40	0.11	0.294	0.337	0.0128	0.0128	0.0028
16	0.10	0.308	0.263	0.0376	0.0376	0.0075	41	0.12	0.290	0.228	0.0118	0.0118	0.0028
17	0.14	0.305	0.216	0.0313	0.0313	0.0088	42	0.16	0.336	0.217	0.0114	0.0114	0.0037
18	0.12	0.261	0.263	0.0135	0.0135	0.0033	43	0.19	0.259	0.256	0.0165	0.0165	0.0063
19	0.19	0.273	0.236	0.0151	0.0150	0.0057	44	0.17	0.327	0.315	0.0109	0.0109	0.0037
20	0.04	0.267	0.261	0.0107	0.0107	0.0009	45	0.09	0.300	0.224	0.0249	0.0249	0.0045
21	0.16	0.295	0.329	0.0154	0.0154	0.0049	46	0.07	0.276	0.262	0.0246	0.0246	0.0035
22	0.19	0.294	0.254	0.0079	0.0079	0.0030	47	0.12	0.351	0.252	0.0167	0.0167	0.0040
23	0.20	0.276	0.248	0.0092	0.0092	0.0037	48	0.23	0.311	0.261	0.0162	0.0162	0.0075
24	0.12	0.301	0.245	0.0363	0.0363	0.0087	49	0.20	0.346	0.263	0.0129	0.0129	0.0052
25	0.13	0.314	0.195	0.0241	0.0241	0.0063	mean	0.14	0.297	0.249	0.0200	0.0200	0.0055

Table 3. Clonal gametic contribution correlation in *Pinus tabulaeformis* seed orchard.

	M10	M13	M18	M21	F10	F13	F18	F21	MF10	MF13	MF18
M13	0.81**										
M18	0.51**	0.59**									
M21	−0.34	−0.24	−0.17								
F10	−0.09	−0.12	−0.19	−0.13							
F13	−0.21	−0.28	−0.11	−0.11	0.61**						
F18	−0.33	−0.36*	−0.52**	0.17	0.27	0.39*					
F21	0.31	0.21	0.17	−0.01	0.13	−0.33	−0.05				
MF10	−0.38*	−0.44*	−0.37*	−0.01	0.60**	0.36*	0.57**	−0.05			
MF13	−0.28	−0.34	−0.20	−0.18	0.69**	0.51**	0.54**	0.00	0.90**		
MF18	−0.30	−0.37*	−0.20	−0.20	0.65**	0.50**	0.50**	0.00	0.91**	0.97**	
MF21	0.33	0.33	0.36*	−0.07	0.02	−0.29	−0.16	0.83**	−0.16	−.06	−0.01

Except for the 21st year, gametic contribution of each clone as male, female, and as all parent in the 10th, 13th, and 18th years has significant or highly significant correlation with each other. Cloanal gametic contribution as male and female parents in the 10th, 13th and 21st years produced correlations higher than 0.9, again indicative of relative stability over the study period (Table 3).

Clonal variation in genetic contribution

There is a significant difference in the estimated genetic contribution of the clones as male, female, and as parents to the orchard's zygote pool (Fig. 6, 7 and 8). At the initial reproductive stage, some clones failed to produce female or male strobili, so they did not contribute genetically to the seed crops, whereas other clones had higher than expected genetic contribution (e.g., clones 1, 15, and 24 as female parents) (Fig. 7 and 8). In the 21st year, all clones contributed to the zygote pool, especially those with high breeding values, thus increasing the seed crop's estimated genetic gain and reflecting the magnitude of genetic gain fluctuation over years (Fig. 6).

In the study years, the average genetic contribution of all the clones was 0.005541, and the variation coefficient reached 57.48%. The average estimated genetic contribution of each clones at different year substantially varied with clones 15 and 20 showing the highest and lowest average genetic contribution, respectively, with as high as 21 fold difference between them (Fig. 6). It demonstrates clonal effect on gametic contribution and subsequently the expected genetic gain. This information could serve as the bases for the seed orchard genetic thinning through the removal of low contributors.

Factors influencing clonal genetic contribution

Parameters related to the gamete contribution of clones such as (flowering synchronization index, gamete contribution, quantity of female and male flowers and number of full seeds per cone) were estimated (Table 2). According to correlation analyses (Table 4), the genetic contribution of each clone has some degrees of correlation with its clonal breeding value. This indicates that these are the most important factors influencing the genetic contribution of each clone.

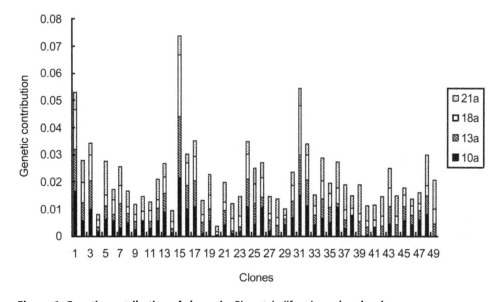

Figure 6. Genetic contribution of clones in *Pinus tabuliformis* seed orchard.

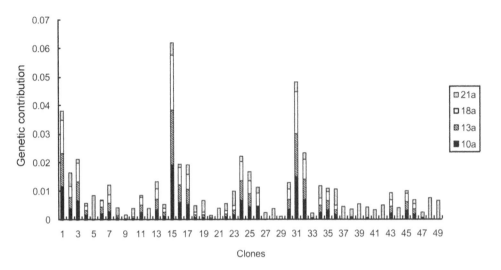

Figure 7. Clonal genetic contribution as female in *Pinus tabuliformis* **seed orchard.**

Discussion

By analysis of reproductive synchrony and its annual stability across clones in a seed orchard from initial seed harvest to stable seed production, clonal female and male contribution can be effectively estimated. Based on this, the gamete and genetic contribution of clones to the seed crop can be calculated. Finally, with generated information, genetic thinning, clonal management measures can be adopted to adjust the genetic composition of orchard pool and to improve the level of genetic gain of the resulting seed crops. Results of this study show that the reproductive synchrony among orchard's clones has some variation among the studied years. The reproductive synchrony index of pollen release has shown to be slightly higher than that of the female receptivity; however, both can be drastically influenced by the ambient climate factors resulting in some changes. For example, the cool and damp weather during the 18th year significantly improved the reproductive synchrony in the studied orchard due to the extended receptivity period. This indirectly indicates that the use of techniques such as those affecting ambient

conditions such as bloom delay can effectively alter reproductive synchrony. The reproductive synchrony index observed over short periods (i.e., few years) cannot reflect the relative difference among clones and thus longer observation periods conducted over several years are needed to effectively evaluate the differences in reproductive synchrony among orchard's clones.

Parental contribution to the seed orchard gametic pool actually means the contribution to the seed crop. Askew and Blush [17] and El-Kassaby and Askew [18] proposed a formula to calculate parental genetic contribution that focuses on the combination probability of opposite genders gametes from the different parents. In the present study, we used the relative gamete yield to estimate the gametic and genetic contribution of each clone as male, female and as parents. Clones in the study orchard have shown extreme differences in terms of the gametic and genetic contribution to the seed crop at the orchard's early production phase. Although it can reflect the relative importance of different clones as male, female and as parents to the genetic composition of the seed crop, the estimation of gametic and genetic contribution of the parents is based on ideal assumptions and actual estimation of genetic contribution of the male parent is

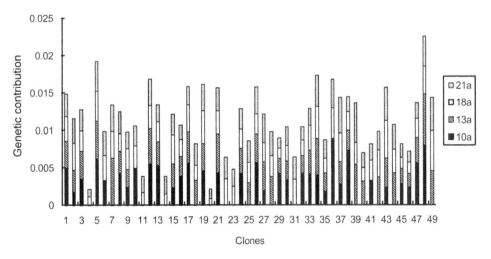

Figure 8. Clonal genetic contribution as male in *Pinus tabuliformis* **seed orchard.**

Table 4. Correlation between clonal gametic contribution and its related parameters in *Pinus tabuliformis* seed orchard.

Correlation efficience	Breedingvaule	Flowering synchronization	Gamete contribution	Number of female flowers	Number of male flowers	Full seeds per cones
Genetic Contribution	0.2666*	0.8589**	0.8589**	0.5840**	−0.3532*	0.6280**
Reliability	0.9560	0.9999	0.9999	0.9999	0.9671	0.9999

uncertain. It subjects to many factors which are influencing the orchard's mating system [21] [22] [23]. Hence, genetic contribution to the seed crop can be finally realized based on the contribution of the female parent. In actual, clonal gamete contribution as male and female often has a negative correlation [23], so the utilization of a clone as pollen, seed or pollen and seed donors should consider the role it should play in the seed crop.

When the clone serves as different parents (i.e., male and/or female), gametic contribution to the seed orchard has a close positive correlation between the observed years. This indicates the presence of certain stability in clonal gametic contribution across the orchard's productive life span. Because genetic contribution is the function of gametic contribution and breeding value, so the genetic contribution of clones is also relatively stable and it can be used to evaluate the genetic composition of seed crops. However, great variation appears in clonal gametic and genetic contribution specifically when they severe as either female or male parents across the study years and those variations are closely related to clonal gender tendency and the features during the orchard's initial reproduction [12]. So genetic composition analysis of seed crops produced from any seed orchard should take the whole clone as the analytic unit of gametic and genetic contribution.

Stability of clonal gametic contribution across years is affected by numerous factors. Generally, gametic contribution of a clone is related to reproductive synchrony between opposite gender, the quantity of female and male strobili production, and based on assumption of no significant difference in the number of valid gametes produced by the female and male strobili (i.e., relationship between reproductive energy and reproductive success) [24]. Moreover, clone arrangement, non-random mating among clones and climate factors may also influence the estimation of gametic contribution, which it is very difficult to reliably estimate the number of viable gametes in the female stobili. Thus, the current study accepts the assumption and replaces the difference between viable female gametes in the clone with the average number of filled seeds produced. Although the estimation result has some variation with that of the actual gametic contribution, it is more convenient operationnally, and basically meets the requirements for seed orchard management. This study has shown that as the number of female and male strobili as well as the average number of filled seeds per cone, have great variances among clones and this variation is associated with relatively high stability during the observed years, and similarly variance in the reproductive synchrony between opposite gender also has its stability. Therefore, gametic contribution of Chinese pine clones in seed orchard demonstrates relatively high stability at the orchard's early reproductive phase. The genetic contribution of clones is the

function of gametic contribution and breeding value, while the breeding value of the clone is relatively independent from its gametic contribution. For example, clones 5, 4, 23, 36, 48, and 50 possess a relative high breeding value (>0.20), but have lower than the average relative gametic contribution. Among clones with a relative higher gametic contribution, 9 clones have breeding values higher than the average, whereas 7 clones have lower than the average. This highlights the relative importance of different clones in the genetic gain of seed crops and the yield of seeds from seed orchard. Genetic contribution comprehensively demonstrates those two values of the clone.

In summary, due to influence of numerous factors on the mating system in seed orchards, genetic contribution of the Chinese pine clone as male parent is uncertain, genetic contribution of the male gamete from a clone has the greatest influence on the genetic composition of the seed orchard during the initial reproductive and seed production phase. Clones with a low breeding value and high contribution of male gametes may significantly reduce the genetic quality of seeds and this can be treated by routine genetic thinning or removal of male strobili carrying branches. Genetic gain from the genetic contribution of female parent to the seed crops has more realistic meanings. Improving yield of female strobili of a clone, which has low female gametic contribution and high breeding values, is an important and effective measure to enhance production and genetic quality of the seed crops. Improved Chinese pine seed orchard can be rebuilt through backward selection based on the general and special combining ability test, and the optimal genetic gain of seed crops from the seed orchard can be reached by optimization with value of general and special combining ability of the clones, reproductive synchrony of opposite genders, male and female gamete contribution to the zygote poll of the seed orchard. Consequently, genetic structure and clone combination can be designed in an improved seed orchard in which clones have the most effective genetic contribution to the seed lots.

Acknowledgments

We are grateful to Prof. Yousry A. El-Kassaby (University of British Columbia, Faculty of Forestry) for his valuable comments, suggestions and reviews during the preparation of this manuscript.

Author Contributions

Conceived and designed the experiments: YL XW. Performed the experiments: YL XW WL. Analyzed the data: YL WL. Contributed reagents/materials/analysis tools: WL. Wrote the paper: WL.

References

1. Kang KS, Lindgren D, Mullin TJ (2001) Prediction of genetic gain and gene diversity in seed orchard crops under alternative management strategies. Theoret Appl Genet 103(6–7): 1099–1107.

2. Gömöry D, Bruchánik R, Longauer R (2003) Fertility variation and flowering asynchrony in *Pinus sylvestris*: consequences for the genetic structure of progeny in seed orchards. For Ecol Manage 174(1–3): 117–126.

3. Koski V (1980) Minimum requirements for seed orchards of Scots pine in Finland. Silv Fenn 14(2): 136–149.

4. Gömöry D, Bruchanik R, Paule L (2000) Effective population number estimation of three Scots pine (*Pinus sylvestris* L.) seed orchards based on an integrated assessment of flowering, floral phenology, and seed orchard design. For Genet 7(1): 65–75.

5. Stoehr M, Webber J, Woods J (2004) Protocol for rating seed orchard seedlots in British Columbia: quantifying genetic gain and diversity. Forestry 77(4): 297–303.

6. Funda T, Lstiburak M, Lachout P, Klapste J, El-Kassaby YA (2009) Optimization of combined genetic gain and diversity for collection and deployment of seed orchard crops. Tree Genet Genom 5(4): 583–593.

7. El-Kassaby YA, Funda T, Lai BSK (2010) Female reproductive success variation in a *Pseudotsuga menziesii* seed orchard as revealed by pedigree reconstruction from a bulk seed collection. J Hered 101(2): 164–168.

8. Zhang DM, Zhang HX, Shen XH, Li Y (2004) Study on temporal and spatial change of the mating system in a seed orchard of *Pinus tabulaeformis* Carr. Sci Silv Sin 40(1): 70–77.

9. Wang SS, Shen XH, Yao LH, Li Y, Li JG, et al. (1985) A comparative study on growth patterns of *Pinus tabulaeformis* clones in Xingcheng seed orchard, Liaoning province. J Beijing For Coll 4: 72–83.

10. Zhang HX, Lu GM, Li ZH (1995) An index of phenology overlap in flowering for *Pinus tabulaeformis* Carr. Seed orchard located in Lushi Henan. Hebei J For Orch Res 10(3): 199–205.

11. Li Y, Wang XR, Li W, Shen XH, Xu JR (2010) Flowering synchronization stabilities of clones among plant ages in a seed orchard of *Pinus tabulaeformis*. J Beijing For Univ 32(5): 88–93.

12. Shen XH, Li Y, Wang XR (1985) Study on flowering habit of *Pinus tabulaeformis* Carr in the seed orchard located in Xingcheng County, Liaoning Province. Journal of Beijing Forestry University 3: 1–13.

13. Li Y, Shen XH (1994) Re-selection of clones and seed orchard roguing of *Pinus tabulaeformis* Carr. Asia–Pacific symposium on forest genetics improvement, IUFRO, Beijing.

14. Matziris DI (1994) Genetic variation in the phenology of flowering in Black pine. Silv Genet 43(5–6): 321–328.

15. Codesido V, Merlo E, Fernandez-Lopez J (2005) Variation in reproductive phenology in a *Pinus radiata* D. Don seed orchard in Northern Spain. Silv Genet 54(4–5): 246–256.

16. Codesido V, Merlo E (2001) Caracterización fenológica del huerto semillero de *Pinus radiata* de Sergude. III Congreso Forestal Español. Actas del Congreso. Tomo III: 69–74.

17. Askew GR, Blush D (1990) An index of phenology overlap in flowering for clonal conifer seed orchards. Silv Genet 39(3–4): 168–171.

18. El-Kassaby YA, Askew GR (1991) The relation between reproductive phenology and output in determining the gametic pool profile in a Douglas-fir seed orchard. For Sci 37(3): 827–835.

19. Askew GR (1988) Estimation of gamete pool compositions in clonal seed orchards. Silv Genet 37(5–6): 227–232.

20. Huang ZH, Chen XY (1993) Research on balance index of gamete contribution of parent in conifer seed orchard. J Beijing For Univ 15(4): 39–43.

21. Gömöry D, Paule L (1992) Inferences on mating system and genetic composition of a seed orchard crop in the European larch (*Larix decidua* Mill.). J Genet and Breed 46(4): 309–314.

22. Hansen OK (2008) Mating patterns, genetic composition and diversity levels in two seed orchards with few clones-Impact on planting crop. For Ecol Manage 256(5): 1167–1177.

23. Torimaru T, Wang XR, Fries A, Andersson B, Lindgren D (2009) Evaluation of pollen contamination in an advanced Scots pine seed orchard. Silv Genat 58(5–6): 262–269.

24. El-Kassaby YA, Cook C (1994) Female reproductive energy and reproductive success in a Douglas-fir seed orchard and its impact on genetic diversity. Silv Genet 43(4): 243–246.

3

Oligo-DNA Custom Macroarray for Monitoring Major Pathogenic and Non-Pathogenic Fungi and Bacteria in the Phyllosphere of Apple Trees

Ying-Hong He, Sayaka Isono, Makoto Shibuya, Masaharu Tsuji, Charith-Raj Adkar Purushothama, Kazuaki Tanaka, Teruo Sano*

Faculty of Agriculture and Life Science, Hirosaki University, Hirosaki, Japan

Abstract

Background: To monitor the richness in microbial inhabitants in the phyllosphere of apple trees cultivated under various cultural and environmental conditions, we developed an oligo-DNA macroarray for major pathogenic and non-pathogenic fungi and bacteria inhabiting the phyllosphere of apple trees.

Methods and Findings: First, we isolated culturable fungi and bacteria from apple orchards by an agar-plate culture method, and detected 32 fungal and 34 bacterial species. *Alternaria, Aureobasidium, Cladosporium, Rhodotorula, Cystofilobasidium,* and *Epicoccum* genera were predominant among the fungi, and *Bacillus, Pseudomonas, Sphingomonas, Methylobacterium,* and *Pantoea* genera were predominant among the bacteria. Based on the data, we selected 29 major non-pathogenic and 12 phytopathogenic fungi and bacteria as the targets of macroarray. Forty-one species-specific 40-base pair long oligo-DNA sequences were selected from the nucleotide sequences of rDNA-internal transcribed spacer region for fungi and 16S rDNA for bacteria. The oligo-DNAs were fixed on nylon membrane and hybridized with digoxigenin-labeled cRNA probes prepared for each species. All arrays except those for *Alternaria, Bacillus,* and their related species, were specifically hybridized. The array was sensitive enough to detect 10^3 CFU for *Aureobasidium pullulans* and *Bacillus cereus.* Nucleotide sequencing of 100 each of independent fungal rDNA-ITS and bacterial 16S-rDNA sequences from apple tree was in agreement with the macroarray data obtained using the same sample. Finally, we analyzed the richness in the microbial inhabitants in the samples collected from apple trees in four orchards. Major apple pathogens that cause scab, Alternaria blotch, and Marssonina blotch were detected along with several non-phytopathogenic fungal and bacterial inhabitants.

Conclusions: The macroarray technique presented here is a strong tool to monitor the major microbial species and the community structures in the phyllosphere of apple trees and identify key species antagonistic, supportive or co-operative to specific pathogens in the orchard managed under different environmental conditions.

Editor: A. Mark Ibekwe, U. S. Salinity Lab, United States of America

Funding: This work was supported in part by Grants-in-Aid for Scientific Research 1865801 and 22658012 from the Ministry of Education, Science, Sports, and Culture of Japan, and a grant from the Ministry of Agriculture, Forestry, and a Fisheries of Japan (Development of mitigation and adaptation techniques to global warming in the sectors of agriculture, forestry and fisheries). The funders had no role in study design, data collection and analysis, decision to publish, or preparation of the manuscript.

Competing Interests: The authors have declared that no competing interests exist.

* E-mail: sano@cc.hirosaki-u.ac.jp

Introduction

The microbial flora of plants, including organisms on the plant exterior as well as those in the interior, plays an important role in shaping the microbial ecosystems in the phyllosphere [1–3]. Both species richness and environmental complexity increase ecosystem functioning, suggesting that detailed knowledge of how individual species interact with complex natural environments is necessary to make reliable predictions about how alterations in the biodiversity affect the functioning of the ecosystem [4]. On the other hand, microbes, especially epiphytic fungi and bacteria surviving on crop plants, have been influenced by environmental changes such as those in temperature, rain, soil, nutrients, and even by agricultural practices, including spraying of chemical pesticides and fertilizer input.

The routine method of spraying chemical pesticides currently employed in modern agriculture has been long believed to have a more or less negative impact on the microbial diversity in the phyllosphere. Actually, on the basis of the findings of a pesticide program on the non-target epiphytic microbial population of apple leaves, it was reported that populations of bacteria, filamentous fungi, yeast, and actinomycetes varied annually and were reduced 10- to 1,000-fold in 1976 and up to 50-fold in 1977 on pesticide-treated leaves [5–6]. A similar study on the agrichemical impact on the growth and survival of non-target microorganisms in the phyllosphere of apple trees revealed that repeated agrichemical applications reduced the *in planta* microbial population 10- to 10,000-fold, suggesting that agrichemicals could affect the non-target, culturable surface microorganisms [7]. A seasonal comparative study of the effect of organic and integrated production systems on the culturable fungi of stored Golden Delicious apples was conducted in Switzerland; the findings revealed that organically produced apples had significantly higher frequencies of filamentous fungi, abundance of total fungi, and higher taxon diversity than the apples produced by integrated systems [8].

Changes in the epiphytic microflora, in turn, are believed to influence the incidence of crop diseases through competitive and/or antagonistic interactions with invading pathogens and pests. Evaluation of the microbial difference in species level in plants is anticipated to become more important because of the promotion of sustainable agriculture and integrated pest management. The traditional methods such as agar-plate culture method are time-consuming, and at times, require additional labor and skills for DNA extraction, polymerase chain reaction (PCR) amplification, and sequencing. Furthermore, potential uncertainty may persist in the differentiation and identification of colonies based on their macroscopic appearance. We also need to consider that not all the microbial inhabitants are culturable; moreover, the growth of some culturable organisms could be underestimated, because of their decreased activity in the medium and culture conditions being used.

To overcome these difficulties, we can utilize the findings from several studies that identified specific pathogens or multiple species by using DNA-based high-throughput techniques to establish risk assessment models or to monitor microbial diversity *in planta* under various environmental conditions; for example, identification and differentiation of the bacterial pathogens of potatoes [9], quantitative assessment of the growth of phytopathogenic fungi on various substrates [10], risk assessment of grapevine powdery mildew [11], diversity pattern of maize leaf epiphytic bacteria in relation to the plants that are genetically resistant to fungal pathogens [12], diagnostic DNA microarray for rapid identification of quarantine bacteria [13], characterization of complex communities of fungi and fungal-like protists [14], proteogenomic analysis of the physiology of phyllosphere bacteria [15], microarray screening of the variability of 16S–23S rRNA internal transcribed spacer region (ITS) in *Pseudomonas syringae* [16], and analysis of molecular battles between the plant and the pathogenic bacteria in the phyllosphere [17].

Here we first present the development of an oligo-DNA custom macroarray technique to detect and monitor the major pathogenic and non-pathogenic fungi and bacteria inhabiting the phyllosphere of apple trees, and then the application of the macroarray to analyze the richness of microbial inhabitants in orchards managed with different disease control measures such as intensive calendar spraying of chemical pesticides, reduced spraying of chemical pesticides, or natural farming practices.

Materials and Methods

Ethics statement

No specific permits were required for the described field studies. No specific permissions were required for these locations/activities, because of the owner's personal kind considerations on our researches. The field studies did not involve endangered or protected species.

Apple orchards used for the analysis

Culturable fungi and bacteria inhabiting the phyllosphere (leaf surface and inner tissue spaces) of *Fuji* apple (*Malus×domestica*) trees were isolated by using an agar-plate culture method 16 times from May 2006 to October 2008 for samples from four apple orchards in Hirosaki City, Aomori Prefecture, Japan. The trees in Orchard A-chemical (intensive spraying of chemical pesticides; owned by

Hirosaki University) were managed under normal cultivation conditions wherein chemical pesticides and fungicides were to be sprayed 11 times in 2009 growing season, according to calendar-based pest management (http://www.applenet.jp/). Chemicals sprayed include fenbuconazole, Score MZ (difenoconazole and mancozeb), ziram, thiram, iminoctadine triacetate, Aliette-C WP (captan), Flint Flowable25 (trifloxystrobin), cyprodinil, Antracol WG (propineb), calcium carbonate, and copper organic compounds as fungicide, and machine oil, organophosphorus compounds (e.g. chlorpyrifos), pyrethroid (e.g. cypermethrin), and neonicotinoid (e.g. clothianidin) as insecticide. In comparison, the trees in Orchard A-organic (Japan Agricultural Standard (JAS) organic; owned by Hirosaki University) in the neighboring field were grown under organic farming conditions for the past several years; *i.e.*, vinegar and acid water were sprayed periodically as alternatives to chemical pesticides. The trees in Orchard B-semi-chemical (reduced spraying of chemicals; owned by Makoto Takeya) were managed under special cultivation (*Tokubetsu Saibai* in Japanese) conditions in which the routine number of chemical pesticide sprays and amount of chemical fertilizer (nitrogen-based) were reduced by half. Chemicals sprayed include iminoctadine triacetate, Score MZ (difenoconazole and mancozeb), copper organic compounds, Stroby (kresoxim-methyl), and captan as fungicide, and machine oil, BT (*Bacillus thuringiensis*) spore, Confuser R (the mating disruptant), organophosphorus compounds (e.g. phenthoate), and neonicotinoid (e.g. acetamiprid) as insecticide. The trees in Orchard B-natural (natural farming; owned by Akinori Kimura) were managed under natural farming conditions without spraying any chemical fertilizers or pesticides, but were sprayed with specially formulated vinegar sprayings for the past 30 years. No specific permissions were required for these locations/activities, because of the owner's personal kind considerations on our researches.

Isolation of the culturable fungi and bacteria inhabiting the phyllosphere of apple trees

Apple leaves (*ca.* 0.5 g) were immersed in 35 mL of distilled water and shaken vigorously for 1 min. The liquid was collected and centrifuged at $4,000 \times g$ for 15 min at room temperature to precipitate the fungi and bacteria, which were then dissolved in 5 mL of distilled water. Aliquots (200-μL each) were spread on multiple 9-cm Petridishes containing *ca.* 30 mL of potato dextrose agar (PDA) and King's B medium and incubated at 20°C and 25°C for the isolation of fungal and bacterial colonies, respectively. The fungal and bacterial colonies thus obtained were further purified by single-colony isolation method. The apple leaves were washed, as described above, sterilized for 1 min by immersion in 70% alcohol, air dried, and homogenized in 1.5 mL of distilled water. Aliquots (200-μL each) were spread, as described above, for isolating the fungi and bacteria present inside the leaves.

Extraction, polymerase chain reaction, sequencing, and identification of fungi and bacteria

After several passages of single-colony isolation, the purified fungal and bacterial isolates were used for DNA isolation by using ISOPLANT II DNA extraction kit (Nippon Gene, Osaka, Japan), according to the manufacturer's instructions. An aliquot of DNA (100 ng in 2 μL) was used for PCR amplification of fungal rDNA-ITS with the primer set ITS1 (5′-TCCGTAGGT-GAACCTGCGG-3′) and ITS4 (5′- TCCTCCGCTTATTGA-TATGC-3′) [18], and of bacterial 16S rDNA region with the primer set Bac16S-27F (5′-AGAGTTTGATCCTGGCTCAG-3′) and Bac16S-1525R (5′-AAAGGAGGTGATCCAGCC-3′). The amplified DNA was sequenced at Macrogen (Seoul, Korea), and the genus or species was determined by performing BLAST analysis of the sequence at the DNA Data Bank of Japan; those showing the sequence homology higher than 98% were identified as species.

Direct nucleotide sequencing analysis of microbial rDNA population in the phyllosphere without the culture step

Apple leaves used for the analysis were the same to those described in details in the latter section on "Sampling of apple leaves and preparation of macroarray probe — in the phyllosphere of the apple orchards". The two methods were used for the direct extraction of microbial DNAs from apple leaves.

Method 1. Apple leaves (0.5 g) were homogenized in 3 volumes (1.5 mL) of distilled water; centrifuged 3 times in a microcentrifuge (MX-150; Tomy Seiko Co., Ltd., Tokyo, Japan), the first at $1,000 \times g$ for 2 min, the second at $1,500 \times g$ for 2 min, and the third at $2,000 \times g$ for 2 min; and left undisturbed for 30 min. The supernatant was then collected and centrifuged at $15,000 \times g$ for 20 min to precipitate microbes inhabiting the phyllosphere of the apple trees. The precipitate was used for DNA extraction by using ISOPLANT II DNA extraction kit, according to the manufacturer's instructions. The extracted DNA was dissolved in 50 μL of distilled water.

Method 2. The preparation by Method 1 includes a large amount of chloroplast and mitochondorial DNAs of host origin. These DNAs could disturb especially bacterial 16S rDNA amplification, because they are also prokaryotic origin. To eliminate the chloroplast and/or mitochondrial from the preparation, apple leaves (0.5 g) were homogenized and fractioned according to Ikeda et al. [19]. Briefly, leaves were homogenized in BCP buffer (5 ml for 1 g leaf), centrifuged at $500 \times g$ for 1 minute, collected supernatant, centrifuged at $5,000 \times g$ for 1 minute, and collected the precipitates. The precipitate was dissolved in 1 ml BCP buffer, vigorously shaken for a second, and re-precipitated by centrifugation at $5,000 \times g$ for 1 minute. BCP buffer (1 ml) treatment was repeated, and the final precipitate was used for DNA extraction by ISOPLANT II.

The rDNA-ITS region for fungi was amplified by PCR using the primer set ITS1 and ITS4, and a part [ca. 500 base pair (bp)] of the 16S-rDNA for bacteria was amplified by the primer set Bac-27F and Bac-519R [20]. The amplified DNA fragments were ligated in pT7blue–T vector (Novagen, Merck KGaA, Darmstadt, Germany) for transformation of *Escherichia coli* (DH5α strain), and the transformant colonies obtained were used for the preparation of recombinant plasmid DNA for sequencing as above.

Macroarray hybridization

Array sequences. On the basis of the nucleotide sequences obtained, 40-bp oligo-DNA sequences specific for each microorganism was selected from fungal rDNA-ITS and bacterial 16S-rDNA sequences as oligo-DNA arrays.

Macroarray preparation. The oligo-DNA arrays specially prepared in this experiment (FASMAC Co., Ltd., Kanagawa, Japan) were dissolved at a concentration of 1 μg/μL in a solution containing 50% (v/v) formamide (Wako), 35% (v/v) formaldehyde (Wako), and 1×saline-sodium citrate (SSC) buffer; denatured at 65°C for 15 min; diluted with 4 volumes of 20× SSC buffer; and then, aliquots (1-μL each) were spotted on a positively charged nylon membrane (Biodyne Plus, Pall Corporation, Mexico). Arrays were fixed on the membranes by ultraviolet (UV) cross-linking (120,000 μJ/cm²).

Preparation of digoxigenin (DIG)-labeled RNA probe for macroarray analysis. A combine of apple leaves (0.5 g) was treated as above in the Method 1 and 2, and a mixture DNA from various microbial inhabitants in the phyllosphere was prepared by using ISOPLANT II DNA extraction kit and dissolved in 50 μL of distilled water. Similarly, the total DNA of each fungal and bacterial isolate for the preliminary analysis was extracted from the cultured preparation.

The microbial DNAs were amplified by PCR in a 25-μL mixture containing 2 μL of total DNA extract, 2.5 μL of each dNTPs at 2.5 mM, 2.5 μL of 10× LA PCR buffer, 2.5 μL of 25 mM MgCl₂, 0.25 μL of LA-Taq DNA polymarase (Takara Bio, Shiga, Japan), and 1 μL of each of the PCR primers (each 20 μM) Bac16S-27F and Bac16S-1525R (5′-AGAG-*TAATAC-GACTCACTATAGGG*-AAAGGAGGTGATCCAGCC-3′) for amplification of bacterial 16S rDNAs, and ITS1 and ITS4-T7 (5′-AGAG-*TAATACGACTCACTATAGGG*-TCCTCCGCTTATTGA-TATGC-3′) for fungal rDNA-ITS region. Bac16S-1525R and ITS4-T7 primers contained the promoter sequence for T7 RNA polymerase at the 5′-end (italicized letters).

Cycle parameters for PCR amplification were heat-denaturation at 94°C for 4 min, followed by 35 cycles of amplification (94°C for 1 min; 55°C, 1 min; and 72°C, 1 min), and a final extension at 72°C for 7 min. The amplified cDNAs were extracted twice by equal volumes of phenol:chloroform (1:1), precipitated by ethanol, and dissolved in 50 μL of distilled water.

A DIG-labeled cRNA probe was prepared in a 5.5-μL transcription mixture containing 2.5 μL of the PCR product (*ca.* 100 ng/μL), 0.5 μL of RNA-labeling mixture (Roche Diagnostics Japan, Tokyo, Japan), 1 μL of 5× T7 buffer (Invitrogen, Life Technologies Japan, Tokyo, Japan), 0.25 μL of 0.1 M dithiothreitol (DTT), 0.125 μL of RNase inhibitor (Wako), and 0.25 μL of T7 RNA polymerase (Invitrogen) by incubating at 37°C for 2 h. The transcription reaction was stopped by adding 0.5 μL of 0.2 M ethylenediaminetetraacetic acid (EDTA) (pH 8.0), 0.625 μL of 4 M LiCl, and 18.75 μL of 99.5% ethanol, and was then stored overnight at −30°C. DIG-labeled cRNAs were collected by centrifugation at $13,000 \times g$ for 10 min, washed in 70% ethanol, air dried, and dissolved in 25 μL of distilled water containing 0.05 μL of RNase inhibitor.

Hybridization. Array membrane (10×10 cm) was placed in a glass hybridization bottle and prehybridized in 5-mL hybridization buffer containing 5× SSC buffer, 1% Denhardt solution, 1% sodium dodecyl sulphate (SDS), and 25 mg of yeast tRNA (Roche Diagnostics) at 58°C for 1.5 h. DIG-labeled cRNA probes were heat-denatured at 95°C for 10 min, and an aliquot (5-μL) was then added to the hybridization solution. Hybridization was carried out at 58°C for at least 18 h. Membranes were washed twice for 15 min each in 70 mL (5 M NaCl, 0.8 M NaH₂PO₄, 0.1 M EDTA) at room temperature, and again twice for 15 min each in 70 mL of 1.5×SSPE (0.5% SDS) buffer at 58°C for 15 min. Membranes were then placed in 5-mL blocking solution containing 1% blocking reagent (Roche Diagnostics) in 0.1 M maleic acid (0.15 M NaCl, pH 7.5), incubated at room temperature for 30 min, and further incubated for 30 min after adding 2 μL of anti-DIG-alkaline phosphatase (AP) (Fab fragment) (Roche Diagnostics). Membranes were washed twice for 15 min each in 70 mL of washing buffer (0.1 M maleic acid (0.15 M NaCl, pH 7.5)) containing 0.3% Tween 20 (v/v) and immersed for 5 min in a 50-mL solution containing 0.1 M Tris-HCl (0.1 M NaCl and 0.05 M MgCl₂, pH 9.5). Membranes were incubated with CSPD star (ready-to-use) (Roche Diagnostics) for 30 min–2 h in ChemiDoc XRS (Bio-Rad Laboratories Japan, Tokyo, Japan) to detect chemiluminescent signals.

Quantification. The signal intensity was quantified by Quantity One (Bio-Rad).

Sampling of apple leaves and preparation of macroarray probe for the analysis of the seasonal changes in richness in the major microbial inhabitants in the phyllosphere of the apple orchards

Apple leaf samples were collected from the four orchards (A-chemical, A-organic, B-semi-chemical, and B-natural), which were the same orchards sampled for agar-plate culture experiments. Apple leaves were collected from these orchards at 2-week intervals from May 29 to October 29, 2009, i.e., 3 leaves from 1 position, 3 positions in 1 tree, and 3 trees in 1 orchard; namely, a total of 27 leaves per orchard. The 27 leaves were crushed into small pieces in the liquid nitrogen, mixed well, and an aliquot (ca. 0.5-g) was used for the harvest of microbial inhabitants in the phyllosphere (Method 1 and 2), followed by extraction of microbial DNAs (ISOPLANT II). Mixtures of fungal rDNA-ITS regions and/or bacterial 16S-rDNAs of the isolates obtained from the phyllosphere of the apple trees were amplified simultaneously or separately by PCR with primer sets specific to the fungal and bacterial species, and DIG-labeled RNA probes were finally transcribed.

Results

Detection and identification of culturable fungi and bacteria in the phyllosphere of apple trees by agar-plate culture method

A total of more than 150 each of independent culturable fungal and bacterial isolates were examined for sequencing, and 112 fungal and 135 bacterial informative sequences were obtained. All the fungal and bacterial species which showed the most high sequence similarity to those isolated from the phyllosphere of apple trees are listed in Table 1 and 2. They are identified at the genus or species level on the basis of the rDNA-ITS nucleotide sequence (ca. 500 bp) for fungi and 16S-rDNA sequence (ca. 1400 bp) for bacteria. A total of 32 different species (or unique sequences) in 31 fungal genera and 34 species in 22 bacterial genera were identified. The genera *Alternaria*, *Aureobasidium*, *Cladosporium*, *Rhodotorula*, *Cystofilobasidium*, and *Eoicoccum* in fungi and *Bacillus*, *Pseudomonas*, *Sphingomonas*, *Methylobacterium*, and *Pantoea* in bacteria were predominant.

Selection and specificity of oligonucleotide array for the detection of microbial species in the phyllosphere of apple trees

On the basis of the results obtained by agar-plate culture method (see Table 1 and 2), we eliminated the species that were detected less than twice in the 16 trials, and selected 11 major non-pathogenic fungi and 18 non-pathogenic bacteria as the targets of macroarray analysis. In addition to these, 11 fungi pathogenic to apple trees and a fire blight pathogen, *Erwinia amylovora*, were added to the list of targets. Consequently, a total of 41 non-pathogenic and pathogenic fungi and bacteria were selected for macroarray analysis, and species-specific 40-bp oligo-DNA sequences were selected as array DNAs from the rDNA-ITS sequence of fungi and 16S-rDNA sequence of bacteria (Table 3 and 4). In case of the four *Bacillus* species selected, four each of 40-bp arrays were designed as described below (Figure 1; array No. 24–27).

Each array DNA was spotted (2 spots/array, except for *Bacillus* No. 24–27) onto a nylon membrane as shown in Figure 1a, *i.e.*, spot nos. 1–11 were arrays for non-pathogenic fungi; 12–22, for major fungal pathogens of apple trees; 23–40, for non-pathogenic bacteria; and 41, for *Erwinia amylovora*.

The specificity of the array was examined by hybridization with individual probes prepared separately from the 40 purified fungal and bacterial species. *E. amylovora*, however, could not be examined because the bacterium was unavailable in our laboratory. As summarized in Figure 2, most of the arrays with the exception of those listed below hybridized specifically with the corresponding species. Cross-hybridization was observed among those targeting *Botrytis elliptica* (or *byssoidea*), *B. cinerea*, *Monilinia fructicola* and *M. mali*, or *Alternaria alternate* and *A. mali*, or the four *Bacillus* spp.

In our preliminary examination, when we used Bacillus arrays No. 24a, 25a, 26a, and 27a, cross-hybridization was observed among the four *Bacillus* species (Fig. 2). Then we have designed supplementary combination of arrays to discriminate four Bacillus species. At present, it seemed difficult to distinguish these species by single DNA array, because the target 16S-rDNA sequences showed very high identity values. In order to distinguish the four *Bacillus* species, we have aligned all the four *Bacillus* 16S-rDNA sequences and selected three each of additional 40-nucleotide sequences unique to each species (Table 4; b, c, and d of array No. 24–27). The four sets of three oligo-DNAs, in addition to the original ones, were quantified, denatured, spotted, and hybridized with probes individually prepared from five *Bacillus* 16S-rDNAs. As the result, all the four arrays in each sets showed positive signals only in the homologous probe–array combinations (Fig. 1b). Consequently, the four *Bacillus* species can be distinguished using the four sets of four oligo-DNAs for each species.

Simultaneous detection of major pathogenic and non-pathogenic fungi inhabiting the apple phyllosphere

Simultaneous detection of major pathogenic and non-pathogenic fungi in the phyllosphere of the apple trees by the macroarray was examined with a probe prepared from apple leaves collected on August 27, 2009, from Orchard A-organic, where Alternaria blotch, scab, and Marssonina blotch were visibly epidemic. As a result, the macroarray allowed us to detect multiple signals ranging from strong to weak in the arrays not only for *Aureobasidium*, *Cladosporium*, *Cryptococcus*, and *Cystofilobasidium* genera of non-pathogenic fungi, but for *A. mali* (the pathogen causing Alternaria blotch), *Venturia inaequalis* (the pathogen causing apple scab), and *Diplocarpon mali* (the pathogen causing Marssonina blotch) of pathogenic fungi (Fig. 3). Notably, major pathogenic fungi such as *A. mali*, *V. inaequalis*, and *D. mali* were simultaneously detected, because these phytopathogenic fungi hardly be detected by agar-plate culture method, due to mainly by their inferior growth rate on the medium, even in the presence of severe disease symptoms on the leaves.

Consequently, the macroarray was able to simultaneously detect multiple species of major fungi, both pathogenic and non-pathogenic, inhabiting the apple phyllosphere. Various intensities of signals, ranging from strong to weak, supported that the data obtained are proportional to those inhabiting in the phyllosphere as we will examine in the next section.

Quantitative nature of the macroarray

We have conducted two experiments to examine whether the signal intensity obtained by the macroarray is proportional to the actual microbial population in the phyllosphere; i.e., dilution kinetics of macroarray probe and direct nucleotide sequencing of microbial rDNAs in the phyllosphere.

First, in order to examine the dilution kinetics and the detection limit of the macroarray probe developed in this study, we selected *A. pullulans* from fungi and *B. cereus* from bacteria. *A. pullulans* is a ubiquitous yeast-like fungus predominating in apple phyllosphere.

Table 1. List of fungi species detected from four apple orchards by agar-plate culturing method in 2006–2008 seasons.

genus	species	identity (%)	accession No. matched	frequency	accession No. deposited
Alternaria	alternata	523/523 (100)	JF835810	12	AB693900
Arthrinium	sacchari	527/528 (99)	HQ914941	6	AB693901
Aureobasidium	pullulans	554/554 (100)	HQ909089	16	AB693902
Biscogniauxia	latirim	417/448 (93)	EF026135	1	AB693903
Botryosphaeria	dothidea	539/539 (100)	HQ730969	1	AB693904
Botrytis	elliptica	486/486 (100)	FJ169671	2	AB693905
Cladosporium	tenuissimum	491/492 (99)	JN689952	16	AB693906
Coprinus	xanthothrix	631/634 (99)	FJ755223	1	AB693907
Curvularia	trifolii	552/562 (98)	AF455446	1	AB693908
Cryptococcus	victoriae	452/453 (99)	AF444645	4	AB693909
Cystofilobasidium	macerans	547/555 (98)	AF444317	5	AB693910
Epicoccum	nigrum	483/484 (99)	DQ981396	5	AB693911
Fusarium	chlamydosporum	495/495 (100)	FJ426391	3	AB693912
Gibberella	avenacea	509/510 (99)	FJ224099	1	AB693913
Hormonema	prunorum	535/536 (99)	AJ244248	1	AB693931
Leptosphaeria	sp.	476/480 (99)	FN394721	1	AB693914
Leptosphaerulina	australis	490/490 (100)	JN712494	2	AB693915
Microdiplodia	sp.	561/564 (99)	EF432267	1	AB693916
Monilinia	sp.	427/430 (99)	AY805571	1	AB693917
Mucor	racemosus	564/580 (97)	AJ271061	1	AB693918
Myrothecium	verrucaria	538/541 (99)	EF211127	1	AB693919
Nigrospora	sp.	455/462 (98)	AM262341	2	AB693920
Paraconiothyrium	variabile	377/403 (93)	HM150642	4	AB693921
Penicillium	mali	526/528 (99)	AF527056	4	AB693922
Phomopsis	sp.	520/524 (99)	AB302248	3	AB693923
Ramularia	pratensis	466/478 (97)	EU019284	1	AB693924
Rhodotorula	glutinis	492/504 (98)	AY188373	6	AB693925
Rhodotorula	laryngis	504/510 (98)	AF444617	6	AB693926
Sclerotinia	sclerotiorum	541/541 (100)	AF455526	1	AB693927
Stemphylium	solani	500/504 (99)	EF104156	1	AB693928
Xylaria	sp.	523/527 (99)	AB255244	1	AB693930

They are identified at the genus or species level on the basis of the rDNA-ITS nucleotide sequence (*ca.* 500 bp) for fungi. "Identity" was shown by the number of nucleotide matched per number of nucleotide compared. "Frequency" indicates the numbers of detection out of 16 trials.

B. cereus is also a ubiquitous bacterium predominating in apple phylloshere. These fungus and bacterium suspension were prepared independently at the concentration of 10^5 CFU/ml in distilled water and diluted serially from 10^1 to 10^4 by 10-fold dilution. According to the method described, we extracted DNA from 1 ml of each dilutions, prepared macroarray probes, and carried out macroarray hybridization. The macroarray analysis was repeated twice, and the relative ratios of microbes in the phyllosphere were estimated based on the average volume (intensities/mm^2) of the four replicates.

As the result, the positive signals were obtained by the probes prepared from 10^3–10^5 CFU in both of *A. pullulans* and *B. cereus*; indicating that the detection limit is 10^3 CFU (Fig. 1c). The signal intensity was proportional to the microbial quantity ranging from 10^3–10^5 CFU.

Next, by using the same field DNA preparation from the leaves corrected from the A-organic orchard in July 10th, 2010, we have conducted a comparative analysis of macroarray and direct nucleotide sequencing analyses of microbial rDNA and rDNA-ITS populations in the phyllosphere without culturing. To minimize the sampling bias as possible, the apple leaves collected from the above orchard (3 leaves from 1 position, 3 positions in 1 tree, and 3 trees in 1 orchard; namely, a total of 27 leaves per orchard), were crushed into small pieces in the liquid nitrogen, mixed well, and an aliquot (ca. 0.5-g) was used for the direct DNA extraction as described. The direct DNA extraction was repeated three times, PCR amplification was repeated three times for each DNA extracts, and finally nine PCR amplicons were combined for cloning and sequencing. After cloning the PCR amplicons, we have picked up 100 for fungal and 65 for bacterial independent clones, sequenced, and obtained 89 and 52 informative nucleotide sequences, respectively. These sequences were analyzed by BLAST and identified the species by the sequence similarity higher than 98%. As the result, fungi was consisted of 44 *A.*

Table 2. List of bacteria species detected from four apple orchards by agar-plate culturing method in 2006–2008 seasons.

genus	species	identity (%)	accession No. matched	frequency	accession No. deposited
Achromobacter	xylosoxidans	1394/1413 (99)	AF511516	2	AB695331
Acinetobacter	sp.	1400/1423 (98)	JN887918	1	AB697151
Agrobacterium	tumefaciens	1373/1378 (99)	AB681363	1	AB695332
Arthrobacter	sp.	1407/1418 (99)	DQ519082	1	AB695333
	cereus or thuringiensis	1437/1438 (99)	JN315893	16	AB697152
Bacillus	megaterium	1414/1419 (99)	HQ202555	16	AB697153
Bacillus	pseudomycoides	1442/1457 (98)	AB681414	5	AB695334
Bacillus	pumilus	1443/1446 (99)	GU125624	8	AB695335
Bacillus	subtilis	1442/1450 (99)	HQ711983	7	AB697154
Bradyrhizobium	elkanii	1360/1364 (99)	AB672634	3	AB695336
Burkholderia	fungorum	1422/1437 (98)	FJ708122	3	AB695337
Curtobacterium	flaccumfaciens	1399/1419 (98)	AM410688	1	AB695338
Dermacoccus	sp.	1378/1386 (99)	JF905611	1	AB697155
Gluconobacter	oxydans	1364/1373 (99)	AB178421	1	AB695339
Methylobacterium	suomiense	1359/1378 (98)	AB175645	1	AB697157
Methylobacterium	radiotolerans	1342/1364 (98)	AY616142	4	AB697158
Microbacterium	foliorum	1414/1422 (99)	EU714371	2	AB695340
Micrococcus	luteus	1395/1403 (99)	HM755622	2	AB697159
Paenibacillus	amylolyticus	1377/1394 (98)	DQ313379	2	AB695341
Paenibacillus	pasadenensis	1395/1427 (97)	AB681404	2	AB697160
Pantoea	agglomerans	1348/1376 (97)	FJ357813	7	AB695342
Pseudomonas	graminis	1418/1426 (99)	Y11150	8	AB695343
Pseudomonas	fluorescens	1386/1392 (99)	JN679853	4	AB695344
Pseudomonas	oryzihabitans	1380/1401 (98)	AB681726	1	AB697161
Pseudomonas	putida	1397/1406 (99)	EU275363	7	AB697162
Pseudomonas	syringae	1420/1446 (98)	AY574914	5	AB697163
Pseudomonas	reactans	1240/1273 (97)	JN411452	5	AB695345
Raoultella	ornithinolytica	1360/1386 (98)	FJ823046	1	AB697156
Rothia	dentocariosa	1321/1330 (99)	CP002280	1	AB697164
Rhodococcus	corynebacterioides	1306/1313 (99)	AY167850	5	AB695346
Sphingomonas	echinoides	1335/1361 (98)	AB680957	5	AB695347
Sphingomonas	yunnanensis	1315/1330 (98)	EU730917	5	AB697165
Stenotrophomonas	maltophilia	1418/1432 (99)	AJ131117	1	AB697167

They are identified at the genus or species level on the basis of the 16S-rDNA sequence (ca. 1400 bp) for bacteria. "Identity" was shown by the number of nucleotide matched per number of nucleotide compared. "Frequency" indicates the numbers of detection out of 16 trials.

pullulans (50%; HQ909089), 31 Cladosporium tenuissimum (35%; JN689952, FQ832794), 8 Cryptococcus victoriae (9%; AF444645), 3 Venturia inaequalis (3%; EU035437), 1 Cryptococcus aff. amylolyticus (1%; EF363151), and 2 unknown species (2%; no significant similarity) (Fig. 4a). Bacteria was consisted of 26 Sphingomonas sp. (S. yunnaensis, S. echinoides, and sp.: 50%; AY336550, AY336556, AM989061, AB649018, AF395038, EU730917), 6 Methylobacterium radiotolerans and sp. (11%; AM989028, AF324201), 3 Pseudomonas syringae (5%; CP000075), 2 each of Actinobacterium sp. (4% AY275506, GU586309), Aggregatibacter aphrophilus (4%; EF605278), Streptococcus sp. (4%; AY518677, EU189961), Neisseria elongate (4%; L06171), Lautropia mirabilis (4%; GU397890), and etc. (AB594202, CP001277) (Fig. 4b).

Since the major species by the direct sequencing of apple leaf extract was completely matched to the data by the 3-year of culture method, it is unlikely that the apple leaves harbor unknown major unculturable species.

In the mean time, a macroarray probe was prepared from the same DNA preparation, and used for the macroarray analysis. The signal intensities of two each of dots per array were quantified by Quantity one. The macroarray analysis was repeated twice, and the relative ratios of microbes in the phyllosphere were estimated based on the average volume (intensities/mm^2) of the four replicates. The result identified several non-pathogenic and pathogenic microbial inhabitants in the phyllosphere; i.e., the fungi of A. pullulans (with the relative ratio in the population of 36.5%), C. tenuissimum (41.7%), V. inaequalis (15.0%), and Cystofilobasidium macerans (6.7%) (Fig. 4c), and bacteria of Sphingomonas yunnaensis (38%), P. syringae (18%), Methylobacterium radiotolerans (14%), Sphingomonas echinoides (13%), P. fluorescens (10%), P. graminis (7%) (Fig. 4d).

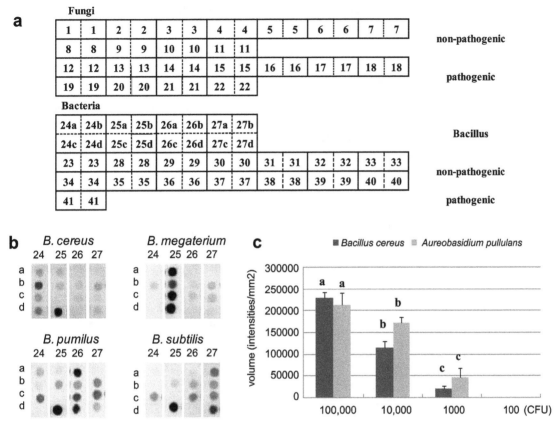

Figure 1. The arrangement, specificity, and quantitative nature of macroarray. (a) The arrangement of macroarray membrane. The numbers are corresponding to those in Table 3 and 4. Each array spots are duplicated except for those targeting four Bacillus species. (b) Four sets of four oligo-DNA arrays to discriminate four Bacillus species in the apple phyllosphere. (c) Quantitative analysis of the major fungus *A. pullulans* and bacterium *B. cereus* by macroarray. Error bars represent the standard deviation (±SD). Mean bars followed by different letters indicate significant differences by Tukey's test (P<0.05). Horizontal axis indicates the amounts (CFU) of *A. pullulans* and *B. cereus*. Vertical axis indicates volume measured by Quantity One.

These macroarray data were in agreement with those obtained by nucleotide sequencing, both in fungi and in bacteria, either in the predominant genera and their relative ratios in the population.

Macroaray analysis of the seasonal changes in the major microbial inhabitants in the phyllosphere of the apple trees in the orchards under the different pest managements

The seasonal changes in the major fungi and bacteria inhabiting the phyllosphere of the apple trees in relation to diseases were examined by using leaf samples collected from the four apple orchards (A-chemical, A-organic, B-semi-chemical, and B-natural) from May 8 to October 29, 2009. All macroarray analyses were performed twice. It should be noted here that, unexpectedly but fortunately, the array "25d" targeting *B. megaterium* reacted stably and strongly to host chloroplast rDNA due to the high sequence homology, so that we used it as internal standard to normalize the signal intensity among the membranes.

On the disease incidence in the four apple orchards, a conspicuous disease epidemic was not observed throughout the growing season in Orchards A-chemical or B-semi-chemical that were managed by normal cultivation with intensive spraying of chemical pesticides and chemical spraying reduced to less than half of the routine, respectively. Only scattered primary spots caused by secondary infections of Marssonina blotch (*D. mali*) were

observed in Orchard B-semi-chemical in mid-October, albeit with minor damage. In Orchard A-organic managed by JAS organic, in contrast, the scab increased from June, and considerable amounts of Alternaria and Marssonina blotches were prevalent in August. In Orchard B-natural managed by natural farming, Monilinia blossom blight was the first epidemic in early May in the blooming period, and the scab started to develop from the end of May. Furthermore, Alternaria and Marssonina blotches started to develop from the end of June. Considerable amount of symptoms of apple scab, Alternaria blotch, and Marssonina blotch persisted at the end of the harvest season (Fig. 5a).

In addition, we analysed seasonal quantitative changes in the population of pathogenic and non-pathogenic fungi and bacteria inhabiting the phyllosphere of apple trees in the four orchards in 2009. First, the quantitative data obtained by fungal macroarray was summarized in Fig. 5a. In Orchard A-chemical, the amount of fungi was maintained at a low level from April to early June. Among the non-pathogenic fungi in the phyllosphere, *A. pullulans* was the most predominant species from September to October, and *C. tenuissimum* and *Cry. victoriae* were also detected in relatively high densities. A considerable amount of scab fungus *V. inaequalis* was first detected in late July; the fungus population decreased to a low level in early-to mid-August and increased again in late August, and was detected in October also.

In Orchard A-organic, at the same location as that in A-chemical, *A. pullulans* was again the most predominant species in

Table 3. List of target fungi species for macroarray and the nucleotide sequences of oligo-DNA arrays.

	array No.	species	oligonucleotide array sequence
non-pathogenic			
	1	Alternaria alternata	ACCCTTGTCTTTTGCGTACTTCTTGTTTCCTTGGTGGGTT
	2	Arthrinium sacchari	AAGCTCGGTTGGAGGCACCTGCAGCTACCCTGTAGTTGCG
	3	Aureobasidium pullulans	AGAATTTATTCGAACGTCTGTCAAAGGAGAGGAACTCTGC
	4	Botrytis byssoidea	GAGTCTATGTCAGTAATGGCAGGCTCTAAAATCAGTGGCG
	5	Cladosporium tenuissimum	TCTAACCACCGGGATGTTCATAACCCTTTGTTGTCCGACT
	6	Cystofilobasidium macerans	CTCTCACCTCCAGCCTTCTTTAATTAGAGGTGTTGGGGCG
	7	Epicoccum nigrum	ATTACCTAGAGTTTGTGGACTTCGGTCTGCTACCTCTTAC
	8	Fusarium equiseti	TTTTTAGTGGAACTTCTGAGTAAAACAAACAAATAAATCA
	9	Mucor racemosus	GGATGACTGAGAGTCTCTTGATCGTCAGATCTCGAACCTC
	10	Rhodotorula laryngis	CACACATTTTAACACTATAGTATAAGAATGTAACAGTCTC
	11	Cryptococcus victoriae	TGAAACCTCACCCCACTTGGGTTTTTGCCTGAGCGGTGGT
pathogenic			
	12	Alternaria mali	AGCGCAGCACAAGTCGCACTCTCTATCAGCAAAGGTCTAG
	13	Botrytis cinerea	GTATTGAGTCTATGTCAGTAATGGCAGGCTCTAAAATCAG
	14	Colletotrichum acutatum	TTTACACGACGTCTCTTCTGAGTGGCACAAGCAAATAATT
	15	Helicobasidium mompa	TAGTCTAAGAATGTAAAGGACCCTTATAATTAATATAAAA
	16	Monilinia fructicola	CTATGTCAGTAATGGCAGGCTCTAAAATCAGTGGCGGCGC
	17	Monilinia mali	GTATTGAGCCCATGTCAGCGATGGCAGGCTCCAAAGTCAG
	18	Penicillium expansum	CCCGAACTCTGCCTGAAGATTGTCGTCTGAGTGAAAATAT
	19	Schizophyllum commune	CGGGCGGCGGTTGACTACGTCTACCTCACACCTTAAAGTA
	20	Valsa ceratosperma	CGCTGGCTGCCCCTCCCGCTCCGGGAGGGGGCCCGCCTCT
	21	Venturia inaequalis	ATTCGGCGCCTGGCGGGGACCACCCCCCGTTCGCGGGGGG
	22	Diplocarpon mali	CCTCGGGGCCGGCGGCTCCGGCTGCTGCGCCCTCGCCAGA

the phyllosphere from late July to early October, and *C. tenuissimum* and *Cry. victoriae* were detected at levels similar to that of *A. pullulans*. A large amount of pathogenic fungi, such as *V. inaequalis*, *A. mali*, and *D. mali*, were detected from late May, late June, and early August, respectively, and continued to be detected until the end of October (the harvest season). It should be noted that the total amount of fungi in A-organic was 3 times that in A-chemical, which resulted from the high number of non-pathogenic fungi, such as *C. tenuissimum* and *Cry. victoriae*, in addition to the pathogenic *V. inaequalis*, *A. mali*, and *D. mali*, inhabiting the phyllosphere of trees in Orchard A-organic.

In Orchard B-semi-chemical, *A. pullulans*, *C. tenuissimum*, and *Cry. victoriae* predominated from late July to the end of the harvest season at levels almost equivalent to those observed in Orchard A-organic. Low levels of *A. mali* and *V. inaequalis* were detectable in early August and early October, respectively, indicating a potential merit of the reduced spraying of chemical fungicides in controlling the major fungal pathogens of apple trees, such as *V. inaequalis*, *A. mali*, and *D. mali*, without a serious negative impact on the major non-pathogenic fungi inhabiting the phyllosphere.

In Orchard B-natural, *V. inaequalis* was the most predominant species in the phyllosphere throughout the growing season, especially from late May to late October. *M. mali* was detectable in May, which is consistent with the observation that Monilinia blight was epidemic in May in the orchard. *A. mali* and *D. mali* were also detected from early August and late September to the end of growing season, respectively. It was noted that in this orchard, the numbers of both non-pathogenic and pathogenic

fungi, with the exception of *V. inaequalis*, were suppressed to levels lower than those in the other orchards throughout the growing season. Although no chemical fungicide was sprayed, the numbers of *A. mali* and *D. mali* in Orchard B-natural were suppressed to levels lower than those in Orchard A-organic, indicating that disease control was more successful in B-natural. Meanwhile, 12 different species of pathogenic and non-pathogenic fungi were detected in this orchard, suggesting that the fungal diversity in the phyllosphere of the trees in Orchard B-natural was richer than that of the other orchards. In contrast, it was noted that the number of fungi inhabiting B-natural decreased to extremely low levels in mid-to late-August (Fig. 5a, red arrow).

Next, the quantitative data obtained by bacterial macroarray was summarized in Fig. 5b. Of all bacterial species, *Bacillus cereus* and *S. yunnaensis* predominated in all the orchards. *Pseudomonas* sp. were also detected in several samples. The variation in bacterial species was maximum in A-organic, *i.e.*, *S. yunnaensis* from late July to late October, *P. fluorescens* from late July to the end of the growing season, and *P. syringae* from early August to the end of the growing season. *P. putida* and *B. subtilis* were also temporarily detected. In Orchard B-semi-chemical, like in Orchard A-organic, *Bacillus cereus*, *S. yunnaensis*, and a trace of *S. echinoids* were detected. In Orchard B-natural, like in Orchard B-semi-chemical, in addition to *Bacillus* and *Sphingomonas*, *Pantoea aggromerans* and *P. graminis* were also detected, meaning that bacterial diversity was a bit richer in Orchard B-natural than in Orchard B-semi-chemical. The bacterial biomass, except for *S. yunnaensis*, was apparently

Table 4. List of target Bacteria species for macroarray and the nucleotide sequences of oligo-DNA arrays.

	array No.	species	oligonucleotide array sequence
non-pathogenic			
	23	*Acinetobacter johnsonii*	GTCGAGCGGGGAAGGGTAGCTTGCTACCTGACCTAGCGGC
	24a	*Bacillus cereus* or *B. thuringiensis*	TGGACCCGCGTCGCATTAGCTAGTTGGTGAGGTAACGGCT
	24b		GACTTTCTGGTCTGTAACTGACACTGAGGCGCGAAAGCGT
	24c		TAACTCCGGGAAACCGGGGCTAATACCGGATAACATTTTG
	24d		GGGGCTAATACCGGATAACATTTTGAACTGCATGGTTCGA
	25a	*B. megaterium*	GGCTTTTTGGTCTGTAACTGACGCTGAGGCGCGAAAGCGT
	25b		TGGGCCCGCGGTGCATTAGCTAGTTGGTGAGGTAACGGCT
	25c		TAACTTCGGGAAACCGAAGCTAATACCGGATAGGATCTTC
	25d		AGGATGAACGCTGGCGGCGTGCCTAATACATGCAAGTCGA
	26a	*B. pumilus*	TAACTCCGGGAAACCGGAGCTAATACCGGATAGTTCCTTG
	26b		TGGACCCGCGGCGCATTAACTAGTTGGTGAGGTAACGGCT
	26c		GACTCTCTGGTCTGTAACTGACGCTGAGGAGCGAAAGCGT
	26d		GGAGCTAATACCGGATAGTTCCTTGAACCGCATGGTTCAA
	27a	*B. subtilis*	GGGGCTAATACCGGATGGTTGTTTGAACCGCATGGTTCAA
	27b		TGGACCCGCGGCGCATTAGCTAGTTGGTGAGGTAACGGCT
	27c		GACTCTCTGGTCTGTAACTGACGCTGAGGAGCGAAAGCGT
	27d		TAACTCCGGGAAACCGGGGCTAATACCGGATGGTTGTTTG
	28	*Methylobacterium radiotolerans*	ACGCCCTTTTGGGGAAAGGTTTACTGCCGGAAGATCGGCC
	29	*Micrococcus luteus*	AACGATGAAGCCCAGCTTGCTGGGTGGATTAGTGGCGAAC
	30	*Paenibacillus amylolyticus*	AAGGAAACTGGAAAGACGGAGCAATCTGTCACTTGGGGAT
	31	*Paenibacillus polymyxa*	CCTGGTAGAGTAACTGCTCTTGAAGTGACGGTACCTGAGA
	32	*Pantoea agglomerans*	GGAAGGCGATGGGGTTAATAACCCTGTCGATTGACGTTAC
	33	*Pseudomonas graminis*	AGGAAGGGCAGTAAGCGAATACCTTGCTGTTTTGACGTTA
	34	*P. fluorescens*	GTTGGGGAGGAAGGGCATTAACCTAATACGTTAGTGTTTTG
	35	*P. putida*	GGGCATTAACCTAATACGTTAGTGTTTTGACGTTACCGAC
	36	*P. syringae*	AGCGGCAGCACGGGTACTTGTACCTGGTGGCGAGCGGCGG
	37	*Rhodococcus corynebacterioides*	GAAAACCAGCAGCTCAACTGTTGGCTTGCAGGCGGATACGG
	38	*Sphingomonas echinoides*	CTCAGGTTCGGAATAACAGCGAGAAATTGCTGCTAATACC
	39	*S. yunnanensis*	TCCAAAGATTTATCGCCAGAGGATGAGCCCGCGTGAGATT
	40	*Staphylococcus epidermidis*	AATATATTGAACCGCATGGTTCAATAGTGAAAGACGGTTT
pathogenic			
	41	*Erwinia amylovora*	GGGGAGGAAGGGTGAGAGGTTAATAACCTCCTGCATTGAC

lower in this orchard than in the other orchards from June, and especially decreased in early August (Fig. 5b, red arrow).

In conclusion, *Bacillus*, *Pseudomonas*, and *Sphingomonas* genera predominated in all the orchards. The numbers of species detected in chemical fungicide-sprayed sites were apparently less than those detected in the organic sites.

Discussion

In a 3-year study (2006–2008) using agar-plate culture method, we detected 32 fungal and 34 bacterial species inhabiting the phyllosphere of apple trees in northern Japan. Because we used PDA and King's B agar for isolating fungi and bacteria, respectively, most of the isolates were non-pathogenic saprophytes with higher growth rates in these media. In contrast, major fungal pathogens of apple trees, including the agents causing scab, Alternaria blotch, and Marssonina blotch, were not detected in these experiments, despite the presence of severe symptoms. *Aureobasidium*, *Cladosporium*, *Alternaria*, *Rhodotorula*, and *Cystofilobasidium* genera were the predominant fungal species that showed extensive growth. This is consistent with the findings of previous studies conducted in New Zealand and Switzerland that showed that *A. arborescens*, *A. pullulans*, *C. tenuissimum*, and *A. mali* were frequently isolated from apple leaves [8,21]. *Bacillus*, *Pseudomonas*, and *Sphingomonas* genera were the predominant bacterial species that showed extensive growth.

On the basis of our results, we selected 41 species of major pathogenic and non-pathogenic fungi and bacteria inhabiting the phyllosphere of apple trees in northern Japan. In the preliminary steps, we examined nearly full-length rDNA-ITS regions for cDNA arrays, but they lacked specificity (data not shown). We also examined 30-bp oligo-DNAs for arrays, but their sensitivities were not high enough (data not shown). Finally, we adapted 40-bp oligo-DNAs specific for each fungal rDNA-ITS region or bacterial

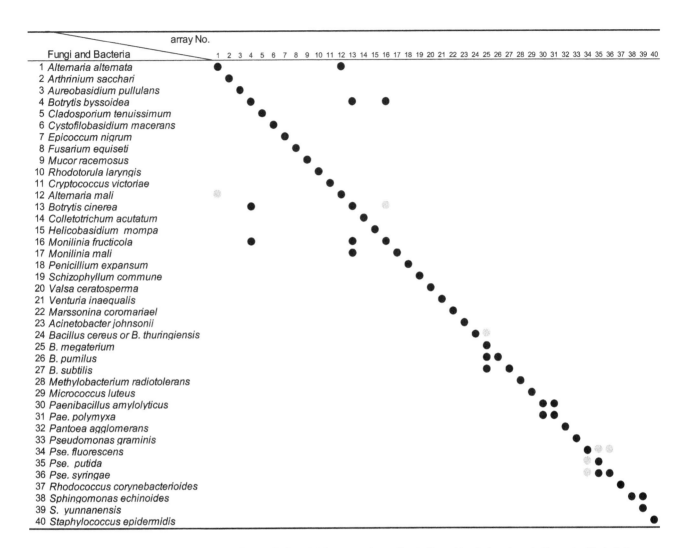

Figure 2. Schematic representation of specificity of oligo-DNA arrays. Arrays No. 1–40 are identical to those in Figure 1a. Black circles mean strong signals and gray ones mean weak non-specific cross-hybridization signals.

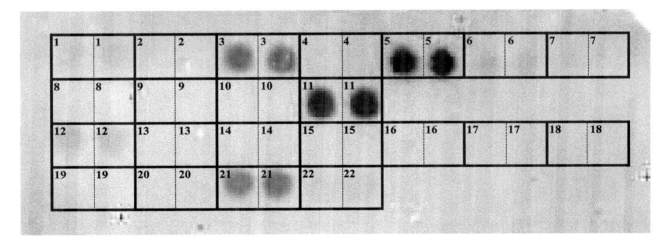

Figure 3. An image of macroarray hybridization for simultaneous detection of major pathogenic and non-pathogenic fungi in the phyllosphere of the apple trees. Arrangement of the arrays was the same to those in Figure 1a (Fungi). The arrays No. 3 (*A. pullulans*), 5 (*Cla. tenuissimum*), 11 (*Cry. victoriae*), and 21 (*V. inaequalis*) showed strong positive, and 6 (*Cys. macerans*) and 12 (*A. mali*) showed weak positive.

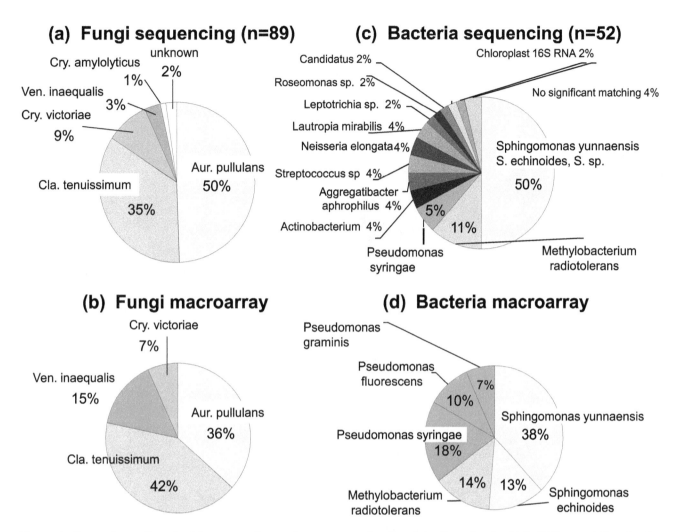

Figure 4. Comparison of nucleotide sequencing and macroarray for the detection of microbial rDNA population in the phyllosphere. Note that both of the fungal species and the ratio obtained by nucleotide sequencing (a) almost completely matched to the data obtained by macroarray (b). Although the minor bacteria species could not be detect by macroarray (d), but the major ones such as *Sphingomonas*, *Methylobacterium*, and *Pseudomonas* were consistent with both methods (c and d).

16S-rDNA and established an oligo-DNA macroarray for analyzing/monitoring richness of the major microbial species in the phyllosphere of apple trees. Most of the arrays specifically identified the target species. However, in some cases, *Alternaria* and related fungal species or *Bacillus* spp. could not be clearly distinguished because of cross-hybridization.

Sholberg et al. [22] used 19- to 25-bp-long oligo-DNA from the ribosomal spacer regions of bacterial and fungal pathogens to identify and monitor economically important apple diseases. The DNA array correctly identified *B. cinerea*, *Penicillium expansum*, *Podosphaera leucotricha*, *V. inaequalis*, and *E. amylovora*, and eliminated closely related species. When the array was used to monitor *V. inaequalis* ascospores collected from spore traps located in orchards, it confirmed the presence of ascospores as predicted by the disease-forecasting model, suggesting that the DNA array can be a useful tool for epidemiological studies. By using the macroarray developed, we have successfully detected pathogens such as *A. mali*, *Valsa ceratosperma*, and *V. inaequalis* even from the orchards where the disease symptoms were virtually invisible, indicating that the macroarray is actually useful for monitoring economically important apple diseases.

Zhang et al. [23] developed macroarray for the detection of solanaceous plant pathogens in the *Fusarium solani* species complex. Thirty-three 17- to 27-bp-long oligonucleotides were designed from the rDNA-ITS sequences of 17 isolates, which belonged to 12 phylogenetically related species. The array was validated by testing inoculated greenhouse samples and diseased field plant samples. Furthermore, Zhang et al. [24] designed 105 17- to 27-bp-long oligonucleotides specific for 25 pathogens of solanaceous crops on the basis of the rRNA-ITS gene sequence. They adapted at least 2 specific oligonucleotides per pathogen to distinguish between closely related species. Although both of the research purpose and target species were totally different, we specifically detected most of 41 major pathogenic and non-pathogenic fungi and bacteria using single array per species by taking the data obtained by preliminary surveillance in consideration. The strategy we employed here is definitely useful for monitoring the richness in the major microbial diversity in a specific host or restricted ecological environment.

The macroarray was adapted to analyze seasonal changes in major epiphytic microbial populations in the phyllosphere of apple trees in the growing season of 2009 in the four representative orchards in northern Japan. The findings were consistent with the

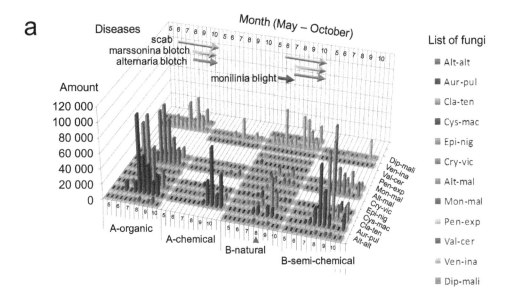

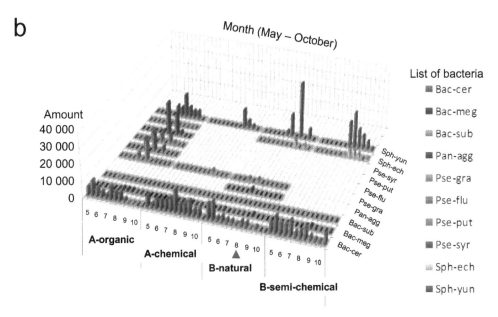

Figure 5. Seasonal changes in pathogenic and non-pathogenic fungi and bacteria inhabiting the apple phyllosphere in the four orchards. Histograms of seasonal changes of pathogenic and non-pathogenic fungi (a) and bacteria (b) detected from four orchards (A-chemical, A-organic, B-semi-chemical, B-natural) in 2009 May–October, by macroarray analysis. Y-axis shows relative amounts (average of two replicate) of each species quantified by QuantiOne software. The fungi and bacteria detected at least once in the orchard were indicated by grey background. The red vertical arrows indicate early-mid August when all the fungi and bacteria decreased to extremely lower levels. The major apple disease epidemics of Monilinia blight, scab, Marssonina blotch, and Alternaria blotch were indicated by horizontal arrows. Abbrebiations for fungi were *Alternaria alternata* (Alt-alt), *Aureobasidum pullulans* (Aur-pil), *Cladosporium tenuissimum* (Cla-ten), *Cystofilobasidium macerans* (Cys-mac), *Epicoccum nigrum* (Epi-nig), *Cryptococcus victoriae* (Cry-vic), *Alternaria mali* (Alt-mal), *Monillinia mali* (Mon-mal), *Penicillium expansum* (Pen-exp), *Valsa ceratosperma* (Val-cer), *Venturia inaequalis* (Ven-ina), and *Diplocarpon mali* (Dip-mal), and for bacteria were *Bacillus cereus* (Bac-cer), *B. megaterium* (Bac-meg), *B. subtilis* (Bac-sub), *Pantoea aggromerans* (Pan-agg), *Pseudomons graminis* (Pse-gra), *P. fluorescens* (Pse-flu), *P. putida* (Pse-put), *P. syringae* (Pse-syr), *Sphingomonas echinoids* (Sph-ech), and *S. yunnanensis* (Sph-yun).

data obtained in the study conducted during 2006–2008 seasons with an agar-plate culture method and with our field observations on the disease epidemics in these orchards. For example, the signal intensities of the arrays for *Aureobasidium*, *Cladosporium*, and *Cryptococcus* fungal genera and *Bacillus*, *Pseudomonas*, and *Sphingomonas* bacterial genera were clearly visible and changed strong to week throughout the growing season in most of the orchards. It should be noted that similar results were independently obtained

by employing culture methods in New Zealand and Switzerland [8,21], suggesting that the predominant species in the phyllosphere of apple trees maybe the same all around the world. Furthermore, as a merit, not only the major saprophytic epiphytes, but also major pathogenic fungi, such as *A. mali*, *V. inaequalis*, and *D. mali*, which were not detected by agar-plate culture method, were certainly detected by this macroarray technique. These phyto-pathogenic fungi were also detectable in the chemical fungicide-sprayed orchards (A-chemical and B-semi-chemical) in the absence of foliar symptoms, suggesting that the macroarray is sensitive enough to monitor changes in the richness of phytopathogenic and non-phytopathogenic fungi and bacteria inhabiting the phyllosphere of apple trees.

Several interesting differences could be seen in the epiphytic microbial diversity in the phyllosphere of apple trees in the orchards with or without employing intensive spraying of chemical fungicides. Firstly, the intensive spraying of chemical fungicides (Orchard A-chemical) reduced the diversity and abundance of both fungi and bacteria in the phyllosphere of apple trees. This is partly consistent with the former observation with Golden Delicious apples in Spain that the fungicide regime on apple trees significantly decreased the total filamentous fungal population; however, bacterial populations were higher on the apples from fungicide-treated plots [25]. Secondary, in the Orchard B-semi-chemical, the development of diseases was successfully controlled without giving adverse impact on the epiphytic microbial diversity in the phyllosphere of apple trees. Finally, a couple of unexpected and interesting findings were obtained in Orchard B-natural, the "natural-farming" orchard; i.e., the abundance of fungal species was the highest in this orchard but the amounts of individual species, with the exception of phytopathogenic *V. inaequalis*, were apparently lower than the others throughout the growing season as represented by *A. pullulans*. Because the orchard has been

maintained to produce apples without relying on any chemical fungicides, it is essential to conduct a more intensive and advanced analysis in relation to microbial diversity and disease control.

The macroarray technique presented here is a strong tool to monitor the complexities of microbial species or the community structures of microbial flora in the phyllosphere of apple trees and identify key species antagonistic, supportive or co-operative to specific pathogens in the orchard managed under different environmental conditions.

Acknowledgments

The authors are grateful to Akinori Kimura and Makoto Takeya, the owners of the orchard B-natural and B-semi-chemical respectively, for their kindness to allow us to survey and collect samples in their orchards. The authors are also grateful to Daiyu Ito for the useful suggestions and kind considerations on sampling apple leaves in the orchard of Hirosaki University.

Author Contributions

Conceived and designed the experiments: YHH KT TS. Performed the experiments: YHH SI MS MT CRAP. Analyzed the data: YHH SI TS. Contributed reagents/materials/analysis tools: YHH SI MS MT CRAP. Wrote the paper: YHH KT TS.

References

1. Kinkel LL (1997) Microbial population dynamics on leaves. Annu Rev Phytopathol 35: 327–347.
2. Lindow SE, Brandl MT (2003) Microbiology of the Phyllosphere. Appl Environ Microbiol 64: 1875–1883.
3. Whipps JM, Hand P, Pink D, Bending GD (2008) Phyllosphere microbiology with special reference to diversity and plant genotype. J Appl Microbiol 105: 1744–1755.
4. Langenheder S, Bulling MT, Solan M, Prosser JI (2010) Bacterial biodiversity-ecosystem functioning relations are modified by environmental complexity. PLoS ONE 5: e10834. doi:10.1371/journal.pone.0010834.
5. Andrews JH, Kenerley CM (1978) The effects of a pesticide program on non-target epiphytic microbial populations of apple leaves. Can J Microbiol 24: 1058–1072.
6. Andrews JH, Kenerley CM (1979) The effects of a pesticide program on microbial populations from apple leaf litter. Can J Microbio l25: 1331–1344.
7. Walter M, Frampton CM, Boyd-Wilson KSH, Harris-Virgin P, Waipara NW (2007) Agrichemical impact on growth and survival of non-target apple phyllosphere microorganisms. Can J Microbiol 53: 45–55.
8. Granado J, Thürig B, Kieffer E, Petrini L, Fliessbach A, et al. (2008) Culturable fungi of stored 'golden delicious' apple fruits: a one-season comparison study of organic and integrated production systems in Switzerland. Microb Ecol 56: 720–732.
9. Fessehaie A, Boer HS, Lévesque AC (2003) An oligonucleotide array for the identification and differentiation of bacteria pathogenic on potato. Phytopathology 93: 262–269.
10. Lievens B, Brouwer M, Vanachter ACRC, Lévesque CA, Cammue BPA, et al. (2005) Quantitative assessment of phytopathogenic fungi in various substrates using a DNA macroarray. Environmental Microbiology 7: 1698–1710.
11. Falacy JS, Grove GG, Mahaffee WF, Galloway H, Glawe DA, et al. (2007) Detection of Erysiphe necator in air samples using the polymerase chain reaction and species-specific primers. Phytopathology 97: 1290–1297.
12. Balint-Kurti P, Simmons SJ, Blum JE, Ballaré CL, Stapleton AE (2010) Maize leaf epiphytic bacteria diversity patterns are genetically correlated with resistance to fungal pathogen infection. MPMI 23: 473–484.
13. Pelludat C, Duffy B, Frey JE (2009) Design and development of a DNA microarray for rapid identification of multiple European quarantine phytopathogenic bacteria. Eur J Plant Pathol 125: 413–423.

14. Izzo AD, Mazzola M (2008) Hybridization of an ITS-based macroarray with ITS community probes for characterization of complex communities of fungi and fungal-like protests. Mycological Research 113: 802–812.
15. Delmotte N, Knief C, Chaffron S, Innerebner G, Roschitzki B, et al. (2009) Community proteogenomics reveals insights into the physiology of phyllosphere bacteria. Proc Natl Acad Sci USA 106: 16428–16433.
16. Lenz O, Beran P, Fousek JI, Mráz I (2010) A microarray for screening the variability of 16S–23S rRNA internal transcribed spacer in Pseudomonas syringae. J Microbiol Methods 82: 90–94.
17. Baker CM, Chitrakar R, Obulareddy N, Panchal S, Williams P, et al. (2010) Molecular battles between plant and pathogenic bacteria in the phyllosphere. Braz J Med Biol Res 43: 698–704.
18. White TJ, Brun TLS, Taylor JW (1990) Amplification and direct sequencing of fungal ribosomal RNA genes for phylogenetics. In: Innis MA, Gelfand DH, Sninsky JJ, White TJ, eds. PCR protocols: a guide to methods and applications, Academic, New York. pp 315–322.
19. Ikeda S, Kaneko T, Ohkubo T, Rallos LE, Eda S, et al. (2009) Development of a bacterial cell enrichment method and its application to the community analysis in soybean stems. Microbiol Ecol 58: 703–714.
20. Lane DJ (1991) 16S/23S rRNA sequencing. In Nucleic acid techniques in bacterial systematics (ed E Stackebrandt & M Goodfellow) 115–175, New York: John Wiley & Sons.
21. Pennycook SR, Newhook FJ (1981) Seasonal changes in the apple phylloplane microflora. New Zealand J Botany 19: 273–283.
22. Sholberg P, O'Gorman D, Bedford K, Lévesque CA (2005) Development of a DNA macroarray for detection and monitoring of economically important apple diseases. Plant Dis 89: 1143–1150.
23. Zhang N, Geiser DM, Smart CD (2007) Macroarray detection of solanaceous plant pathogens in the Fusarium solani species complex. Plant Dis 91: 1612–1620.
24. Zhang N, McCarthy ML, Smart CD (2008) A macroarray system for the detection of fungal and oomycete pathogens of solanaceous crops. Plant Dis 92: 953–960.
25. Teixido N, Usall J, Magan N, Viras I (1999) Microbial population dynamics on Golden Delicious apples from bud to harvest and effect of fungicide applications. Ann ApplBiol 134: 109–116.

4

Weak Spatial and Temporal Population Genetic Structure in the Rosy Apple Aphid, *Dysaphis plantaginea*, in French Apple Orchards

Thomas Guillemaud[1]*, **Aurélie Blin**[1], **Sylvaine Simon**[3], **Karine Morel**[3], **Pierre Franck**[2]

1 Equipe "Biologie des Populations en Interaction", UMR 1301 I.B.S.V. INRA-UNSA-CNRS, Sophia Antipolis, France, 2 UR1115 Plantes et Systèmes de Culture Horticoles, INRA, Avignon, France, 3 UE695 Recherche Intégrée, INRA, Domaine de Gotheron, Saint-Marcel-lès-Valence, France

Abstract

We used eight microsatellite loci and a set of 20 aphid samples to investigate the spatial and temporal genetic structure of rosy apple aphid populations from 13 apple orchards situated in four different regions in France. Genetic variability was very similar between orchard populations and between winged populations collected before sexual reproduction in the fall and populations collected from colonies in the spring. A very small proportion of individuals (~2%) had identical multilocus genotypes. Genetic differentiation between orchards was low ($F_{ST} < 0.026$), with significant differentiation observed only between orchards from different regions, but no isolation by distance was detected. These results are consistent with high levels of genetic mixing in holocyclic *Dysaphis plantaginae* populations (host alternation through migration and sexual reproduction). These findings concerning the adaptation of the rosy apple aphid have potential consequences for pest management.

Editor: Marco Salemi, University of Florida, United States of America

Funding: This work was funded by the French Program ECOGER AO 2005. The funders had no role in study design, data collection and analysis, decision to publish, or preparation of the manuscript.

Competing Interests: The authors have declared that no competing interests exist.

* E-mail: guillem@sophia.inra.fr

Introduction

The rosy apple aphid *Dysaphis plantaginea* (Hemiptera: Aphididae) is one of the most serious pests of apple trees in Europe [1] and North America [2]. It causes fruit deformation and severe leaf-curling [3], distorts shoots, reduces flower formation and slows tree growth [4].

In commercial apple tree orchards, the damage caused by even very low densities of aphids may decrease the commercial value of the crop. This economic loss justifies aphid management techniques, based principally on pesticide use. Recommendations generally suggest the use of several pesticide treatments in apple orchards: in early spring, before flowering and after flowering or in late summer [5]. The intensive use of chemical insecticides against *D. plantaginea* has resulted in an intense selection regime and the development of mechanisms of insecticide resistance in the field [6]. Alternative control strategies, such as the application of organic pesticides (neem extract or potassium soap [5]), the use of repellent or barrier-effect products (kaolin [7,8,9]), biological control [10, 11,12], and plant resistance [13,14,15,16], are being developed and tested.

Whatever the pest management strategy applied, the likelihood of developing resistance to management depends on the ecological characteristics of the target species: its migration capability, sexual reproduction and clonal multiplication determine, at least in part, its genetic variability and, thus, its capacity to adapt to control measures. An analysis of genetic variation in the *D. plantaginea* population may therefore provide essential information about these crucial ecological parameters.

The life cycle of *D. plantaginea* almost certainly has profound consequences for its genetic variability. Like many aphid species, *D. plantaginea* has a cyclic parthenogenetic (or holocyclic) life cycle [17,18]. In late summer and fall, cyclically parthenogenetic aphids give birth to gynoparae (precursor forms of sexual females), followed by winged males. Both fly from the herbaceous secondary host plant, *Plantago*, to the primary host, apple trees, where the gynoparae give birth to sexual females [19]. Mating occurs on apple and sexual females lay eggs that hatch by the beginning of spring. During late spring and early summer, after 3 to 4 (maximum 6) parthenogenetic generations, winged morphs are produced that migrate from the primary to the secondary host on which about 3 to 8 successive parthenogenetic generations occur [19]. Thus, due to the annual host alternation, two large migration events take place in biological cycle of *D. plantaginea*, in the fall and spring.

In many species, cyclic parthenogenetic populations coexist with obligate parthenogenetic populations [20,21]. In such populations, the aphids have lost the ability to reproduce sexually and remain on herbaceous plants throughout the year. According to Lathrop [22], the rosy apple aphid does not occur on plantain during winter in colder parts of the USA. However, "in the mild climate of western Oregon, overwintering on plantain as well as apple is the rule" [22]. This suggests that this species displays variation in reproductive modes, with cyclic parthenogenetic populations coexisting with obligate parthenogenetic populations. However, we are not aware of any other study demonstrating such a polymorphism in *D. plantaginea*.

Cyclic parthenogenetic aphids would be expected to display high levels of genotypic variability, due to the recombination occurring during sexual reproduction [21,23,24]. However, drift and/or selection may strongly decrease neutral genetic variability during successive parthenogenetic generations after egg hatching on apple and on secondary hosts, due to the absence of recombination and the rapid rate of increase during clonal reproduction as shown in the peach-potato aphid *Myzus persicae* [24,25]. During this clonal phase, genetic signs of parthenogenesis may accumulate: linkage disequilibrium (LD), Hardy-Weinberg (HW) disequilibrium, and decrease in multilocus genotype diversity [23].

Little is known about the genetic diversity of the *D. plantaginea* species. The only data available are the preliminary results obtained by Salomon *et al.* [26], who reported high levels of genetic variability in a single apple orchard, based on an analysis of microsatellite genetic markers previously developed by Harvey *et al.* [27] for this species. We therefore know little about the effects of the succession of sexual and asexual reproduction on the genetic variability of this species or those of the major migration events occurring during host shift.

The aim of this study was to determine whether the complex mode of reproduction, with a single sexual generation and successive clonal generations, and host shift-related migration events affected genetic variation in this species. In other words, we evaluated the geographic scale over which *D. plantaginea* populations function and possible decreases in the genetic variation of *D. plantaginea* on apple due to cyclic parthenogenesis.

More specifically, we used a geographic and temporal sampling scheme and highly polymorphic genetic markers (microsatellite) data to address the following questions: (i) What degree of genetic variability does the rosy apple aphid display at the national scale (over the whole of France)? (ii) Is there any genetic differentiation between populations of *D. plantaginea* and at what level (regions, orchards, apple cultivars) can this differentiation be detected? (iii) Are the genetic diversity and geographic population structure of *D. plantaginea* stable at different parts of the life cycle and in different years?

Materials and Methods

Sample collection

Samples were collected according to a geographic and temporal scheme in experimental apple orchards belonging to INRA institute. Here an orchard is defined as a field of apple trees with a given management strategy and a specific tree cultivar. The term "sample" refers to as a group of aphids collected during a specific season and at a specific position in a given orchard. No specific permission was required to sample aphids in these orchards. They were collected at one location in north-western France (near Angers), one location in south-western France (near Agen), and two locations in southern France (near Avignon and Valence) (Figure 1). Depending on the location, aphids were sampled at one (Agen), two (Avignon, Angers) or three (Valence) different periods of the aphid life cycle, in fall 2006 and 2007, and in spring 2007 (see Table S1). Furthermore, at Avignon, Valence and Angers, samples were taken from different orchards at the same time (Table S1). The distances between these orchards were as follows. At Valence, the various orchards that were sampled were located from within a circle with a radius of 250 m. The Smoothee1 orchard sample was located about 350 to 450 m from the other orchard samples, the Conventional Ariane orchard sample was about 300–450 m from the other samples, and the remaining orchard samples were located about 10 to 100 meters apart. At Valence, samples were collected on different apple cultivars (Smoothee, Melrose and Ariane) under organic management, but also from different plants of the same cultivar (Ariane) grown under organic, low-input and conventional pest management

regimes (i.e. organic-registered for the organic system, minimized for the low-input system and supervised for the conventional system). Two locations (center and border) in Smoothee1 orchard in Valence were sampled in autumn 2006 to test for micro-geographic genetic structure that would not depend on tree cultivars and management strategies. At Angers, the two orchards sampled, P32 and D1, were located 500 meters apart. Finally, at Avignon, orchards 65 and 157 were located 2.5 km apart, each about 12 to 15 km from the INRA orchard. In the fall, winged gynoparae were sampled manually by branch tapping. In spring, individuals were collected by hand, with a small brush, with no more than one individual collected per colony and per tree on two sampling dates (May 8 and 23). Aphids were stored in absolute ethanol for DNA extraction.

DNA extraction and microsatellite analysis

Template material for the amplification of microsatellites by PCR was prepared from individual aphids with the "salting out" rapid extraction protocol [28] and resuspended in 50 µl H_2O. Eight microsatellite loci for *D. plantaginea* (DpL4, DpB10) [27], *Sitobion* species (S24, Sa4Σ, S3.43, S16b) [29], *Rhopalosiphon padi* (R5.29B) [29] and *Aphis fabae* (AF93) [30] were amplified in two separate multiplex PCRs. The first reaction amplified *DpL4, DpB10, S24* and *Sa4Σ*, and the second amplified *S3.43, AF93, R5.29B* and *S16b*. Both multiplex reactions were carried out with Qiagen multiplex PCR kits (Qiagen, Hilden, Germany), according to the manufacturer's instructions, in a final volume of 10 µl containing 1 µl of DNA template. The forward primer for each microsatellite was labeled with a fluorescent dye, to allow the detection of PCR products on an ABI 3100 DNA sequencer (Applied Biosystems, Foster City, CA). We used the following PCR program for both reactions: 95°C for 15 minutes, followed by 35 cycles of 30 s at 94°C, 90 s at 56°C, 1 min at 70°C, and 30 s at 60°C.

Data analysis

Within-population genetic diversity was estimated by calculating the number of alleles per locus, and observed and expected heterozygosities calculated with GENEPOP ver. 4.0 [31,32]. Exact tests for deviation from Hardy-Weinberg (HW) expectations, linkage disequilibrium and population differentiation were carried out with GENEPOP. A Mantel test of isolation by distance was also carried out with Genepop ver. 3.1 [31]. MICROCHECKER was used to detect the presence of null alleles at each microsatellite locus [33] and genotypic differentiation between pairs of populations (F_{ST}) was corrected for null alleles as described by Chapuis *et al.* [34]. We compared the number of alleles per locus between population samples, by estimating allelic richness (AR) on the basis of minimum sample size, with the rarefaction method [35] implemented in FSTAT 2.9.3 [36].

If more than one copy of the same multilocus genotype (MLG) was observed, the null hypothesis of the same MLG being obtained repeatedly by chance through sexual reproduction was tested with Genclone ver. 2.0 [37]. This test is based on calculation of the probabilities of obtaining MLGs from sexual events, taking into account the estimated F_{IS} for the population.

Finally, the number of distinct populations (K) present in the set of samples was estimated with STRUCTURE [38]. This software was used to estimate $Pr(X|K)$, the probability of the observed set of genotypes (X), conditional on the number of genetically distinct populations, K, for values of K between 1 and the number of samples. The program was run for 10^5 iterations, preceded by an initial burn-in period of 2×10^4 iterations. Three runs were performed for each value of K, to check that estimates of $Pr(X|K)$ were consistent between runs. The posterior probabilities, $Pr(K|X)$, were then calculated as described by Pritchard *et al.* [38].

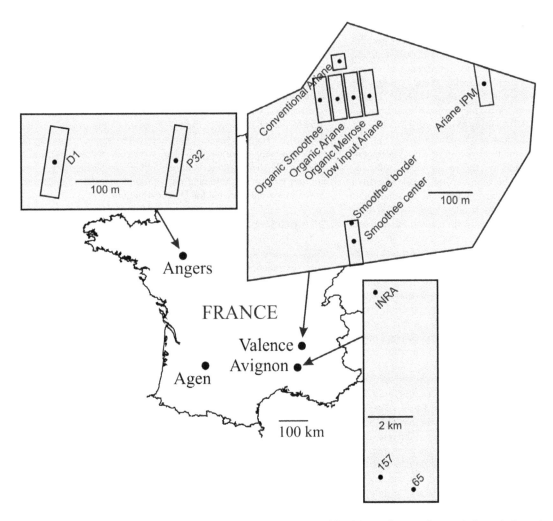

Figure 1. Locations of the samples of *Dysaphis plantaginea* **used in this study.** Sampling periods are indicated.

For multiple tests of a single hypothesis and non orthogonal comparisons, we used Benjamini & Hochberg [39] and sequential Bonferroni [40] correction procedures, respectively, to correct significance levels.

Results

Within-population variability

We genotyped 532 individuals in total and found the level of genetic variation to be high. There were seven (locus *S3.43*) to 34 (locus *S24*) alleles per microsatellite locus. Within-population genetic variability was high, with mean numbers of alleles per locus (Na) of more than seven for samples with more than 15 individuals. Allelic richness (*AR*), calculated on a sample of at least 15 individuals for inter-population comparisons, was between 3.9 and 4.3 (mean *AR* = 4.14, SEM = 0.13), and revealed no difference in population variability between samples and between spring and fall (Friedman analysis of variance and Wilcoxon's signed rank test, *p* > 0.05). Consistent with this, no heterogeneity of Nei's heterozygosity was detected (mean *He* = 0.63, SE = 0.03; Friedman analysis of variance and Wilcoxon's signed rank test, *p* = 0.41 and *p* = 0.32, for between-sample and between spring and fall comparisons, respectively). All samples displayed a heterozygote deficiency, with many genotypic compositions showing departure from HW equilibrium (Table S1). The instances of HW departure identified frequently involved the same three loci (*DPL4*, *DPB10*

and *AF93*), suggesting the presence of null alleles at these loci. Loci *DPL4*, *DPB10* and *AF93* displayed departure from HW equilibrium eight, seven and five times, respectively, in a total of 26 significant per locus and per sample tests. MICROCHECKER suggested the existence of null alleles for *DPB10* and *DPB4*. No heterogeneity in the proportion of significant HW tests was found between samples or between spring and fall samples (Fisher's exact test on RxC contingency tables, *p* > 0.05 for both tests). Accordingly, no heterogeneity in F_{IS} value was detected between samples or between spring and fall samples (Friedman analysis of variance and Wilcoxon's signed rank test on mean F_{IS} value per locus, p = 0.51 and p = 0.33, respectively). After removal of the *DPL4*, *DPB10* and *AF93* loci from the analysis, the general heterozygote deficiency remained and no heterogeneity was apparent between samples or between spring and fall (Friedman analysis of variance and Wilcoxon's signed rank test on mean F_{IS} value per locus, *p* = 0.72 and 0.89 respectively).

A very high level of multilocus genotypic variability was found. PCR amplification was unsuccessful in some cases. In total, 342 individuals were genotyped with no missing data, and 336 different multilocus genotypes (MLG) were detected in these individuals (ratio of the number of multilocus genotypes over the total number of individuals, $N_{MLG}/N = 0.98$). Six MLGs were found in multiple copies. Each of these repeated MLGs was found in two individuals sampled from the same orchard on the same date: orchards 65 and 157 in fall 2006, orchards Bio Smoothee and Bio

Melrose in Valence in spring 2007, and Agen in fall 2007. These repeated MLGs were probably generated by clonal rather than sexual reproduction (test of the null hypothesis of sexual recombination, $p < 8 \times 10^{-4}$). Consistent with the extensive multilocus genotypic variability observed, an analysis of the genotypic disequilibrium between each pair of loci in each sample revealed very few cases of significant linkage. No heterogeneity in the number of significant LD was found either between samples, or between spring and fall (Fisher's exact test on RxC contingency tables, $p > 0.05$ for both tests).

Population differentiation

As most samples displayed heterozygote deficiency, we carried out exact tests of genotypic differentiation between samples only. All comparisons between parts of orchards or between orchards at the same location or at the same period were characterized by small F_{ST} values (<1%) and non significant differentiation tests ($p = 0.078$ and 0.25 at Angers, $p = 0.15$ and 0.47 at Avignon in fall 2006 and 2007 respectively, and $p = 0.43$ at Valence in fall 2007). Parts of orchards and orchards at the same location were therefore pooled by period for analyses of regional genetic differentiation (Table 1).

As null alleles were suspected for several loci, we also performed an analysis taking these null alleles into account [34]. We found the same absence of differentiation between samples from the same location, with the exception of two orchards in Avignon sampled in 2006 (165 and 57, $p = 10^{-3}$). As the level of genetic differentiation was very low ($F_{ST} = 4.3 \times 10^{-3}$) we decided to pool the samples from each location.

Significant, but weak ($F_{ST} < 1\%$) genotypic differentiation was detected between Angers, Avignon and Valence in fall 2006 (Table 1). In fall 2007, significant moderate levels of differentiation were observed between Avignon and other locations ($F_{ST} \sim 2\%$). A low level of differentiation was found between Agen and Angers or Valence ($F_{ST} \sim 1\%$) and no differentiation was detected between Angers and Valence. The same overall pattern was observed if null alleles were taken into account: significant, but low to moderate levels of differentiation between locations.

Only low to very low levels of differentiation were found between samples from the same location collected at different time periods. Almost no difference was found between samples collected at Valence in fall 2006, spring 2007 and fall 2007 (although

the differentiation between fall 2006 and spring 2007 was of borderline significance, $p = 0.023$, $F_{ST} = 0.002$). Comparisons between fall 2006 and 2007 for each location revealed significant but weak (in the case of Angers and Avignon, $p < 4 \times 10^{-3}$, $F_{ST} = 0.005$ and 0.008, respectively) and non significant (in the case of Valence, $p = 0.3$, $F_{ST} = -0.002$) differentiation.

Very similar results were obtained when null alleles were taken into account. In this case, significant differentiation was detected in all comparisons other than that between fall 2006 and fall 2007 at Valence. No isolation by distance was detected between the 16 samples with more than 15 individuals (Mantel test, $p = 0.153$).

A Bayesian analysis of population structure grouped all individuals together in a single population, regardless of their location and sampling period ($P(K = 1 | X) = 1$). This was true for the default model (admixture and correlated allele frequency), but also for the admixture and independent allele frequency model. Models without admixture gave inconsistent results ($P(K = 2 | X) = 1$ and $P(K = 14 | X) = 1$ for the correlated and independent allele frequency models, respectively). Evanno's ΔK [41] also gave inconsistent results for the models without admixture (K = 5 and K = 2 for the correlated and independent allele frequency models, respectively).

Discussion

Considerable variability and no evidence for obligate parthenogenesis

In this study, we analyzed the genetic structure of populations of the rosy apple aphid, *D. plantaginae*, collected from its primary host. The goal was to characterize, for the first time, the genetic variability of this aphid, and to evaluate the impact of three evolutionary forces potentially affecting this variation: drift, migration and selection. Rosy apple aphid populations collected from apple trees in four regions of France displayed extensive genetic variation. In particular, a very high degree of genotypic diversity was observed, with almost all individuals genetically different from each other. This was true for all locations and sampling periods. This result confirms and extends the findings of Solomon *et al.* [26], who were the first to report high levels of genetic variability in *D. plantaginea* sampled from apple orchards.

The rosy apple aphid is thought to be a cyclic parthenogenetic species, with a single sexual generation and many asexual generations. It is unknown whether this species displays polymorphism

Table 1. Regional and temporal differentiation of *Dysaphis plantaginea* samples in France.

		Fall 2006			Spring 2007		Fall 2007		
		Angers	Avignon	Valence	Valence	Agen	Angers	Avignon	Valence
Fall 2006	Angers	-	0.002	0.008			0.005		
	Avignon	0.018*	-	0.009				0.008	
	Valence	0.001**	3×10^{-4}**	-	0.002				−0.002
Spring 2007	Valence		0.023		-				0.001
Fall 2007	Agen					-	0.006	0.026	0.012
	Angers	0.004*				0.006*	-	0.016	0.003
	Avignon		8×10^{-4}**			10^{-5}**	3×10^{-4}**	-	0.024
	Valence			0.3	0.56	0.035*	0.223	10^{-5}**	-

Pairwise estimates of F_{ST} are above the diagonal and the p-values of genotypic differentiation exact tests are shown below the diagonal. * and ** after p-values indicate that the tests were significant before and after Bonferroni correction, respectively. Only pertinent comparisons (i.e. between periods at the same sites or between sites during the same period) are shown.

in its mode of reproduction, with the coexistence of obligate parthenogenetic and parthenogenetic individuals, as in many other aphid species [42]. The mode of reproduction has consequences for the genetic variation of populations [43], and this topic has been particularly well studied in aphids [44]. In the case of holocycly, two antagonistic effects occur. Asexual generations (reproducing by mitotic parthenogenesis in this species) are expected to generate individuals with an identical genetic background, with mutations as the only source of variation. The occurrence of such asexual generations also leads to systematic linkage disequilibrium (LD) and departure from HW equilibrium. By contrast, (panmictic) sexual generation disrupts inter-locus associations, resulting in each individual being genetically different from all others. It also re-establishes HW equilibrium within a single generation and decreases LD. Note that, in the long term, obligate parthenogenesis (parthogenesis as the only form of reproduction) tends to lead to excess heterozygosity due to the accumulation of mutations without recombination [44].

In French populations of the rosy apple aphid collected from its primary host we found neither general LD, nor a global excess of heterozygotes. We found extensive multilocus genotypic variability. These genetic signals provide evidence of sexual reproduction, supporting the hypothesis that the populations collected from apple trees in the spring and fall are holocyclic. This is consistent with what is known of the lifecycle of D. plantaginea, and with the observation of eggs on apple trees during the winter [17,22]. We found no evidence for the existence of obligate parthenogenesis in D. plantaginea, at least on apple trees in the fall and spring. However, it remains possible that anholocyclic lineages exist during these periods of the year on secondary hosts, as reported for many aphid species displaying host alternation [42].

The populations sampled in the fall, before the occurrence of recombination, were produced by lineages that had gone through several parthenogenetic generations since the last sexual event. We therefore expected to find genetic signs of clonal reproduction (repeated multilocus genotypes, LD, systematic HW disequilibrium) in the samples collected in the fall. However, no such signs were observed. This suggests that a single yearly sexual reproduction event is sufficient to generate a high level of genetic variability and to cancel out the genetic signs of clonality, even in the fall, before the occurrence of sexual reproduction. The almost entire absence of individuals with identical multilocus genotypes in samples collected in the fall suggests that the number of individuals from an individual clone of D. plantaginea present on apple trees in France in the fall is not large. This may be due to 1) the limited size of the clonal populations sharing the same genotype on secondary hosts compared to the number of different clonal genotypes present on these plants and/or 2) an extensive geographic redistribution of the aphids during their return flight to their primary hosts (but see below), leading the dilution of repeated clonal genotypes.

The high level of genetic variability found in D. plantaginea on its primary host is similar to that found in other cyclic parthenogenetic aphids, such as M. persicae in France [24] and Australia [45], S. avenae [46] or R. padi [23] and other cyclic parthenogenetic animals, such as rotifers (e.g. Brachionus plicatilis (Müller), [47]), which display high levels of genotypic diversity despite going through numerous parthenogenetic generations each year.

We frequently observed heterozygote deficits associated with HW disequilibrium. Possible explanations based on previous findings for aphids include a Wahlund effect, null alleles, inbreeding and selection [23,24,48,49,50].

The Wahlund effect is the unintentional pooling of differentiated populations into a single sample, resulting in excess homozygosity

[43]. A Wahlund effect may occur in the fall, due to the co-occurrence on the primary host of migrants originating from populations that were genetically differentiated on secondary hosts. Such genetic differentiation may result from genetic drift or selection (e.g. adaptation to various secondary host plants). Panmictic sexual reproduction leads to HW equilibrium in only one generation [43]. Thus, assuming panmictic sexual reproduction, heterozygote deficits in the spring (i.e. after sexual reproduction) cannot be accounted for by a Wahlund effect.

Null alleles were suspected for three loci, and specific statistical treatments were carried out to take this possibility into account. A specific statistical analysis was carried out to detect loci with null alleles, but we cannot rule out the possibility that this problem occurred at a larger number of loci.

Inbreeding and selection are often proposed as explanations for heterozygote deficits in sexual aphid populations [23,48,49, 50], but we found no evidence to support this hypothesis in this study.

Spatial genetic differentiation

Another key finding of this study was the very weak spatial genetic differentiation between D. plantaginae populations. We detected no population genetic differentiation at the regional scale or at the intra-orchard or inter-orchard level, for samples located less than 20 km apart. Classically, spatial genetic differentiation results from the balance between migration and genetic drift [51]. In species with mitotic parthenogenesis, selection at one or a few loci affects allele frequency not only at these loci, but throughout the genome, because there is no recombination. Therefore, in a species like D. plantaginae, the use of microsatellites to assess spatial population genetic structure also provides information about selection (until sexual reproduction takes place). Our results therefore suggest that the effect of local drift or selection is largely compensated by migration. The fall and spring flights of the aphids mediating host shift are thus sufficient to homogenize genetic variability at a local and regional scale. However, we observed significant levels of population genetic differentiation at the scale of the entire country (France), between different apple-growing areas, with differences observed between Avignon, Agen, Valence and Angers. This genetic differentiation was weak (F_{ST} generally below 1%) and no isolation by distance was observed, but these results nonetheless suggest that the emigration and return flights of D. plantaginea are limited by geographic distance, at regional scale at least, in France. D. plantaginea has only one winter host-plant, apple, and this species has a patchy distribution in France. This may account for the spatial limitation of migration. We also found evidence for a local dispersion component in the fall and spring. The sharing of the same multilocus genotype by a pair of individuals on the primary host in the fall, before sexual reproduction, was rare, but nonetheless observed in three instances. In each case, the two individuals sharing the same MLG were found in the same orchard. This strongly suggests that the return flight was local. In other words, this migration may connect secondary and primary hosts located close together, rather than reflecting global geographic homogenization.

The situation in spring was similar to that in the fall and provides information about dispersal between primary hosts after sexual reproduction: three repeated MLGs, each shared by a single pair of individuals, were observed in three different orchards, among 118 colonies. One of the repeated MLGs corresponded to individuals collected from the same tree on two different dates and, thus, probably reflected sampling from the same aphid colony. However, in the two other cases of aphids sharing other repeated MLGs, individuals were collected from non contiguous trees, probably reflecting the dispersion of aphids between different trees in the spring. It is unknown whether such dissemination between distant trees is passive (through wind or cropping practices) or active. Overall, these findings suggest that, at the time of sampling in May, i) aphid dispersal between primary hosts occurred but was not frequent and/or ii) dispersion may have been frequent but only a small proportion of the total number of colonies was sampled. A rough estimate of the sampling effort in spring would be one colony sampled per five actual colonies, so the probability of sampling the same MLG twice or more was low.

Overall, spatial genetic differentiation in *D. plantaginea* was very weak or null over short distances and weak but significant over large distances, suggesting that local migration occurs in *D. plantaginea*. This situation is similar to that reported for other aphid species. For instance, in *R. padi*, no genetic differentiation was found between populations located less than 1000 km apart [23]. Weak population differentiation was found between both close (<100 km) and distant (>500 km) populations of the cereal aphid, *S. avenae* [48,52,53]. This work provides an additional demonstration that genetic differentiation is not rare in aphids and that aphid migration probably therefore occurs over limited spatial areas [24,48,53,54,55,56].

Temporal genetic differentiation

The third key result of this study is the almost complete temporal genetic homogeneity among samples. Only very low levels of genetic differentiation were observed between samples collected in fall 2006, spring 2007 and fall 2007. There was thus no decrease in genetic variability between the sampling periods. Between the two first sampling periods, one phase of sexual reproduction occurred and a few clonal generations were produced on the primary host. After sexual reproduction on apple trees, *D. plantaginea* is frequently subject to strong demographic bottlenecks, due to pest management practices (e.g. insecticide treatments [5,6]). In our study system, eight of the 13 orchards were treated conventionally with pesticides. If resistance genes are present in the treated populations, then such pesticide treatments may generate strong selection pressure, increasing the frequency of resistance genes in the clonal aphid population during spring. As recombination is absent during this part of the life cycle, we would expect (i) a change in microsatellite allelic frequencies due to the complete linkage between neutral genetic markers and genes subject to selection and (ii) a decrease in genetic variability due to the increase in frequency of some adapted MLGs. No such change was observed. Moreover, almost no repeated multilocus genotypes potentially resulting from the selection of a few adapted clones were observed in spring. Conventional apple orchards undergo a large number of pesticide treatments (up to 10 treatments are commonly applied in apple orchards when *D. plantaginea* is present, in France [57], and elsewhere see e.g. Blommers *et al.* [18]). Thus, the selection pressure resulting from pesticide treatments is likely to be very intense. Our observation is therefore more consistent with an absence of adaptive gene polymorphism, particularly for insecticide resistance genes, in the populations sampled, the resistance alleles being either fixed or absent. No failure of insecticide treatment was reported in spring 2007, suggesting that the mechanisms of insecticide resistance mechanisms documented by Delorme [58] did not occur.

Using a similar temporal sampling scheme for the peach potato aphid, *Myzus persicae*, Guillemaud *et al.* [24] detected a change in insecticide resistance allele frequency in holocyclic populations in southern France. The *kdr* mutation, which confers resistance to pyrethroid insecticides, increased in frequency between autumn and spring, probably because of insecticide treatments. Conversely,

the *rdl* mutation, which confers resistance to cyclodiene insecticides, decreased in frequency over the same period, probably because of the negative pleiotropic effects of the mutation [24].

We also found almost no differentiation between spring 2007 and autumn 2007, a period of time spanning a few clonal generations on the primary host, the emigration flight to secondary hosts followed by a sequence of several clonal generations and the return flight to the apple tree. Again, no decrease in genetic variability was observed between the two sampling points, suggesting that selection and/or drift during the asexual phase of the life cycle has little or no effect on the genetic structure of *D. plantaginea* . This contrasts sharply with what was reported for *M. persicae* by Vorburger [25] and by Guillemaud *et al.* [24], who analyzed changes in population genetic structure during the asexual phase. Vorburger [25] followed the temporal dynamics of *M. persicae* clones on secondary hosts in detail over a period of one year, and Guillemaud *et al.* [24] measured the differentiation between aphids collected during emigration and the return flight. Both studies revealed significant temporal variation of the structure of the population, interpreted in both cases as a result of selection rather than genetic drift. Selection in aphids is now well documented, and it appears that host plant [59,60,61,62] and pesticide treatment [62,63] are among the most important selective factors to be taken into account when trying to understand the population genetic structure of aphid species acting as crop pests.

No such selective forces appear to shape the population genetic structure of *D. plantaginea* during the asexual phase, which occurs mostly on secondary hosts. The known secondary hosts of *D. plantaginea* are herbaceous plants of the genus *Plantago* [18]. Little is known about possible environmental selection on these plants. No control treatments (such as pesticide applications) are used against *D. plantaginea* when feeding on *Plantago* spp. because these plants are of neither economic nor ornamental value. However, we cannot exclude the possibility that, during the summer, *D. plantaginea* is exposed to pesticides applied to crops or vegetation stands in which their *Plantago* spp. host plants are common (e.g. as weeds). We tried to sample *D. plantaginea* on *Plantago* close to the primary host sampling locations at Valence, without success. This may be because (i) the populations of *D. plantaginea* on the secondary host are small, (ii) secondary host colonization is restricted to particular *Plantago* populations or to plants growing under specific favorable conditions or (iii) *Plantago* is not the only secondary host of *D. plantaginae*. It may be important to identify the entire set of actual secondary host plants of *D. plantaginea* and their distribution, to determine which processes may occur during the asexual phase on the secondary host plant (currently seen as a "black-box").

Practical aspects of aphid management

Our results concerning the genetic structure of the rosy apple aphid population have practical implications for the management of this aphid. We found no genetic differences between samples collected from orchards planted with different cultivars (Ariane, Smoothee and Melrose; unfortunately we could not test for an effect of pesticide treatments in Valence in spring 2007 because the sample size was too small for low-input and conventional orchards). There are three possible explanations for this result: (i) None of the three apple cultivars was thought to be resistant to the rosy apple aphid, so there is probably no adaptation to these cultivars in *D. plantaginea*. (ii) Determination of the genetic structure of the population with microsatellites does not reveal genetic structure due to selection, because recombination during sexual reproduction breaks the linkage between adaptive alleles and microsatellite markers. (iii) Migration homogenizes genotypic frequencies, so it is not possible to determine the genetic structure of the population linked to selective forces. The balance between migration and selection was in favor of migration, as discussed below.

We found that migration had a larger effect than drift and selection in shaping the population genetic structure of this species at various geographic scales. The imbalance in favor of migration was found within orchards, between orchards separated by tens of meters at the same site and between sites separated by one to several hundreds of kilometers. This imbalance has two consequences: local adaptation [64] probably cannot occur, and adaptations to control practices may spread rapidly over large geographic areas. Local adaptation may occur when the environment is heterogeneous for selection (e.g. with or without pesticide treatment) and when a there is cost associated with adaptation (e.g. a cost to pesticide resistance). It occurs when a mutated genotype (e.g. a pesticide-resistant genotype) is better adapted to certain local conditions (e.g. pesticide application) but less well adapted to other environmental conditions (e.g. absence of pesticide treatment) than the wild-type genotypes (e.g. pesticide-susceptible genotypes). Management strategies, such as treatment applications limited to small geographic pockets (the stable zone strategy in [65]), based on local adaptations may therefore not be applicable for the rosy apple aphid on apple trees in France. The second consequence of the apparently extensive migration of the rosy apple aphid is that a monogenic or oligogenic genotype adapted to control strategies (e.g. pesticide-resistant genotypes or genotypes circumventing plant resistance) may invade large areas very rapidly after its emergence. This is a potential Achilles heel of control strategies against *D. plantaginea*, because adaptation at any one site may lead to the failure of control everywhere. Resistance to carbamate and organophosphate insecticides has recently been found in a *D. plantaginea* clone collected in Avignon (Southern France) [58]. This resistance is probably oligenic and based on a small number of biochemical mechanisms. Our results suggest that it is likely to increase rapidly in frequency and spread geographically, leading to the failure of pest control over large areas if no other pesticides (such as pyrethroids) are used.

Acknowledgments

We thank Delphine Racofier (FREDON Aquitaine), Isabelle Lafargue (DRAF Aquitaine), Arnaud Lemarquand, René Rieux, and Hubert Defrance for their help with aphid sampling. We also thank Armelle Coeur d'Acier for her help with aphid species identification.

Author Contributions

Conceived and designed the experiments: TG SS KM PF. Performed the experiments: AB SS KM. Analyzed the data: TG. Wrote the paper: TG PF.

References

1. Hill D (1987) Agricultural insect pests of temperate regions and their control. Cambridge, UK: Cambridge University Press.
2. Hull LA, Starner VR (1983) Effectiveness of insecticide applications timed to correspond with the development of rosy apple aphid (Homoptera, Aphididae) on apple. Journal of Economic Entomology 76: 594–598.
3. Forrest JMS, Dixon AFG (1975) Induction of leaf-roll galls by apple aphids Dysaphis devecta and Dysaphis plantaginea. Annals of Applied Biology 81: 281.
4. Lyth M (1985) Hypersensitivity in apple to feeding by Dysaphis plantaginea - effects on aphid biology. Annals of Applied Biology 107: 155–161.
5. Cross JV, Cubison S, Harris A, Harrington R (2007) Autumn control of rosy apple aphid, Dysaphis plantaginea (Passerini), with aphicides. Crop Protection 26: 1140–1149.
6. Delorme R, Ayala V, Touton P, Auge D, Vergnet C (1999) Le puceron cendré du pommier (Dysaphis plantaginea) : Etude des mécanismes de résistance à divers insecticides. In: ANPP, editor; 7–8–9 décembre 1999; Montpellier. pp 89–96.
7. Burgel K, Daniel C, Wyss E (2005) Effects of autumn kaolin treatments on the rosy apple aphid, Dysaphis plantaginea (Pass.) and possible modes of action. Journal of Applied Entomology 129: 311–314.
8. Marko V, Blommers LHM, Bogya S, Helsen H (2008) Kaolin particle films suppress many apple pests, disrupt natural enemies and promote woolly apple aphid. Journal of Applied Entomology 132: 26–35.
9. Wyss E, Daniel C (2004) Effects of autumn kaolin and pyrethrin treatments on the spring population of Dysaphis plantaginea in apple orchards. Journal of Applied Entomology 128: 147–149.
10. Brown MW, Mathews CR (2007) Conservation biological control of rosy apple aphid, Dysaphis plantaginea (Passerini), in Eastern North America. Environmental Entomology 36: 1131–1139.
11. Minarro M, Hemptinne JL, Dapena E (2005) Colonization of apple orchards by predators of Dysaphis plantaginea: sequential arrival, response to prey abundance and consequences for biological control. Biocontrol 50: 403–414.
12. Wyss E, Villiger M, Hemptinne JL, Muller-Scharer H (1999) Effects of augmentative releases of eggs and larvae of the ladybird beetle, Adalia bipunctata, on the abundance of the rosy apple aphid, Dysaphis plantaginea, in organic apple orchards. Entomologia Experimentalis Et Applicata 90: 167–173.
13. Angeli G, Simoni S (2006) Apple cultivars acceptance by Dysaphis plantaginea Passerini (Homoptera: Aphididae). Journal of Pest Science 79: 175–179.
14. Minarro M, Dapena E (2007) Resistance of apple cultivars to Dysaphis plantaginea (Hemiptera: Aphididae): Role of tree phenology in infestation avoidance. Environmental Entomology 36: 1206–1211.
15. Minarro M, Dapena E (2008) Tolerance of some scab-resistant apple cultivars to the rosy apple aphid, Dysaphis plantaginea. Crop Protection 27: 391–395.
16. Qubbaj T, Reineke A, Zebitz CPW (2005) Molecular interactions between rosy apple aphids, Dysaphis plantaginea, and resistant and susceptible cultivars of its primary host Malus domestica. Entomologia Experimentalis Et Applicata 115: 145–152.
17. Bonnemaison L (1959) Le puceron cendré du pommier (Dysaphis plantaginae Pass.) – Morphologie et biologie – Méthode de lutte. Annales de l'Institut National de la Recherche Agronomique, Série C, Epiphyties 10: 257–322.
18. Blommers LHM, Helsen HHM, Vaal F (2004) Life history data of the rosy apple aphid Dysaphis plantaginea (Pass.) (Homopt., Aphididae) on plantain and as migrant to apple. Journal of Pest Science 77: 155–163.
19. Bonnemaison L (1961) Les Ennemis Annimaux des Plantes Cultivées et des Forêts. Paris: Paris 1er.
20. Blackman RL (1981) Species, sex and parthenogenesis in aphids. In: Forey PL, ed. The evolving biosphere. Cambridge: Cambridge University Press. pp 75–85.
21. Simon J-C, Rispe C, Sunnucks P (2002) Ecology and evolution of sex in aphids. Trends in Ecology and Evolution 17: 34–39.
22. Lathrop FH (1928) The biology of apple aphids. The Ohio Journal of Science 28: 177–204.
23. Delmotte F, Leterme N, Gauthier JP, Rispe C, Simon JC (2002) Genetic architecture of sexual and asexual populations of the aphid Rhopalosiphum padi based on allozyme and microsatellite markers. Molecular Ecology 11: 711–723.
24. Guillemaud T, Mieuzet L, Simon JC (2003) Spatial and temporal genetic variability in French populations of the peach-potato aphid, Myzus persicae. Heredity 91: 143–152.
25. Vorburger C (2006) Temporal dynamics of genotypic diversity reveal strong clonal selection in the aphid Myzus persicae. Journal of Evolutionary Biology 19: 97–107.
26. Solomon MG, Harvey N, Fitzgerald J (2003) Molecular approaches to population dynamics of Dysaphis plantaginea. In: Cross JV, Solomon MG, eds. 10–14 March 2002 Vienna, Austria: International Organization for Biological and Integrated Control of Noxious Animals and Plants (OIBC/OILB), West Palaearctic Regional Section (WPRS/SROP). pp 79–81.
27. Harvey NG, Fitz Gerald JD, James CM, Solomon MG (2003) Isolation of microsatellite markers from the rosy apple aphid Dysaphis plantaginea. Molecular Ecology Notes 3: 111–112.
28. Sunnucks P, England P, Taylor AC, Hales DF (1996) Microsatellite and chromosome evolution of parthenogenetic Sitobion aphids in Australia. Genetics 144: 747–756.
29. Wilson ACC, Massonnet B, Simon JC, Prunier-Leterme N, Dolatti L, et al. (2004) Cross-species amplification of microsatellite loci in aphids: assessment and application. Molecular Ecology Notes 4: 104–109.
30. Gauffre B, Coeur d'Acier A (2006) New polymorphic microsatellite loci, cross-species amplification and PCR multiplexing in the black aphid, Aphis fabae Scopoli. Molecular Ecology Notes 6: 440–442.
31. Raymond M, Rousset F (1995) Genepop (version. 1.2), a population genetics software for exact tests and ecumenicism. Journal of Heredity 86: 248–249.
32. Rousset F (2008) GENEPOP ' 007: a complete re-implementation of the GENEPOP software for Windows and Linux. Molecular Ecology Resources 8: 103–106.

33. Van Oosterhout C, Hutchinson WF, Wills DPM, Shipley P (2004) MICRO-CHECKER: software for identifying and correcting genotyping errors in microsatellite data. Molecular Ecology Notes 4: 535–538.

34. Chapuis MP, Estoup A (2007) Microsatellite null alleles and estimation of population differentiation. Molecular Biology and Evolution 24: 621–631.

35. Petit RJ, El Mousadik A, Pons O (1998) Identifying populations for conservation on the basis of genetic markers. Conservation Biology 12: 844–855.

36. Goudet J (2001) FSTAT, a program to estimate and test gene diversities and fixation indices (version 2.9.3). Updated from Goudet (1995).

37. Arnaud-Haond S, Belkhir K (2007) GENCLONE: a computer program to analyse genotypic data, test for clonality and describe spatial clonal organization. Molecular Ecology Notes 7: 15–17.

38. Pritchard JK, Stephens M, Donnelly P (2000) Inference of population structure using multilocus genotype data. Genetics 155: 945–959.

39. Benjamini Y, Hochberg Y (1995) Controlling the False Discovery Rate - a Practical and Powerful Approach to Multiple Testing. Journal of the Royal Statistical Society Series B-Methodological 57: 289–300.

40. Sokal RR, Rolf FJ (1995) Biometry. The Principles and Practice of Statistics in Biological Research. New York: W.H. Freeman and Company.

41. Evanno G, Regnaut S, Goudet J (2005) Detecting the number of clusters of individuals using the software STRUCTURE: a simulation study. Molecular Ecology 14: 2611–2620.

42. Simon JC, Rispe C, Sunnucks P (2002) Ecology and evolution of sex in aphids. Trends in Ecology & Evolution 17: 34–39.

43. Hartl DL, Clark AG (1997) Principles of Population Genetics. SunderlandMA, , U.S.A.: Sinauer Associates, Inc.

44. Halkett F, Simon JC, Balloux F (2005) Tackling the population genetics of clonal and partially clonal organisms. Trends in Ecology & Evolution 20: 194–201.

45. Wilson ACC, Sunnucks P, Blackman RL, Hales DF (2002) Microsatellite variation in cyclically parthenogenetic populations of Myzus persicae in south-eastern Australia. Heredity 88: 258–266.

46. Jensen AB, Hansen LM, Eilenberg J (2008) Grain aphid population structure: no effect of fungal infections in a 2-year field study in Denmark. Agricultural and Forest Entomology 10: 279–290.

47. Gomez A, Carvalho GR (2000) Sex, parthenogenesis and genetic structure of rotifers: microsatellite analysis of contemporary and resting egg bank populations, 9, 203–214. Molecular Ecology 9: 203–214.

48. Simon JC, Baumann S, Sunnucks P, Hebert PDN, Pierre JS, et al. (1999) Reproductive mode and population genetic structure of the cereal aphid Sitobion avenae studied using phenotypic and microsatellite markers. Molecular Ecology 8: 531–545.

49. Papura D, Simon JC, Halkett F, Delmotte F, Le Gallic JF, et al. (2003) Predominance of sexual reproduction in, Romanian populations of the aphid Sitobion avenae inferred from phenotypic and genetic structure. Heredity 90: 397–404.

50. Massonnet B, Weisser WW (2004) Patterns of genetic differention between populations of the specialized herbivore Macrosiphoniella tanacetaria (Homoptera, Aphididae). Heredity 93: 577–584.

51. Wright S (1969) The Theory of Gene Frequencies: The University of Chicago Press, Chicago. 511 p.

52. De Barro PJ, Sherratt TN, Brookes CP, David O, MacLean N (1995) Spatial and temporal genetic variation in British field populations of the grain aphid Sitobion avenae (F.) (Hemiptera: Aphididae) studied using RAPD-PCR. Proceedings of the Royal Society of London, B 262: 321–327.

53. Sunnucks P, DeBarro PJ, Lushai G, Maclean N, Hales D (1997) Genetic structure of an aphid studied using microsatellites: Cyclic parthenogenesis, differentiated lineages and host specialization. Molecular Ecology 6: 1059–1073.

54. Loxdale HD, Brookes CP (1990) Temporal genetic stability within and restricted migration (gene flow) between local populations of the blackberry-grain aphid Sitobion fragariae in South-East England. J anim Ecol 59: 497–514.

55. Loxdale HD, Hardie J, Halbert S, Foottit R, Kidd NAC, et al. (1993) The relative importance of short-range and long-range movement of flying aphids. Biological Reviews of the Cambridge Philosophical Society 68: 291–311.

56. Martinez-Torres D, Carrio R, Latorre A, Simon JC, Hermoso A, et al. (1997) Assessing the nucleotide diversity of three aphid species by RAPD. Journal of Evolutionary Biology 10: 459–477.

57. Butault J, Dedryver C, Gary C, Guichard L, Jacquet F, et al. (2010) Ecophyto R&D. Quelles voies pour réduire l'usage des pesticides? Synthèse du rapport d'étude. 90 p.

58. Delorme R, Ayala V, P T, Auge D, Vergnet C (1999) Le puceron cendré du pommier (Dysaphis plantaginea): étude des mécanismes de résistance à divers insecticides; Montpellier. ANPP.

59. Carletto J, Lombaert E, Chavigny P, Brevault T, Lapchin L, et al. (2009) Ecological specialization of the aphid Aphis gossypii Glover on cultivated host plants. Molecular Ecology 18: 2198–2212.

60. Peccoud J, Ollivier A, Plantegenest M, Simon JC (2009) A continuum of genetic divergence from sympatric host races to species in the pea aphid complex. Proceedings of the National Academy of Sciences of the United States of America 106: 7495–7500.

61. Simon JC, Carre S, Boutin M, Prunier-Leterme N, Sabater-Munoz B, et al. (2003) Host-based divergence in populations of the pea aphid: insights from nuclear markers and the prevalence of facultative symbionts. Proceedings of the Royal Society B-Biological Sciences 270: 1703–1712.

62. Zamoum T, Simon JC, Crochard D, Ballanger Y, Lapchin L, et al. (2005) Does insecticide resistance alone account for the low genetic variability of asexually reproducing populations of the peach-potato aphid Myzus persicae? Heredity 94: 630–639.

63. Carletto J, Martin T, Vanlerberghe-Masutti F, Brevault T (2010) Insecticide resistance traits differ among and within host races in Aphis gossypii. Pest Management Science 66: 301–307.

64. Roughgarden J (1996) Theory of population genetics and evolutionary ecology: an introduction. Upper Saddle River: Prentice-Hall, Inc. 612 p.

65. Lenormand T, Raymond M (1998) Resistance management: the stable zone strategy. The Proceedings of the Royal Society of London, B 265: 1985–1990.

5

West Nile Virus Prevalence across Landscapes Is Mediated by Local Effects of Agriculture on Vector and Host Communities

David W. Crowder[1]*, **Elizabeth A. Dykstra**[2], **Jo Marie Brauner**[2], **Anne Duffy**[2], **Caitlin Reed**[2], **Emily Martin**[1], **Wade Peterson**[1], **Yves Carrière**[3], **Pierre Dutilleul**[4], **Jeb P. Owen**[1]

1 Department of Entomology, Washington State University, Pullman, Washington, United States of America, 2 Washington State Department of Health, Olympia, Washington, United States of America, 3 Department of Entomology, University of Arizona, Tucson, Arizona, United States of America, 4 Department of Plant Science, McGill University, Macdonald Campus, Ste-Anne-de-Bellevue, Quebec, Canada

Abstract

Arthropod-borne viruses (arboviruses) threaten the health of humans, livestock, and wildlife. West Nile virus (WNV), the world's most widespread arbovirus, invaded the United States in 1999 and rapidly spread across the county. Although the ecology of vectors and hosts are key determinants of WNV prevalence across landscapes, the factors shaping local vector and host populations remain unclear. Here, we used spatially-explicit models to evaluate how three land-use types (orchards, vegetable/forage crops, natural) and two climatic variables (temperature, precipitation) influence the prevalence of WNV infections and vector/host distributions at landscape and local spatial scales. Across landscapes, we show that orchard habitats were associated with greater prevalence of WNV infections in reservoirs (birds) and incidental hosts (horses), while increased precipitation was associated with fewer infections. At local scales, orchard habitats increased the prevalence of WNV infections in vectors (mosquitoes) and the abundance of mosquitoes and two key reservoir species, the American robin and the house sparrow. Thus, orchard habitats benefitted WNV vectors and reservoir hosts locally, creating focal points for the transmission of WNV at landscape scales in the presence of suitable climatic conditions.

Editor: Tian Wang, University of Texas Medical Branch, United States of America

Funding: This study was supported by United States Department of Agriculture (USDA) Agriculture and Food Research Initiative Project 2011-67012-30718 and USDA Risk Avoidance and Mitigation Program Project 2006-0207436. The funders had no role in study design, data collection and analysis, decision to publish, or preparation of the manuscript.

Competing Interests: The authors have declared that no competing interests exist.

* E-mail: dcrowder@wsu.edu

Introduction

Viruses transmitted by arthropods (arboviruses) threaten the health of humans, livestock, and wildlife worldwide [1,2]. Most arboviruses cycle primarily between blood-feeding arthropod vectors and wild vertebrates, and can subsequently be spread to incidental hosts such as humans or livestock [1,2]. Diseases associated with arboviruses include dengue fever, yellow fever, West Nile encephalitis, and Chikungunya disease, all of which can cause severe symptoms and/or fatality in humans and other hosts.

Arbovirus transmission is governed by the ecological interactions between vectors, hosts, and pathogens across landscapes [1]. Focal points of infection develop in areas where populations of competent vectors, reservoir hosts, and susceptible recipient hosts interact. These focal points are often ephemeral, leading to dramatic fluctuations in the prevalence of some diseases over space and time [1,2]. Understanding the complex set of factors that lead to the formation of focal points of infection, and subsequent disease spread across landscapes, is therefore essential for predicting and mitigating disease outbreaks [1].

As the causal agent of West Nile encephalitis, West Nile virus (WNV) is the most widespread arbovirus in the world [3]. In the United States, WNV was detected in New York State in 1999 and rapidly spread across the country. The invasion of the United States by WNV has caused regional declines of multiple bird species [4] and thousands of infections and deaths in humans and horses [3]. At a landscape scale, where infections per county have been analyzed, urbanization and agricultural intensification appear to increase the prevalence of WNV infection in humans and horses [5–7]. Increased temperatures and decreased precipitation have also been linked to increased infections [6,8,9]. However, although the prevalence of West Nile virus is assumed to be strongly affected by vector and host distributions [10], it remains unclear how local (i.e., sub-county level) interactions between WNV vectors and hosts are affected by land-use and climate to create focal points for WNV transmission across landscapes.

The work reported here had two objectives. First, we used spatially-explicit models to test whether land-use and climate affected the prevalence of WNV infection at landscape and local spatial scales. Second, to examine the mechanisms by which land-use and climate affect focal points of infection for WNV spread, we assessed the distributions of mosquitoes and birds involved in the transmission cycle. Thus, we first determined if factors associated with WNV infection were different, or not, depending on spatial scale. By subsequently analyzing communities of WNV vectors and reservoirs, we linked local vector and host distributions with the prevalence of WNV infection across landscapes.

Methods

Prevalence of West Nile Virus Infection

We examined the effects of land-use and climate on the prevalence of WNV infection at two main spatial scales: 1) landscape: the prevalence of WNV infections per county in humans, horses, and birds over Idaho (ID), Oregon (OR), and Washington (WA) states during 2007–2010 and 2) local: the prevalence of WNV in mosquitoes at field locations over eight counties in WA during 2009–2010. No specific permits were required for the described field studies. Data at the landscape scale were collected from the Center for Disease Control ArboNet database and the WA Department of Health database [11,12].

At the local scale, data on the prevalence of WNV in mosquitoes (*Culex pipiens* and *Cx. tarsalis*) were collected at 101 and 108 field locations in 2009 and 2010, respectively. The locations of these sites were determined by respective mosquito control districts based on public input (complaints about mosquitoes) or their assessment of risk. All field sites were located on public land, and no specific permissions were required for these sampling activities. At each site, mosquitoes were collected using Encephalitis virus surveillance (EVS) traps baited with dry ice. The location of each trap was recorded with a Global Positioning System unit. Mosquitoes were trapped from 21 April to 7 October in 2009, and from 15 April to 22 September in 2010. The number of traps collected varied among locations (range 1–90). Variation in trap density was not associated with the prevalence of WNV at any particular location, but instead was based on methodology of the respective mosquito control districts and the accessibility of the field locations. For data analysis, we only included the locations with at least 5 traps, a condition met at 54 and 69 field locations in 2009 and 2010, respectively.

Collected bags of mosquitoes from EVS traps were kept in coolers until they were processed. For processing, mosquitoes were knocked down with dry ice and then sorted on ice or a chill table. The total number of female mosquitoes collected was recorded, and all female mosquitoes were pooled according to species (12–50 specimens per pool) prior to testing for WNV. We only included females in mosquito pools as only females blood-feed and are responsible for WNV transmission. Female mosquitoes were identified to species using a clear dichotomous key [13] by trained technicians at respective mosquito control districts. Identification of mosquitoes to species was necessary so that tested pools only contained *Cx. tarsalis* and *Cx. pipiens*, as these species account for the majority of WNV infections in the Pacific Northwestern United States [3]. Other mosquito species are not important vectors of WNV in our sampled region, and were therefore excluded from analyses. These data were also used in the analysis of mosquito abundance (see Vector and host distributions). When traps had more than 50 mosquitoes, a random subsample of 50 was used for WNV testing.

Mosquito pools were examined for the presence/absence of WNV RNA with the Rapid Analyte Measurement Platform (RAMP®) WNV test (Response Biomedical Corp., Burnaby, Canada), following the manufacturer's instructions or by reverse transcription-polymerase chain reaction (RT-PCR). RAMP test results with a value of ≥300.0 RAMP units were considered positive. Test results with values between 50.1 and 299.9 RAMP units were considered negative, unless confirmed positive by PCR testing (these samples were shipped to Oregon State University for confirmatory testing). Samples with values <50 RAMP units were

considered negative and were not tested by PCR. Mosquito samples tested by PCR only were shipped to the Center for Vector-borne Diseases at the University of California, Davis. The detection of WNV RNA was conducted with real-time -PCR, using TaqMan Fast Virus 1-Step Master Mix (Applied Biosystems, Carlsbad, USA) with WNV specific primers [14] on an ABI MagMax instrument.

Vector and Host Distributions

To link the prevalence of WNV infections with vector and host distributions, we examined local factors affecting mosquito and bird communities. Data on mosquito abundance were taken from the WNV survey sites. Data on bird abundance and species composition were obtained from 136 Breeding Bird Survey (BBS) sites from 2007 to 2010 [15] (Fig. S1). The BBS is a United States Geographical Survey funded project that examines bird communities throughout the United States. The BBS follows a standard protocol, with observers driving along a 39.4-km roadside route. Every 0.8 km, observers record the total number and species of birds seen or heard during a 3-min observation period. The total area sampled per route is 25.4 km². Survey routes are sampled once per year. At each BBS site, we calculated the total bird abundance, the number of bird species, and the abundance of two common enzootic amplification hosts of WNV: the American robin and the house sparrow [16–18].

Land Cover and Climatic Data

We obtained land cover and climatic data to relate land-use and climate with the prevalence of WNV infections and the distributions of vectors and hosts. We determined land cover using USDA Cropland Datalayer (CDL) maps, which provide remotely sensed data on land-use throughout the United States [19]. From 2007–2009, CDL maps were produced at a 56-m resolution; 2010 maps were produced at a 30-m resolution. These differing resolutions, however, did not affect how land cover was evaluated.

To determine land cover for the landscape analysis, we imported the CDL maps into ArcGIS [20] and then calculated the abundance of three habitat types in each county over WA, OR, and ID: vegetable/forage crops, orchards, and natural (Table S1). Vegetable/forage crops were considered differently from orchards because they are typically grown under central-pivot irrigation, while orchards are not. To determine land cover surrounding each BBS site, we followed the methods of Meehan et al. [21]. Briefly, we extracted the area for each habitat type from rectangular buffers around survey routes, with buffers extending 0.4 km from the route to reflect the observation distance at which bird species were surveyed. To scale habitat area derived from the rectangular buffer (31.5 km²) to the sum of circular buffers sampled by the BBS (25.4 km², see above), we multiplied land cover areas by the factor 0.81 (25.4 km²/31.5 km²). This scaling assumes that land-use contained within the rectangular buffers is the same as in the area sampled by the BBS observer [21].

To determine land cover surrounding mosquito trapping locations, we drew 10 concentric rings in GIS around each field location over a scale from 0.05 to 0.5 km (each ring was 0.05 km). The maximum radii was based on usual mosquito dispersal, which typically occurs over distances <0.5 km [22]. Furthermore, the use of distances >0.5 km did not improve the fit of models during data analysis, indicating that variation in habitat structure did affect mosquito over these distances. In each ring, we measured the acreage of each of the three habitat types (vegetable/forage, orchard, natural) using ArcGIS.

Climatic data (temperature and precipitation) were collected from the Western Regional Climate Center [23]. To obtain climatic data for the landscape analysis, we randomly selected three weather stations from each county where WNV had been detected. From each station, we obtained the annual average temperature and precipitation across the sampling period (2007 to 2010) to determine the average climatic conditions over the period where WNV was surveyed. The obtained climatic data were averaged over these stations to produce county averages; in cases where only one or two stations were located in a county, data from those stations were used.

To assess climatic factors associated with our mosquito and bird sites, we used data from these same weather stations. In these analyses, we estimated the temperature and precipitation at field locations where mosquitoes where trapped or at BBS sites (at the center of BBS survey rectangles), using inverse distance weighting (IDW) interpolation in ArcGIS [20]. Here, IDW was used to interpolate climatic data between spatially discontinuous weather stations. For each field location, the temperature and precipitation for each year (2007–2010) was calculated by averaging the weighted sums of temperature and precipitation data from the 12 nearest weather stations, the stations farther away influencing the climate estimates less than those closer to the site (decay component = 2).

Data Analyses

At the landscape scale we used multiple regression models to evaluate associations between the three habitat type acreages (vegetable/forage crops, orchards, and natural), two climatic variables (temperature and precipitation), and all two-way interactions, on the prevalence of WNV infections in humans, horses, and birds. Prior to data analysis, the number of WNV infections in humans was standardized by county population; infections in horses and birds were standardized by county area. WNV prevalence in each of the three groups (human, horse, bird) was highly non-normal, so we used rank-based statistics which do not require the normality assumption [24]. Each county served as one sampling unit in analyses performed at the landscape scale. For each model, we first used stepwise regression [24] to select a subset of explanatory variables that minimized the Akaike's Information Criterion (AIC); these models were subsequently used in all further analyses of factors affecting the prevalence of WNV infection. Results obtained with the Bayesian Information Criterion (BIC) were very similar and therefore are not reported. These models were fit in JMP [25].

We used logistic regression to evaluate relationships between habitat type acreages, climatic variables, and the local prevalence of WNV infection in mosquito pools. WNV infection counts were binomial, with each field location where mosquitoes were trapped providing one observation. The number of mosquitoes tested per pool was included as a covariate. Models were fitted separately in 2009 and 2010, at each of the 10 spatial scales (0.05–0.5 km). In each year, we used the AIC and the corresponding R^2 value to determine the scale at which the most variation was explained, and used these models for further analyses. To assess the presence of spatial autocorrelation in the residuals from the logistic regression models selected in each year analyzing the prevalence of WNV in mosquitoes, we used the co-regionalization analysis with a drift method [26–28], and Pearson's chi-square residuals and deviance residuals were evaluated before and after applying a Box-Cox transformation-. In this context (i.e., geostatistical analysis of spatial data at small vs. large scale), it is possible to model a 'drift' (representing large-scale heterogeneity of the mean) globally or using a moving window. These models were fit in Matlab [29].

From the 16 spatial autocorrelation analyses (2 years × 2 types of residuals × 2 data transformations × 2 drift models), only one (2010, Pearson residuals, no Box-Cox transformation, local drift model) revealed significant spatial autocorrelation at relatively small distances (up to 6 km). There was thus no need for adjusting the statistical tests of significance of the logistic regression models and estimated slopes. Resulting logistic regression models were fit in JMP [25].

For analysis of mosquito and bird abundance, and bird species richness, we used multiple regression models. Each field location where mosquitoes were trapped or each BBS site provided one observation. In these models, we used rank-based statistics that did not require the normality assumption [24]. For mosquito abundance, models were analyzed at each of 10 spatial scales (from 0.05 to 0.5 km), with land-use at each scale as an explanatory variable (climate was the same at all scales). For birds, abundance and species richness at each BBS site were averaged over sampling years, and a single value of land-use from GIS models was used in the analyses. Semivariograms were computed to quantify and analyze spatial autocorrelation [30] in the abundance of mosquitoes and the abundances and species richness of bird hosts. Spatial autocorrelation was accounted for in tests of significance following the approach of Carrière et al. [30], by using effective sample sizes and degrees of freedom in modified t tests designed for multiple regression analyses with spatial data [28]. These analyses were performed in Matlab [29].

Results

Prevalence of West Nile Virus Infection

At the landscape scale, the numbers of WNV infections in horses and birds were significantly positively associated with the acreage of orchard habitats, but were not significantly affected by other habitats (Fig. 1a,b, Table S2). The number of human infections was not significantly associated with any habitat (Fig. 1c, Table S2). The number of infections in humans, horses, and birds was significantly negatively associated with precipitation (Fig. 2, Table S2), and the interaction between orchard acreage and precipitation was significantly negative (Table S2). Temperature was not significantly associated with the number of infections in humans, horses, or birds (Table S2).

At a local scale, in both 2009 and 2010 the prevalence of WNV infections in mosquitoes was significantly positively associated with the acreage of orchard habitats (Fig. 3, Table S3). These results were based on a total of 22,141 and 28,504 Cx. pipiens collected and tested in 2009 and 2010, respectively; a total of 25,461 and 49,293 Cx. tarsalis were collected and tested in both years, respectively. The prevalence of WNV infection was similar in both species, with 14.5 or 13.5% of mosquito pools containing Cx. pipiens or Cx. tarsalis testing positive for WNV, respectively. In 2010, the prevalence of WNV in mosquitoes was also positively associated with the acreage of vegetable/forage and natural habitats, and negatively associated with temperature (Table S3). The prevalence of WNV in mosquitoes was not significantly associated with precipitation in either year (Table S3). In both years, the strength of these effects varied from 0.1–0.5 km (Figs. S2, S3).

Effects of land-use and climate on mosquito and bird distributions. The abundance of both Cx. pipiens and Cx. tarsalis, and the combined abundance of these two species, increased locally with greater acreage planted to orchards, but were unaffected by other habitat types or climatic variables (Table 1). Similarly, the abundance of American robins and house sparrows increased at sites with greater orchard acreages (Table 1).

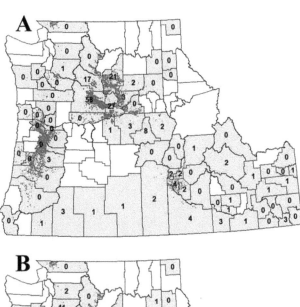

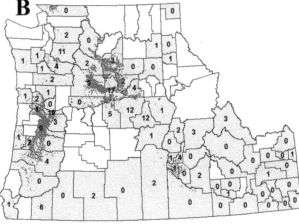

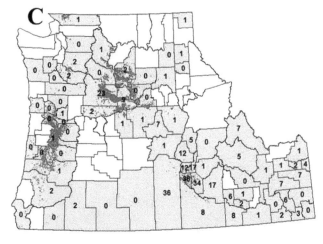

Figure 1. Land-use and the prevalence of West Nile virus infection. Number of West Nile virus (WNV) infections in (A) horses, (B) birds, and (C) humans from 2007 to 2010 over Idaho, Oregon, and Washington States. Counties shaded gray had at least one WNV case detected in any species over this period, with the numbers indicating the number of infections in each organism(s). Land covered by orchard habitats is shown in green.

While American robins and house sparrows benefited from orchard habitats, we found no evidence that land-use or climate affected total bird abundance or species richness (Table 1). Thus,

American robins and house sparrows increased in relative as well as absolute abundance in orchards (Table 1).

Discussion

The prevalence of WNV infections increased in birds (reservoir hosts) and horses (incidental hosts) across landscapes modified for orchard production and with reduced precipitation (Fig. 1). Similarly, the prevalence of WNV infections in humans was negatively associated with precipitation, although human infections were not associated with land-use. Human cases, however, may not accurately reflect counties where infections actually occurred due to traveling; the same could be true for migratory birds that travel outside of the county where they were infected.

Our results show that decreased precipitation was associated with higher prevalence of WNV infection (Fig. 2). Dry conditions can reduce the abundance of mosquito predators and competitors, leading to increased mosquito abundance and disease prevalence [31]. Dry conditions can also promote congregations of mosquitoes and birds in refuges where water is present, and dispersal from such refuges can promote disease spread [9,32]. Conversely, high levels of precipitation can decrease adult mosquito activity and larval survival [33,34]. Our results show that climate did not significantly affect mosquito or bird distributions (Table 1). This suggests that the effects of climate on the prevalence of WNV infections could be due to altered mosquito or bird activity, rather than a quantitative effect on vector or host abundances.

Our results show that in addition to climatic factors, land-use strongly affected the prevalence of WNV across landscapes (Fig. 1). Results seen here are generally in accordance with studies on the prevalence of WNV infection at landscape scales [8,9]. For example, intensification of agriculture has been shown to promote the prevalence of WNV infections, measured on a per-county basis, in humans and horses [7–9]. Here we show that orchard habitats, but not vegetable/forage crops, were associated with a greater prevalence of WNV. Thus, our results show that not all forms of agriculture should be considered equal in terms of their suitability for promoting or limiting the spread of WNV. As agricultural management practices differ widely among regions, states, counties, and crop types, future models should explicitly test how specific types of agriculture (or other land-use factors) affect disease prevalence.

While our results suggest that land-use and climate strongly affect the prevalence of WNV infections across landscapes, infections per county may fail to reveal underlying ecological processes that operate over smaller spatial scales [35]. To address this uncertainty, we examined the prevalence of WNV infection in mosquitoes at a local scale (field locations within a county). This allowed us to determine whether greater prevalence of infection in incidental hosts (mammals) and reservoir hosts (birds) was linked with greater prevalence of WNV in mosquito vectors. Furthermore, this allowed us to increase the resolution of our estimates for infection risk at a sub-county level scale, a commonly overlooked component in epidemiological studies of arboviruses [35,36].

Our results reveal that orchard habitats, which promoted infections in mammals and birds across landscapes, were also associated with the local prevalence of WNV infection in mosquitoes (Fig. 3). By performing spatially-explicit analyses at two spatial scales, our results linked landscape patterns of infection with local ecological factors that support pathogen transmission. For example, while natural habitats and vegetable/forage crops were associated with more WNV infections in mosquitoes in 2010, they did not affect infections in humans, horses, or birds at the landscape scale (Table 1). This suggests that these habitats did not

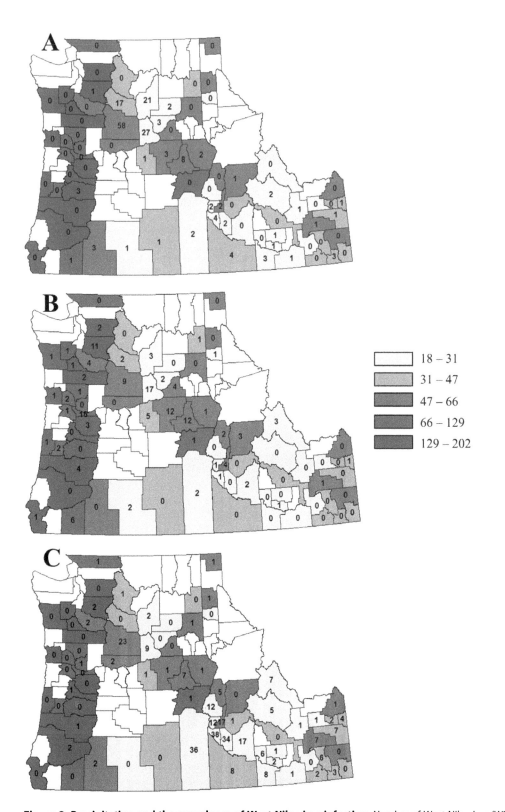

Figure 2. Precipitation and the prevalence of West Nile virus infection. Number of West Nile virus (WNV) infections in (A) horses, (B) birds, and (C) humans from 2007 to 2010 over Idaho, Oregon, and Washington States. Counties are shaded according to their annual average precipitation (mm, only counties where WNV was detected in any species are shaded), with the numbers indicating the number of infections in each organism(s).

produce focal points for WNV infection. In contrast, more abundant orchard habitats were associated with greater WNV prevalence in mosquitoes in both 2009 and 2010, and greater prevalence of WNV across landscapes (Fig. 3). This suggests that orchard habitats produced local focal points that promoted vector-host contact and subsequent pathogen transmission across landscapes, supporting the hypothesis that local processes strongly impact WNV infections across landscapes [10].

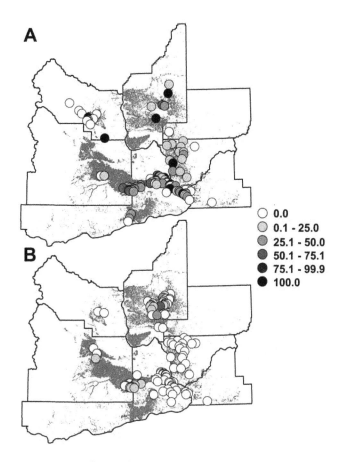

Figure 3. Prevalence of West Nile virus infection in mosquitoes. Proportion of mosquito pools that tested positive for West Nile virus (WNV) infection at each mosquito trap location in (A) 2009 and (B) 2010 across eight counties in Washington State. Each mosquito trapping location is indicated with a circle, with different levels of shading indicating a different class of values for the proportion of positive mosquito pools. Land covered by orchards in each year is shown in green.

Legend:
- 0.0
- 0.1 – 25.0
- 25.1 – 50.0
- 50.1 – 75.1
- 75.1 – 99.9
- 100.0

The anthropophilic nature of mosquito vectors and bird reservoirs is often assumed to be associated with the strong effects of agricultural intensification on WNV prevalence [10]. To test this hypothesis, and to determine the mechanism by which orchard habitats promoted the prevalence of WNV infection in mosquitoes, birds, and incidental hosts, we analyzed the impacts of land-use and climate on mosquito and bird communities. Our analysis focused on the two most prevalent WNV vectors in the Pacific Northwestern United States, *Cx. pipiens* and *Cx. tarsalis*, and two key bird species, the American robin and the house sparrow. Kilpatrick et al. [16] showed that even when relatively uncommon, American robins were likely associated with nearly 60% of WNV infections in mosquitoes from several locations in the eastern United States. Similarly, Hamer et al. [18] found that >95% of infectious *Cx. pipiens* mosquitoes had fed on house sparrows and/or robins; Kent et al. [17] found that in Colorado, American robins were a source of WNV infected mosquitoes early in the season and house sparrows were a key host later in the season. These seasonal dynamics are important for WNV transmission to incidental hosts, as mosquitoes often shift their feeding from bird to non-bird hosts when preferred bird hosts become less abundant [37].

Not surprisingly, orchards were associated with greater abundance of mosquitoes, American robins, and house sparrows across sites (Table 1). Orchards provide a readily available source of plant nectar during flowering, which is essential for the survival of adult mosquitoes [38]. Both American robins and house sparrows also use orchards as nesting and feeding sites [39,40]. Thus, orchards promoted the abundance of three species that are critical components of the WNV transmission cycle. In other words, orchard habitats likely amplified vectorial capacity by promoting host-vector contact and supporting mosquito survival [41,42]. By amplifying vectorial capacity, these habitats promoted the spread of WNV from focal points of infection across landscapes. These relationships were only identified by linking local and landscape-level processes.

Although American robins and house sparrows benefitted from orchards, we found no evidence for effects of land-use or climate on bird richness or total abundance (Table 1). The spread of infection is often reduced when host diversity is high [43,44]. This "dilution effect" occurs because the number of host species that are unsuitable blood-meal sources, or are poor reservoir hosts for the pathogen, increases in diverse communities. This makes it less likely that vectors will feed on suitable blood-meal hosts or

Table 1. Results of multiple regression analysis.

Response	Explanatory Variable									
	Temperature		Precipitation		Vegetable/forage		Orchard		Natural	
	Slope	P	Slope	P	Slope	P	Slope	P	Slope	P
Cx. pipiens abundance	−0.40	0.34	−0.24	0.51	−0.16	0.91	0.58	0.0002	0.082	0.51
Cx. tarsalis abundance	0.017	0.95	0.16	0.58	0.27	0.17	0.34	0.034	0.23	0.26
Cx pipiens+Cx tarsalis abundance	0.028	0.91	0.24	0.30	0.17	0.12	0.51	<0.0001	0.088	0.39
American robin abundance	0.039	0.45	0.10	0.41	−0.17	0.26	0.22	0.053	−0.063	0.69
House sparrow abundance	0.011	0.89	−1.02	0.38	−0.11	0.42	0.30	0.0036	−0.12	0.40
Robin+sparrow abundance	0.062	0.50	0.034	0.79	−0.15	0.33	0.22	0.046	−0.092	0.56
Proportion robins+sparrows	0.13	0.17	0.029	0.81	−0.045	0.77	0.21	0.064	−0.0098	0.95
Total bird abundance	−0.037	0.68	−0.023	0.86	−0.26	0.096	0.18	0.11	−0.26	0.11
Bird species richness	0.13	0.17	0.077	0.54	−0.19	0.21	0.078	0.49	−0.044	0.78

Results from a multiple regression model on effects of land-use and climate on mosquito and bird abundance and community composition.

pathogen reservoirs [43]. However, our results suggest that the dilution effect did not occur in our study regions. In contrast, American robins and house sparrows increased in both relative and absolute abundance in orchards (Table 1). This might lead to a greater proportion of feedings on these highly-suitable reservoirs, amplifying WNV spread, which suggests these species are key for WNV transmission [16–18].

It is clear that predicting the spread of arboviruses requires a system-based approach that explores ecological interactions between vectors, hosts, and pathogens across landscapes [1,10]. While the spread of many pathogens are often well characterized at landscape scales, the complex ecological factors driving these patterns at local scales are often poorly understood [35]. Here, we showed that combining spatially-explicit models with an assessment of vector, host, and pathogen distributions allows for a robust examination of the processes driving arbovirus transmission across multiple spatial scales. Linking local and landscape-level epidemiological studies in this way can form the basis for management strategies to predict and reduce the spread of arboviruses and other pathogens.

Acknowledgments

We thank the Benton, Yakima, Franklin, Columbia and Grant #1 Mosquito Control Districts, and the Kittitas County Local Health Department, for sharing their mosquito surveillance data. * Disclaimer. The contents of this manuscript are solely the responsibility of the authors and do not necessarily represent the official views or policies of the Washington State Department of Health.

Author Contributions

Conceived and designed the experiments: DWC EAD JPO. Performed the experiments: DWC EAD JMB AD CR EM WP. Analyzed the data: DWC YC PD. Wrote the paper: DWC EAD YC PD JPO.

References

1. Reisen WK (2010) Landscape epidemiology of vector-borne diseases. Annu Rev Entomol 55: 461–483.
2. Weaver SC, Reisen WK (2010) Present and future arboviral threats. Antiviral Res 85: 328–345.
3. Kramer LD, Styer LM, Ebel GD (2008) A global perspective on the epidemiology of West Nile virus. Annu Rev Entomol 53: 61–81.
4. LaDeau SL, Kilpatrick AM, Marra PP (2007) West Nile virus emergence and large-scale declines of North American bird populations. Nature 447: 710–713.
5. Brown HE, Childs JE, Diuk-Wasser M, Fish D (2008) Ecological factors associated with West Nile virus transmission, Northeastern United States. Emerg Infect Dis 14: 1539–1545.
6. Ward MP, Wittich CA, Fosgate G, Srinivasan R (2009) Environmental risk factors for equine West Nile virus disease cases in Texas. Vet Res Commun 33: 461–471.
7. Bowden SE, Magori K, Drake JM (2011) Regional differences in the association between land cover and West Nile virus disease prevalence in humans in the United States. Am J Trop Med Hyg 84: 234–238.
8. Wimberly MC, Hildreth MB, Boyte SP, Lindquist E, Kightlinger L (2008) Ecological niche of the 2003 West Nile virus epidemic in the Northern Great Plains of the United States. PLoS One 3: e3744.
9. Wang G, Minnis RB, Belant JL, Wax CL (2010) Dry weather induces outbreaks of human West Nile virus infections. BMC Infect Dis 10: 38.
10. Kilpatrick AM (2011) Globalization, land use, and the invasion of West Nile virus. Science 334: 323–327.
11. Center for Disease Control. ArboNet. Available: http://diseasemaps.usgs.gov/wnv_historical.html. Accessed 2012 October 29.
12. Washington State Department of Health. West Nile virus surveillance maps and statistics. Available: http://www.doh.wa.gov/DataandStatisticalReports/DiseasesandChronicConditions/WestNileVirus.aspx. Accessed 2012 October 29.
13. Darsie RF, Ward RA (2004) Identification and geographical distribution of the mosquitoes of North America, north of Mexico. Gainesville: University of Florida Press. 400 pp.
14. Kesavaraju B, Farajollahi A, Lampman RL, Hutchinson M, Krasavin NM, et al. (2012) Evaluation of a rapid analyte measurement platform for West Nile virus detection based on United States mosquito control programs. Am J Trop Med Hyg 87: 359–363.
15. US Geological Survey Patuxent Wildlife Research Center. North American Breeding Bird Survey. Available: http://www.pwrc.usgs.gov/bbs/. Accessed 2012 October 29.
16. Kilpatrick AM, Daszak P, Jones MJ, Marra PP, Kramer LD (2006) Host heterogeneity dominates West Nile virus transmission. Proc Roy Soc B 273: 2327–2333.
17. Kent R, Juliusson L, Weissmann M, Evans S, Komar N (2009) Seasonal Blood-Feeding Behavior of Culex tarsalis (Diptera: Culicidae) in Weld County, Colorado, 2007. J Med Entomol 46: 380–390.
18. Hamer GL, Chaves LF, Anderson TK, Kitron UD, Brawn JD, et al. (2011) Fine-scale variation in vector host use and force of infection drive localized patterns of West Nile virus transmission. PLoS One 6: e23767.
19. US Department of Agriculture National Agriculture Statistics Service Spatial Analysis Research Section. Cropland Data Layer. Available: http://www.nass.usda.gov/research/Cropland/SARS1a.htm. Accessed 2012 October 29.
20. ESRI (2010) ArcGIS Desktop: Release 10. Redlands, CA.
21. Meehan TD, Hurlbert AH, Gratton C (2010) Bird communities in future bioenergy landscapes of the Upper Midwest. Proc Nat Acad Sci USA 107: 18533–18538.
22. Service MW (1997) Mosquito dispersal – the long and short of it. J Med Entomol 34: 578–588.
23. Western regional climate center. Cooperative climatological data summaries. Available: http://www.wrcc.dri.edu/climatedata/climsum/. Accessed 2012 October 29.
24. Carrière Y, Dutilleul P, Ellers-Kirk C, Pedersen B, Haller S, et al. (2004) Sources, sinks, and the zone of influence of refuges for managing insect resistance to Bt crops. Ecol Appl 14: 1615–1623.
25. SAS Institute (2010) JMP 10.0. Cary, NC.
26. Pelletier B, Dutilleul P, Larocque G, Fyles JW (2009) Coregionalization analysis with a drift for multi-scale assessment of spatial relationships between ecological variables 1. Estimation of drift and random components. Environ Ecol Stat 16: 439–466.
27. Pelletier B, Dutilleul P, Larocque G, Fyles JW (2009) Coregionalization analysis with a drift for multi-scale assessment of spatial relationships between ecological variables 2. Estimation of correlations and coefficients of determination. Environ Ecol Stat 16: 467–494.
28. Dutilleul P (2011) Spatio-Temporal Heterogeneity: Concepts and Analyses. Cambridge: Cambridge University Press. 393 pp.
29. The Mathworks Inc (2008) MATLAB Version R2008a. Natick, MA.
30. Carrière Y, Ellers-Kirk C, Hartfield K, Larocque G, Degan B, et al. (2012) Large-scale, spatially-explicit test of the refuge strategy for delaying insecticide resistance. Proc Natl Acad Sci USA 109: 775–780.
31. Chase JM, Knight TM (2003) Drought-induced mosquito outbreaks in wetlands. Ecol Lett 6: 1017–1024.
32. Shaman J, Day JF, Stieglitz M (2002) Drought-induced amplification of Saint Louis encephalitis virus, Florida. Emerg Infect Dis 8: 575–580.
33. Paaijmans KP, Wandago MO, Githeko AK, Takken W (2007) Unexpected high losses of Anopheles gambiae larvae due to rainfall. PLoS ONE 2: e1146.

34. Jones CJ, Lounibos LP, Marra PP, Kilpatrick AM (2012) Rainfall influences survival of *Culex pipiens* mosquitoes in a residential neighborhood in the mid-Atlantic USA. J Med Entomol 49: 467–473.

35. Eisen RJ, Eisen L (2008) Spatial modeling of human risk of exposure to vector-borne pathogens based on epidemiological versus arthropod vector data. J Med Entomol 45: 181–192.

36. Rochlin I, Ginsberg HS, Campbell SR (2009) Distribution and abundance of host-seeking *Culex* species at three proximate locations with different levels of West Nile virus activity. Am J Trop Med Hyg 80: 661–668.

37. Kilpatrick AM, Kramer LD, Jones MJ, Marra PP, Daszak P (2006) West Nile Virus Epidemics in North America Are Driven by Shifts in Mosquito Feeding Behavior. PLoS Biology 4: e82.

38. Clements AN (2010) The Biology of Mosquitoes. Oxfordshire: CABI. 752 pp.

39. Fluetsch KM, Sparling DW (1994) Bird nesting success and diversity in conventionally and organically managed apple orchards. Environ Toxicol Chem 13: 1651–1659.

40. Lothrop HD, Reisen WK (2001) Landscape affects the host-seeking patterns of *Culex tarsalis* (Diptera: Culicidae) in the Coachella Valley of California. J Med Entomol 38: 325–332.

41. Reisen WK (1989) Estimation of vectorial capacity: relationship to disease transmission by malaria and arbovirus vectors. Bull Soc Vector Ecol 14: 39–40.

42. Marra PP, Griffing S, Caffrey C, Kilpatrick AM, McLean R, et al. (2004) West Nile virus and wildlife. Bioscience 54 : 393–402.

43. Keesing F, Holt RD, Ostfeld RS (2006) Effects of species diversity on disease risk. Ecol Lett 9: 485–498.

44. Allan BF, Langerhans RB, Ryberg WA, Landesman WJ, Griffin NW, et al. (2009) Ecological correlates of risk and prevalence of West Nile virus in the United States. Oecologia 158: 699–708.

6

Subterranean, Herbivore-Induced Plant Volatile Increases Biological Control Activity of Multiple Beneficial Nematode Species in Distinct Habitats

Jared G. Ali[1¤], **Hans T. Alborn**[2], **Raquel Campos-Herrera**[1,3], **Fatma Kaplan**[2], **Larry W. Duncan**[1], **Cesar Rodriguez-Saona**[4], **Albrecht M. Koppenhöfer**[4], **Lukasz L. Stelinski**[1]*

1 Entomology and Nematology Department, Citrus Research and Education Center, University of Florida, Lake Alfred, Florida, United States of America, 2 Center for Medical, Agricultural, and Veterinary Entomology, Agricultural Research Service, U.S. Department of Agriculture, Gainesville, Florida, United States of America, 3 Departamento de Contaminación Ambiental, Instituto de Ciencias Agrarias, Centro de Ciencias Medioambientales, Madrid, Spain, 4 Department of Entomology, Rutgers University, New Brunswick, New Jersey, United States of America

Abstract

While the role of herbivore-induced volatiles in plant-herbivore-natural enemy interactions is well documented aboveground, new evidence suggests that belowground volatile emissions can protect plants by attracting entomopathogenic nematodes (EPNs). However, due to methodological limitations, no study has previously detected belowground herbivore-induced volatiles in the field or quantified their impact on attraction of diverse EPN species. Here we show how a belowground herbivore-induced volatile can enhance mortality of agriculturally significant root pests. First, in real time, we identified pregeijerene (1,5-dimethylcyclodeca-1,5,7-triene) from citrus roots 9–12 hours after initiation of larval *Diaprepes abbreviatus* feeding. This compound was also detected in the root zone of mature citrus trees in the field. Application of collected volatiles from weevil-damaged citrus roots attracted native EPNs and increased mortality of beetle larvae (*D. abbreviatus*) compared to controls in a citrus orchard. In addition, field applications of isolated pregeijerene caused similar results. Quantitative real-time PCR revealed that pregeijerene increased pest mortality by attracting four species of naturally occurring EPNs in the field. Finally, we tested the generality of this root-zone signal by application of pregeijerene in blueberry fields; mortality of larvae (*Galleria mellonella* and *Anomala orientalis*) again increased by attracting naturally occurring populations of an EPN. Thus, this specific belowground signal attracts natural enemies of widespread root pests in distinct agricultural systems and may have broad potential in biological control of root pests.

Editor: Martin Heil, Centro de Investigación y de Estudios Avanzados, Mexico

Funding: This research was partially supported by grants from the USDA-CSREES, Citrus Research and Development Foundation and start up funding provided by the University of Florida to LLS. RCH was supported by the Seventh Framework Programme of the European Union (FP7-PEOLE-2009-10F-252980). Funders had no role in study design, data collection and analysis, decision to publish, or preparation of the manuscript.

Competing Interests: The authors have declared that no competing interests exist.

* E-mail: stelinski@ufl.edu

¤ Current address: Department of Ecology and Evolutionary Biology, Cornell University, Ithaca, New York, United States of America

Introduction

Natural enemies of herbivorous pests use flexible foraging strategies that often incorporate environmental cues emitted by the herbivore's host plant. While the role of herbivore-induced volatiles in plant-herbivore-natural enemy interactions is well-documented aboveground [1–5], new evidence from several systems, including strawberry, maize and, most recently, citrus, indicates that induced root volatiles may protect plants by attracting entomopathogenic nematodes (EPNs) [6–12]. However, to date, only one root-induced attractant has been described and shown to enhance the effectiveness of EPNs in the field: (*E*)-β-caryophyllene from the roots of maize (*Zea Mays* L.) [11,13]. The disparity in the number of aboveground investigations versus analogous belowground research on indirect defense is largely due to technical limitations rather than a lack of ecological or agricultural relevance [9,14]. No previous studies have detected a belowground herbivore-induced volatile from intact plants in the field or measured the effectiveness of belowground attractants for recruiting populations of naturally occurring EPNs in the soil. Depending on the specificity of interactions, the identification and manipulation of a root signal in the field could well enhance biological control of diverse root pests in agroecosystems.

Larvae of the weevil *Diaprepes abbreviatus* (L.), introduced into Florida in 1964 [15], feed on the roots of more than 290 plant species including citrus, sugarcane, potatoes, strawberries, sweet potatoes, papaya, and non-cultivated wild plants [16]. Over the past 40 years, the weevil has significantly contributed to the damage and spread of disease in agricultural plants [17]. Because pesticides are expensive, environmentally hazardous and often ineffective [18,19], currently the most effective alternative method of root-pest control is the application of EPNs from the genera *Heterorhabditis* and *Steinernema* [20]. EPNs are obligate parasites that kill their host with the aid of a symbiotic bacterium [21,22]. Over its 20 years of use, the efficacy of mass release of EPNs as a biopesticide for *D. abbreviatus* has been reported as varying and unpredictable, ranging anywhere between 0 to >90% [23]. Promoting plant attractiveness to natural enemies is a novel alternative to traditional broad-spectrum pesticides, which indiscriminately kill predators and parasitoids and often lead to subsequent pest resurgence [24–27]. Deploying herbivore-induced plant volatiles (HIPVs) aboveground by controlled release dispensers has been shown to increase recruitment and retention of beneficial natural enemies to plants [27–29]. In an analogous belowground investigation, EPN infection of western corn rootworm (*Diabrotica virgifera virgifera* LeConte) larvae was increased by spiking soil surrounding maize roots with the HIPV, (*E*)-β-caryophyllene [11]. Herein, we investigated the mechanisms by which a novel HIPV affects naturally occurring EPN species and their consequence on belowground herbivores in two distinct agroecosystems.

We have recently shown that a citrus root stock (*Citrus paradisi* Macf. × *Poncirus trifoliata* L. Raf.) releases HIPVs in response to larval feeding by the weevil, *D. abbreviatus*, and that these HIPVs attract EPN species in lab bioassays [7,8]. Here we characterize the specific HIPV attractant as 1,5-dimethylcyclodeca-1,5,7-triene (pregeijerene) and show its real-time release in response to herbivory. We also demonstrate that field application of this volatile increases mortality of belowground root weevils by attracting naturally occurring nematodes. We used recently developed qPCR primers and probes to detect and enumerate cryptic species of EPNs allowing for species-specific quantification of nematode response to attractants belowground. Given the broad effect of pregeijerene on EPN species, we also tested and demonstrated its efficacy in a non-citrus, temperate climate agroecosystem. The use of plant-produced signals, such as the damage-induced release of pregeijerene, along with conservation biological control strategies, could extend the usefulness of EPNs in crops damaged by belowground herbivores.

Results

Induction of Root Volatiles

Volatiles were non-destructively sampled every three h from the root zone of citrus seedlings in glass chambers (Analytical Research Systems, Gainesville, FL) with sandy soil [8]. Gas chromatography-mass spectrometry (GC-MS) revealed 1,5-dimethylcyclodeca-1,5,7-triene (pregeijerene) as the dominating volatile, reaching a maximum release between nine and 12 h after initiation of larval feeding (Figure 1). There was an effect of treatment ($F_{1,4} = 26.4$, $P = 0.0005$) and volatile release over time ($F_{3,2} = 2812$, $P = 0.0005$). A stainless steel probe (Figure S3) was designed to collect volatiles in the field from the soil beds surrounding citrus trees in an unmanaged orchard. Here, GC-MS analyses again revealed pregeijerene in the root zone, as the most abundant volatile at one m away from the trunks of trees and still at detectable levels at 10 m from trees (Figure S4).

Nematode Attraction in the Field using HIPVs

We next conducted field tests to determine whether application of volatiles collected from infested roots would impact EPN-inflicted mortality of sentinel *D. abbreviatus* larvae. Commercially available EPN had been applied to the test orchard at numerous occasions; however, their persistence was not monitored. Cylindrical mesh cages containing a single *D. abbreviatus* larva in autoclaved sandy soil [30] were treated with: (i) volatiles collected from weevil-infested roots or (ii) a blank solvent control (Figure S5). Larval mortality was 74±6.9% in the presence of volatiles from infested roots, but only 41±7.5% in the solvent alone treatment (N = 10, $t_{18} = 2.75$, $P = 0.013$).

Nematode Attraction in the Field using Isolated Pregeijerene

A second experiment tested whether pregeijerene alone would increase mortality of larvae by attracting EPNs. For these experiments, a sufficient amount of pregeijerene was first extracted and purified from the roots of Common Rue (*Ruta graveolens* L.). The structure of extracted pregeijerene was confirmed by nuclear magnetic resonance (NMR) (Table S1, Table S2). To test for the attractiveness of pregeijerene, we first used serial dilutions of purified compound in dichloromethane in two-choice, sand-filled olfactometers [7,8,31]; 8 ng/μl (in 30 μl aliquots) was found to be the optimally attractive dosage to EPNs (*S. riobrave* and *H. indica*) (Figure 2). We used real-time qPCR to quantify the attraction of naturally occurring EPNs in the field and identified them to species. Our approach was to use species-specific primers and probes to identify EPN species known to either naturally occur in Florida: *S. diaprepesi*, *H. indica*, *H. zealandica*, and *Steinernema* sp. LWD1 (an undescribed species in the *S. glaseri* group); those which were applied to citrus orchards in the form of commercial biopesticides (*S. riobrave*); or those which might be introduced from natural long-distance spread from pastures and golf courses to manage mole crickets (*S. scapterisci*) [32,33]. Mortality of larvae buried with purified pregeijerene was > 3-fold higher than that of larvae buried with the solvent control (Figure 3A). The number of EPNs detected within (Figure 3B) and around (Figure 3C) cages containing purified pregeijerene was significantly higher than that from cages with the solvent control. Tukey HSD test indicated *H. indica* and *H. zealandica* were more abundant than *Steinernema* sp. LWD1 and *S. diaprepesi* ($P < 0.0001$ in all comparisons); however, there were no differences in the relative representation of species between treated and control samples. Neither *S. riobrave*, nor *S. scapterisci* were detected in any of the samples.

Manipulation of Nematode Behavior with Pregeijerene in Blueberry Plantings

Finally, we tested the generality of pregeijerene as an EPN attractant by deploying the compound in a geographically distant, non-citrus agricultural system: commercial highbush blueberry, *Vaccinium corymbosum* L., in Chatsworth, NJ. No pregeijerene was detected in volatiles collected from soil surrounding blueberry roots. Cages (described above) containing either a third-instar oriental beetle, *Anomala orientalis* Waterhouse, a scarab blueberry root pest or a late instar greater wax moth, *Galleria mellonella* L., larva (a widely used EPN sentinel), were deployed in blueberry fields. As described above, cages were treated with either blank solvent or pregeijerene. EPN-inflicted larval mortality (combined *A. orientalis* and *G. mellonella*) was nearly 2-fold greater in treatments with pregeijerene (55%) than those with solvent alone (30%) ($P = 0.009$). The increase was highly significant (from 40% to 80%) for *G. mellonella* ($P = 0.003$), but not statistically significant (from 20% to 30%) for *A. orientalis* ($P = 0.552$). Emerging EPNs were identified as *S. glaseri* with real-time PCR. On average, there were more *S. glaseri* nematodes surrounding the treatment (Mean±SE, 7.96±2.91) than in the control (4.43±2.56); however, this difference was not significant (N = 10, $t_{18} = 0.909$, $P = 0.38$).

Discussion

The obstacles of investigating belowground chemically mediated interactions between plants and animals are being overcome gradually, opening opportunities for manipulating these interactions for enhanced biological control [34–36]. At least half of all plant biomass is attacked by underground herbivores and

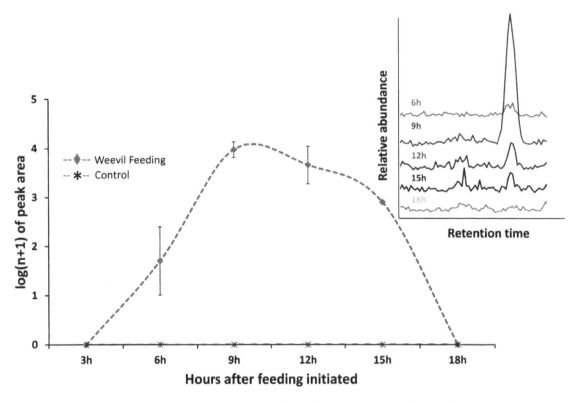

Figure 1. Time course of pregeijerene (1,5-dimethylcyclodeca-1,5,7-triene) release following initiation of weevil (*Diaprepes abbreviatus*) feeding on citrus roots. Insert in the upper right displays chromatogram of volatile abundance at each interval.

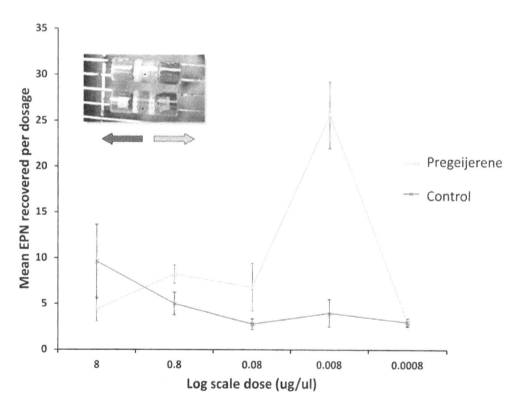

Figure 2. Optimal dosage of pregeijerene (1,5-dimethylcyclodeca-1,5,7-triene) for attracting entomopathogenic nematodes (*Steinernema riobrave* and *Heterorhabditis indica*) based on the log scale dilution of purified compound. Picture in upper left displays sand-filled two-choice olfactometers used for nematode bioassays.

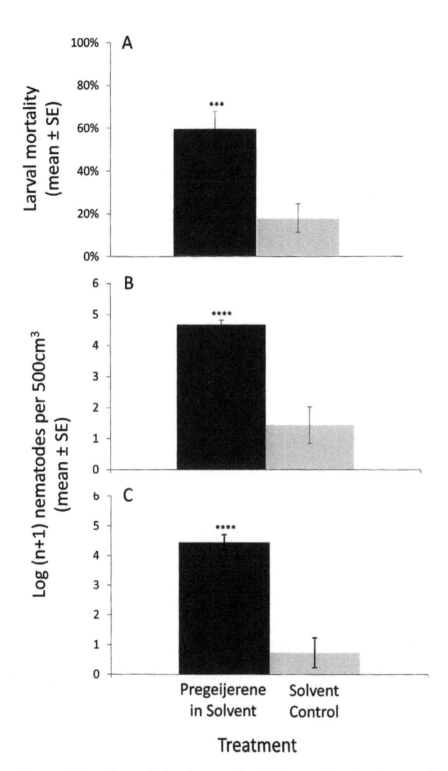

Figure 3. Effect of pregeijerene on mortality of *Diaprepes abbreviatus* larvae and associated attraction of entomopathogenic nematode infective juveniles (IJs) (all species combined). A) Average mortality of larvae buried with purified pregeijerene compared with the solvent control (N = 10, *t* = 4.01, *P* = 0.0008). B) Mean number of IJs recovered from cages containing purified pregeijerene compared with cages containing the solvent control (N = 10, *t* = 5.33, *P* = 0.00005). C) Mean number of IJs recovered from soil samples surrounding cages containing pregeijerene compared with the solvent control (N = 10, *t* = 5.67, *P* = 0.00003).

pathogens, living in a complex ecological food web in the soil [37]. Although induced plant responses were originally postulated as a potential novel approach to pest management in agricultural systems [38] for insect herbivore population regulation [39], few studies [40–43] of induced responses (particularly volatiles) have

addressed their practical application beyond fundamental concepts in ecology and evolutionary biology [35,37,43], with particularly few studies for belowground systems. HIPVs are likely important mediators of tritrophic interactions that afford indirect plant defense within the root zone. Our study not only shows this approach in the field, but also provides the first description of an ecological role for the C_{12} terpene, pregeijerene.

To evaluate applied volatiles for the attraction of belowground natural enemies in the field, studies usually quantify mortality of a target pest by trapping adults emerging from soil [11]. This technique frequently results in low recovery and also gives no confirmation of the specific cause of mortality. In addition, it can be difficult to quantify populations of naturally occurring EPNs, which may be abundant in soil, but remain cryptic. We used real-time qPCR as an efficient method for describing EPN diversity and quantifying their abundance [32,44–46]. Moreover, we showed that pregeijerene was directly responsible for attracting five species of native EPNs in the soil so as to enhance pest mortality. Given the efficacy of this compound, there may be little need for exogenous application of non-native EPNs in systems with a rich fauna of native EPNs. In orchards with established EPN populations, large-scale introduction of non-native species may temporarily reduce native populations due to trophic cascades that increase predatory fungi and attenuate net efficacy of biological control [19,47]. Although it is known that artificially reared and commercially formulated EPNs can persist, it is possible that natives have advantages associated with habitat acclimation and response to HIPVs [9]; thus, further investigation of enhancing conservation biological control of belowground pests in concert with behavioral modification via HIPVs is warranted.

The results of the experiment conducted in blueberries, an agricultural setting vastly different from citrus, demonstrate the potential broad applicability of pregeijerene on diverse species of EPNs. Timing application of pregeijerene to target the most susceptible instar of *A. orientalis* should optimize its efficacy (depending on EPN species, final (third)-instar *A. orientalis* may be less susceptible to EPN infection than earlier larval instars [48]).

Our previous research suggests that volatile production in response to herbivore feeding differs between citrus species [8]. Thus, our current findings could have broad impacts not only for rootstock selection in commercial citriculture, but also for use of attractants in other agroecosystems as demonstrated in blueberry fields. Here, we identified an additional naturally occurring species of EPN responsive to pregeijerene that was not found in Florida. Pregeijerene may thus have extensive application for enhancing native biological control of root feeding insects, including those, which attack a wide range of crops. However, we recognize that plants should benefit from the proposed function of herbivore-induced responses. Our experiments were not designed to assess improved crop fitness as a result of attracting beneficial natural enemies by application of pregeijerene. However, there is strong evidence that plants benefit from the cascading effects caused by EPN-induced suppression of herbivores [49], and our future work will determine whether crop yield is affected by HIPV-mediated manipulation of EPN behavior.

Where most aboveground studies have identified blends of volatiles as being responsible for the attraction of natural enemies [50], analogous belowground studies that identify an attractant in an agricultural system have demonstrated that a single compound can elicit natural enemy responses. This certainly makes application potentially less complex, but also points to an interesting potential property of belowground cues and natural enemy response. Future work should evaluate the complexity of belowground cues and the range of volatiles that cause belowground natural enemies, like EPNs, to respond. Only recently have new methodologies been employed to investigate chemicals stimulating changes in EPN behavior [51] and much more progress is necessary to understand these relationships [52].

Materials and Methods

Insect Larvae

D. abbreviatus larvae were obtained from a culture maintained at University of Florida's Citrus Research and Education Center (CREC) in Lake Alfred, FL. This culture was periodically supplemented from a larger culture maintained at the Division of Plant Industry Sterile Fly Facility in Gainesville, FL. Larvae were reared on a commercially prepared diet (Bio-Serv, Inc., Frenchtown, NJ) using procedures described by Lapointe and Shapiro [53]. Larvae used in experiments were from third to sixth instars.

Third-instar *A. orientalis* were collected from untreated turf areas at the Rutgers University Horticultural Research Farm (North Brunswick, NJ) in late April. The larvae were stored individually in the cells of 24-well plates in sandy loam at 10°C for two weeks and returned to room temperatures (21–24°C) for 24 h before use in experiments. Late instar *G. mellonella* larvae were obtained from Big Apple Herpetological (Hauppauge, NY).

Plants

'Swingle citrumelo' (*C. paradisi* Macf. × *P. trifoliata* L. Raf.) rootstock is very prominent in commercial citrus production [54]. The extensive use of this rootstock in commercial citrus production justified its use in this investigation. All plants were grown and maintained at the CREC in Lake Alfred, FL in a greenhouse at 26±3°C and 60–80% RH. Citrus seedlings used in the experiments were 25–35 cm long. *R. graveolens* was purchased as full grown plants 46–61 cm in height. The plants were immediately bare rooted and rinsed to remove as much soil material as possible; only roots were placed into vials containing dichloromethane for further extractions and purification.

Nematodes Used for Laboratory qPCR and Bioassays

The entomopathogenic nematodes, *S. diaprepesi* HK31, *S. riobrave* Btw1, *Steinernema sp.* LWD1, *H. indica* Ker1 and *H. zealandica* Btw1 were isolated from *D. abbreviatus* larvae buried in commercial citrus orchards in Florida. *S. riobrave* and *S. carpocapsae* isolates were descendants of commercial formulations intended for field application to manage *D. abbreviatus*. Other EPN species included in this study were *S. scapterisci* (provided by Dr. J.H. Frank, University of Florida, FL). All EPN species were cultured in last instar larvae of the greater wax moth, *G. mellonella larvae*, at approximately 25°C according to procedures described in Kaya and Stock [55]. Infective juveniles (IJs) that emerged from insect cadavers into emergence traps were stored in shallow water in tissue culture flasks at 15°C for up to two weeks prior to use.

In situ Volatile Collection from Infested Roots in the Greenhouse

Six 'Swingle citrumelo' plants were initially placed in glass root-zone chambers (300 ml volume capacity) (Analytical Research Systems, Gainesville, FL) filled with sand that had been autoclaved for one hour at 121°C and then adjusted to 10% moisture as described in Ali et al. [7,8] and El-Borai et al. [31]. All seedlings were given three days to adjust to their sand filled chambers. Three of the plants were subjected to feeding by weevil larvae for three days; the remaining three served as undamaged controls. During this period, each of the six root-zone chambers were connected to a vacuum pump (Analytical Research Systems, Gainesville, FL) with a suction flow of 80 ml/min [7]. Compounds emitted from chambers were collected on adsorbent traps filled with 50 mg Super-Q (800–1000 mesh, Alltech, Deerfield, IL) held in glass fittings between the chamber and vacuum pump [7].

Super-Q traps were replaced every three h for a 72-h period to track the time course of volatile release. The removed Super-Q traps were subsequently eluted with 150 µl of dichloromethane into individual 2.0 ml clear glass vials (Varian, Part Number 392611549, equipped with 500 µl glass inserts) [7]. The amount of pregeijerene detected over time for all treatments was subjected to multiple analysis of variance (MANOVA) to determine differences between, as well as, within treatments.

It was a challenge to remove sufficient pregeijerene from infested roots for bioassays and field testing. However, it was previously established [56] that a hydrodistillate of Common Rue (*Ruta graveolens*) roots contained the related terpene, geijerene, as a major constituent (67% of the total volatile compounds). Pregeijerene easily converts to geijerene at temperatures exceeding 120°C [57] (Figure S1). On-column GC-MS analyses confirmed pregeijerene as the main naturally occurring terpene in roots of Common Rue that could be easily extracted and purified from crushed roots using a series of solid phase extractions (Figure S2).

In situ Volatile Collection from Infested Roots in the Field

Volatiles were collected from the soil beds surrounding citrus trees in the field. A soil probe (Figure S3) was used to sample soil volatiles at a depth of 20 cm and at distances of one and 10 m from the trunks of citrus trees. A vacuum pump was used to pull air at a rate of 200 ml/min for a total of 30 min. Compounds were collected on adsorbent traps filled with 50 mg of Super-Q attached to the top of the soil probe (Figure S3). The Super-Q traps were subsequently eluted as described in the previous section.

Identification of Pregeijerene

Pregeijerene isolated from Common Rue and that from citrus roots after herbivore feeding was identified by electron impact (EI) and chemical ionization (CI) GC-MS analyses on DB1, DB5, and DB35 GC columns. Although the EI mass spectra matched pregeijerene in the Adams 2 library, the lack of a standard made it necessary to confirm the structure by nuclear magnetic resonance (NMR) (described in Supporting Information Materials and Methods S1).

Two-Choice Bioassay to Determine Optimal Dosage to Attract EPNs

The behavioral responses of EPNs to collected pregeijerene were quantified in a two-choice, sand-filled olfactometer [7,31]. Briefly, the olfactometer consists of three detachable sections: two opposing 16-ml glass jars which contained treatments and a central connecting tube three cm in length with an apical hole into which EPN were applied. Dilutions from the purified *R. graveolens* root extract were placed on filter paper, which was allowed to dry for 30 s to allow solvent evaporation. Thereafter, filter papers were placed on the bottoms of each glass jar, which were then filled with moist (10% w/v) sterilized sand. The central chamber connecting the two arms of the olfactometer was also filled with sterilized and moistened sand. EPNs (*ca.* 200 IJs) were applied into the central orifice of the connecting tube and given eight h to respond. Following the incubation period, the column was disassembled and the IJs from the two collection jars were extracted using Baermann funnels. The experiment was replicated five times for each dilution and separately tested with two EPN species: *S. riobrave* Btw1 and *H. indica* Ker1.

A student's *t*-test was used to compare nematode response in the two-choice olfactometer. Since responses of both species to pregeijerene *versus* the solvent controls were identical, data for both species were combined prior to analysis ($df = 18$). The dosage

at which a significant proportion of EPNs were attracted to the treatment arm was selected for our field trial.

Application of HIPVs in the Field

An experiment was conducted in a sandy soil (97:2:1, sand:silt:clay; pH 7.1; 0.1% OM) citrus orchard at the CREC (28 07 26.84 N, 81 42 55.31 W). The experiment was placed within a section of mature orange trees spaced (without beds) 4.5 m within and 8.1 m between rows that was irrigated with microsprinklers. A randomized design was used to place treatments between trees in eight adjacent rows. Cylindrical wire-mesh cages containing autoclaved sandy soil (10% moisture) and a single *D. abbreviatus* larva (reared on artificial diet for 3 to 5 weeks) were buried 20 cm deep in the soil beneath the tree canopies. Cages were made of 225-mesh stainless-steel cylinders (7 length × 3 cm diam) secured at each end with polypropylene snap-on caps. A replicate consisted of six cages placed equidistantly from one another in a circle pattern (48 cm diam) for each treatment. All cages contained a single *D. abbreviatus* larva and were baited with one of two treatments per replicate: (i) volatiles from roots fed upon by a *D. abbreviatus* larva or (ii) blank solvent control. There were 10 replicates of six cages per treatment. Treatments were applied as 30 µl aliquots to three cm diameter filter paper discs (Whatman). Solvent was allowed to evaporate for 30 s prior to insertion of filter papers at the base of each cage. The cages were left buried for 72 h. Eight soil core samples (2.5 cm diam × 30 cm deep) were taken from soil surrounding the treatment arena before the cages were removed to measure the number of EPNs attracted to the surrounding treatment arena. Recovered larvae were rinsed and placed on moistened filter paper within individual Petri dishes to confirm EPN infection by subsequent infective juvenile emergence from cadavers. Mortality of the larvae caused by EPNs was recorded from 0 to 72 h after removal from soil.

The effect of isolated pregeijerene on larval mortality was investigated in two additional experiments (one of which was conducted in a blueberry planting in Chatsworth, NJ, using *A. oientalis* and *G. mellonella*). The methods for these experiments were similar to those described above, except that the soil remaining within the six cages from each replication was placed in a container and homogenized for later nematode DNA extraction (n = 10). Soil cores taken from the surrounding treatment arenas were also combined and stored for nematode DNA extraction (n = 10). Fisher's exact test was used to compare larval mortality between the treatment and control. Only soil samples from the citrus experiment were analyzed for DNA quantification.

Detection, Identification, and Quantification of Entomopathogenic Nematodes Using Real-Time qPCR

Real-time qPCR was used to quantify attraction of naturally occurring EPN species to volatiles applied in the field and to identify nematodes to species. This technique targeted 11 EPN species [32,33,44]. In the citrus experiment, we surveyed the natural occurrence of six species (*S. diaprepesi*, *Steinernema* sp. *glaseri* group, *S. riobrave*, *S. scapterisci*, *H. indica*, and *H. zealandica*); in the blueberry experiment, nine species were surveyed (*S. carpocapsae*, *S. feltiae*, *S. glaseri*, *S. kraussei*, *S. scapterisci*, *Steinernema* sp. *glaseri* group, *H. indica*, *H. zealandica*, and *H. bacteriophora*). Briefly, species-specific primers and TaqMan® probes were designed from the ITS rDNA region using sequences of the target species as well as closely related species recovered from the NCBI database or generated by the authors in that study. Multiple alignments of the corresponding sequences were performed [58] to select areas of variability in the ITS region. The designed primers and probes provided no non-specific amplification when they were tested using other EPN species. Standard curve points were obtained from DNA dilution. Four independent DNA extractions were performed from Eppendorf tubes containing 300 IJs in 100 μl of the corresponding nematode species (Ultra Clean SoilTM DNA kit, MO BIO) to generate a standard curve [32,44]. Dilutions corresponding to 100, 30, 10, 3, and 1 IJs were prepared using serial dilution of the appropriate DNA.

Nematodes from soil samples were extracted by sucrose centrifugation [59] from aliquots of 500 cm^3 from the mixed composite sample. Each nematode community was concentrated in a 1.5 ml Eppendorf tube. DNA was processed using the UltraCleanTM soil DNA extraction kit and quantification was performed for each DNA extraction using the nanodrop system with the control program (ND-1000 v3.3.0). All DNA samples were adjusted to 0.2 ng/μl that is required for nematode quantification [32]. The resulting real values were analyzed with analysis of variance (ANOVA) for the EPN species recovered ($F = 41$, $df = 5$, 204). Where ANOVA showed significant differences, Tukey's HSD test ($P < 0.05$) was conducted to separate means in the software R (R Development Core Team 2004).

Supporting Information

Figure S1   Conversion of pregeijerene (A) to geijerene (B).

Figure S2   Chromatograms showing the initial crude extract prior to purification and final purified pregeijerene. The Y-axis represents relative abundance or the ratio of a mass peak profile area to that of another. These were calculated as ratios of the sums of areas used to plot the profiles ×100%.

Figure S3   Soil probe design used to sample volatiles belowground. Probe is inserted into soil and connected to a vacuum pump.

Figure S4   Chromatograms of volatiles taken from intact citrus roots in the field at one and 10 m distances from the trunk of the tree. The Y-axis represents relative abundance or the ratio of a mass peak profile area to that of another. These were calculated as ratios of the sums of areas used to plot the profiles ×100%.

Figure S5   Schematic diagram of the deployment and sampling procedure for field experiments in which sentinel traps with root weevils were deployed with or without HIPVs. One treatment replicate is depicted.

Table S1 ^{1}H (600 MHz), ^{13}C (151 MHz), HMBC and NOESY NMR spectroscopic data for pregeijerene in C_6D_6. ^{13}C was also detected directly (126 MHz) using a 5 mm Cryoprobe. Chemical shifts referenced to residual proton signal in C_6D_6 benzene $\delta(^1H) = 7.16$ ppm for ^{1}H and $\delta(C_6D_6H) = 128.2$ ppm for ^{13}C.

Table S2 ^{1}H (600 MHz), ^{13}C (151 MHz), HMBC and NOESY NMR spectroscopic data for geijerene in C_6D_6. ^{13}C was also detected directly (126 MHz) using a 5 mm Cryoprobe. Chemical shifts referenced to residual proton signal in C_6D_6 benzene $\delta(^1H) = 7.16$ ppm for ^{1}H and $\delta(C_6D_6H) = 128.2$ ppm for ^{13}C. For convenience, the pregeijerene numbering is retained after cope rearrangement to geijerene.

Acknowledgments

We thank Dr. Peter Teal (USDA-ARS-CMAVE) for facilitating access to laboratory space and equipment. Comments from Dr. Anurag Agrawal and the Plant-Interactions Group at Cornell University improved a previous version of the manuscript. Dr. Inna Kuzovkina and Monique Rivera also made valuable contributions to this project. The authors also acknowledge the National Science Foundation through the User Program of the National High Magnetic Field Laboratory (NHMFL) at the Advanced Magnetic Resonance Imaging and Spectroscopy (AMRIS) facility in the McKnight Brain Institute of the University of Florida and Jim Rocca for his assistance.

Author Contributions

Conceived and designed the experiments: LLS JGA HTA. Performed the experiments: JGA HTA LLS RC-H FK. Analyzed the data: JGA HTA FK LLS. Contributed reagents/materials/analysis tools: HTA CR-S AMK LWD LLS. Wrote the paper: JGA LLS HTA LWD RC-H FK CR-S AMK.

References

1. Turlings TCJ, Tumlinson JH, Lewis WJ (1990) Exploitation of herbivore-induced plant odors by host-seeking parasitic wasps. Science 250: 1251–1253.
2. Takabayashi J, Dicke M (1996) Plant-carnivore mutualism through herbivore-induced carnivore attractants. Trends Plant Sci 1: 109–113.
3. Tumlinson J, Turlings T, Lewis W (1993) Semiochemically mediated foraging behavior in beneficial parasitic insects. Arch Insect Biochem Physiol 22: 385–391.
4. Turlings TCJ, Wäckers F (2004) Recruitment of predators and parasitoids by herbivore-injured plants. In: Cardé RT, Millar JG, editors. Advances in insect chemical ecology. Cambridge University Press. pp 21–75.
5. Kant MR, Bleeker PM, Wijk MV, Schuurink RC, Haring MA (2009) Plant volatiles in defense. In: Loon LCV, editor. Advances in botanical research. Academic Press. pp 613–666.
6. Boff MIC, Zoon FC, Smits PH (2001) Orientation of Heterorhabditis megidis to insect hosts and plant roots in a Y-tube sand olfactometer. Entomol Exp Appl 98: 329–337.
7. Ali JG, Alborn HT, Stelinski LL (2010) Subterranean herbivore-induced volatiles released by citrus roots upon feeding by Diaprepes abbreviatus recruit entomopathogenic nematodes. J Chem Ecol 36: 361–368.
8. Ali JG, Alborn HT, Stelinski LL (2011) Constitutive and induced subterranean plant volatiles attract both entomopathogenic and plant parasitic nematodes. J Ecol 99: 26–35.
9. Rasmann S, Agrawal AA (2008) In defense of roots: A research agenda for studying plant resistance to belowground herbivory. Plant Physiol 146: 875–880.
10. van Tol RWHM, van der Sommen ATC, Boff MIC, van Bezooijen J, Sabelis MW, et al. (2001) Plants protect their roots by alerting the enemies of grubs. Ecol Lett 4: 292–294.
11. Rasmann S, Köllner TG, Degenhardt J, Hitpold I, Töpfer S, et al. (2005) Recruitment of entomopathogenic nematodes by insect-damaged maize roots. Nature 434: 732–737.
12. Hiltpold I, Töpfer S, Kuhlmann U, Turlings TCJ (2010) How maize root volatiles affect the efficacy of entomopathogenic nematodes in controlling the western corn rootworm? Chemoecology 20: 155–162.
13. Degenhardt J, Hiltpolt I, Köllner TG, Frey M, Gierl A, et al. (2009) Restoring a maize root signal that attracts insect-killing nematodes to control a major pest. Proc Natl Acad Sci USA 106: 13213–13218.
14. Hunter MD (2001) Out of sight, out of mind: the impacts of root-feeding insects in natural and managed systems. Agr Forest Entomol 3: 3–9.
15. Beavers J (1982) Biology of Diaprepes abbreviatus (Coleoptera: Curculionidae) reared on an artificial diet. Fla Entomol 65: 263–269.
16. Simpson SE, Nigg HN, Coile NC, Adair RA (1996) Diaprepes abbreviatus (Coleoptera: Curculionidae): host plant associations. Environ Entomol 25: 333–349.
17. Graham J, Bright D, McCoy C (2003) Phytophthora-Diaprepes weevil complex: Phytophthora spp. relationship with citrus rootstocks. Plant Dis 87: 85–90.
18. Bullock RC, Pelosi RR, Killer EE (1999) Management of citrus root weevils (Coleoptera: Curculionidae) on Florida citrus with soil-applied entomopathogenic nematodes (Nematoda : Rhabditida). Fla Entomol 82: 1–7.

19. Duncan L, Graham J, Zellers J (2007) Food web responses to augmenting the entomopathogenic nematodes in bare and animal manure-mulched soil. J Nematol 39: 176–189.

20. Schroeder WJ (1994) Comparison of two steinernematid species for control of the root weevil *Diaprepes abbreviatus*. J Nematol 26: 360–362.

21. Kaya H, Gaugler R (1993) Entomopathogenic nematodes. Annu Rev Entomol 38: 181–206.

22. Gaugler R (2002) Entomopathogenic nematology. New York: CABI. 388 p.

23. Georgis R, Koppenhöfer AM, Lacey LA, Bélair G, Duncan LW, et al. (2006) Successes and failures in the use of parasitic nematodes for pest control. Biol Control 38: 103–123.

24. Turlings TCJ, Ton J (2006) Exploiting scents of distress: the prospect of manipulating herbivore-induced plant odours to enhance the control of agricultural pests. Curr Opin Plant Biol 9: 421–427.

25. Degenhardt J, Gershenzon J, Baldwin IT, Kessler A (2003) Attracting friends to feast on foes: engineering terpene emission to make crop plants more attractive to herbivore enemies. Curr Opin Biotech 14: 169–176.

26. Bruce T (2010) Exploiting plant signals in sustainable agriculture. In: Baluška F, Ninkovic V, editors. Plant communication from an ecological perspective. Springer. pp 215–227.

27. Thaler JS, Stout MJ, Karban R, Duffey SS (1996) Exogenous jasmonates simulate insect wounding in tomato plants (*Lycopersicon esculentum*) in the laboratory and field. J Chem Ecol 22: 1767–1781.

28. James D, Grasswitz T (2005) Synthetic herbivore-induced plant volatiles increase field captures of parasitic wasps. BioControl 50: 871–880.

29. Kaplan I (2011) Attracting carnivorous arthropods with plant volatiles: The future of biocontrol or playing with fire? Biol Control 60: 77–89.

30. McCoy C (2000) Entomopathogenic nematodes and other natural enemies as mortality factors for larvae of *Diaprepes abbreviatus* (Coleoptera: Curculionidae). Biol Control 19: 182–190.

31. El-Borai FE, Campos-Herrera R, Stuart RJ, Duncan LW (2011) Substrate modulation, group effects and the behavioral responses of entomopathogenic nematodes to nematophagous fungi. J Invertebr Pathol 106: 347–356.

32. Campos-Herrera R, Johnson EG, El-Borai FE, Stuart RJ, Graham JH, et al. (2011) Long-term stability of entomopathogenic nematode spatial patterns in soil as measured by sentinel insects and real-time PCR assays. Ann Appl Biol 158: 55–68.

33. Campos-Herrera R, El-Borai FE, Stuart RJ, Graham JH, Duncan LW (2011) Entomopathogenic nematodes, phoretic *Paenibacillus* spp., and the use of real time quantitative PCR to explore soil food webs in Florida citrus groves. J Invertebr Pathol 108: 30–39.

34. Johnson SN, Murray PJ (2008) Root feeders. Cambridge, MA: CABI. 226 p.

35. van Dam NM, Heil M (2011) Multitrophic interactions below and above ground: en route to the next level. J Ecol 99: 77–88.

36. van Dam NM (2009) Belowground herbivory and plant defenses. Annu Rev Ecol Evol Syst 40: 373–391.

37. De Deyn GB, van Ruijven J, Raaijmakers CE, de Ruiter PC, van den Putten WH (2007) Above- and belowground insect herbivores differentially affect soil nematode communities in species-rich plant communities. Oikos 116: 923–930.

38. Green TR, Ryan CA (1972) Wound induced proteinase inhibitor in plant leaves: a possible defense mechanism against insects. Science 175: 776–777.

39. Haukioja E, Hakala T (1975) Herbivore cycles and periodic outbreaks. Formation of a general hypothesis. Rep Kevo Subarctic Res Stat 12: 1–9.

40. Khan ZR, Ampog-Nyarko K, Chilishwa P, Hassanali P, Kimani S, et al. (1997) Intercropping increases parasitism of pests. Nature 388: 631–632.

41. Birkett MA, Campbell CAM, Chamberlain K, Guerrieri E, Kick AJ, et al. (2000) New roles for cis-jasmone as an insect semiochemical and in plant defense. Proc Natl Acad Sci USA 97: 9329–9334.

42. Ockroy M, Turlings TCJ, Edwards P (2001) Response of natural populations of predators and parasitoids to artificially induced volatile emissions in maize plants (*Zea mays* L.). Agr Forest Entomol 3: 201–209.

43. Hunter M (2002) A breath of fresh air: beyond laboratory studies of plant volatile–natural enemy interactions. Agr Forest Entomol 4: 81–86.

44. Torr P, Spiridonov SE, Heritage S, Wilson MJ (2007) Habitat associations of two entomopathogenic nematodes: a quantitative study using real-time quantitative polymerase chain reactions. J Anim Ecol 76: 238–245.

45. Holeva R, Phillips MS, Neilson R, Brown DJF, Young V, et al. (2006) Real-time PCR detection and quantification of vector trichodorid nematodes and Tobacco rattle virus. Mol Cell Probes 20: 203–211.

46. Kolb S, Knief C, Stubner S, Conrad R (2003) Quantitative detection of methanotrophs in soil by novel pmoA-targeted real-time PCR assays. Appl Environ Microb 69: 2423–2429.

47. El-Borai F, Brentu C, Duncan LW (2007) Augmenting entomopathogenic nematodes in soil from a Florida citrus orchard: non-target effects of a trophic cascade. J Nematol 39: 203–210.

48. Koppenhöfer AM, Fuzy EM (2004) Effect of white grub developmental stage on susceptibility to entomopathogenic nematodes. J Econ Entomol 97: 1842–1849.

49. Denno RF, Gruner DS, Kaplan I (2008) Potential for entomopathogenic nematodes in biological control: A meta-analytical synthesis and insights from trophic cascade theory. J Nematol 40: 61–72.

50. Hare J (2011) Ecological role of volatiles produced by plants in response to damage by herbivorous insects. Annu Rev Entomol 56: 161–180.

51. Hallem EA, Dilman AR, Hong AV, Zhang Y, Yano JM, et al. (2011) A sensory code for host seeking in parasitic nematodes. Curr Biol 21: 377–383.

52. Rasmann S, Ali JG, Helder J, van der Putten W (2012) Ecology and evolution of soil nematode chemotaxis. J Chem Ecol In press.

53. Lapointe SL, Shapiro JPT (1999) Effect of soil moisture on development of *Diaprepes abbreviatus* (Coleoptera: Curculionidae). Fla Entomol 82: 291–299.

54. Stover E, Castle W (2002) Citrus rootstock usage, characteristics, and selection in the Florida Indian River region. HortTechnology 12: 143–147.

55. Kaya HK, Stock SP (1997) Techniques in insect nematology. In: Lacey, LA, editor. Manual of techniques in insect pathology. London, UK: Academic Press. pp 281–324.

56. Kuzovkina I, Szarka S Héthelyi, É (2009) Composition of essential oil in genetically transformed roots of *Ruta graveolens*. Russ J Plant Physl 56: 846–851.

57. Kubeczka K, Ullmann I (1980) Occurrence of 1, 5-dimethylcyclodeca-1, 5, 7-triene (pregeijerene) in *Pimpinella* species and chemosystematic implications. Biochem Syst Ecol 8: 39–41.

58. Larkin MA, Blackshields G, Brown PN, Chenna R, McGettigan PA, et al. (2007) Clustal W and Clustal X version 2.0. Bioinformatics 23: 2947–2948.

59. Jenkins W (1964) A rapid centrifugal-flotation technique for separating nematodes from soil. Plant Dis Rep 48: 692.

7

Transcriptome Profiling of Citrus Fruit Response to Huanglongbing Disease

Federico Martinelli[1,6]**, Sandra L. Uratsu**[1]**, Ute Albrecht**[2]**, Russell L. Reagan**[1]**, My L. Phu**[1]**, Monica Britton**[5]**, Vincent Buffalo**[5]**, Joseph Fass**[5]**, Elizabeth Leicht**[3,4]**, Weixiang Zhao**[3]**, Dawei Lin**[5]**, Raissa D'Souza**[3,4]**, Cristina E. Davis**[3]**, Kim D. Bowman**[2]**, Abhaya M. Dandekar**[1]*

1 Plant Sciences Department, University of California Davis, Davis, California, United States of America, **2** U.S. Horticultural Research Laboratory, United States Department of Agriculture, Agricultural Research Service, Fort Pierce, Florida, United States of America, **3** Mechanical and Aerospace Engineering Department, University of California Davis, Davis, California, United States of America, **4** Center for Computational Science and Engineering, University of California Davis, Davis, California, United States of America, **5** Bioinformatics Core, Genome Center, University of California Davis, Davis, California, United States of America, **6** Dipartimento di Sistemi Agro-Ambientali, Università degli Studi di Palermo, Viale delle Scienze, Palermo, Italy

Abstract

Huanglongbing (HLB) or "citrus greening" is the most destructive citrus disease worldwide. In this work, we studied host responses of citrus to infection with *Candidatus* Liberibacter *asiaticus* (CaLas) using next-generation sequencing technologies. A deep mRNA profile was obtained from peel of healthy and HLB-affected fruit. It was followed by pathway and protein-protein network analysis and quantitative real time PCR analysis of highly regulated genes. We identified differentially regulated pathways and constructed networks that provide a deep insight into the metabolism of affected fruit. Data mining revealed that HLB enhanced transcription of genes involved in the light reactions of photosynthesis and in ATP synthesis. Activation of protein degradation and misfolding processes were observed at the transcriptomic level. Transcripts for heat shock proteins were down-regulated at all disease stages, resulting in further protein misfolding. HLB strongly affected pathways involved in source-sink communication, including sucrose and starch metabolism and hormone synthesis and signaling. Transcription of several genes involved in the synthesis and signal transduction of cytokinins and gibberellins was repressed while that of genes involved in ethylene pathways was induced. CaLas infection triggered a response via both the salicylic acid and jasmonic acid pathways and increased the transcript abundance of several members of the WRKY family of transcription factors. Findings focused on the fruit provide valuable insight to understanding the mechanisms of the HLB-induced fruit disorder and eventually developing methods based on small molecule applications to mitigate its devastating effects on fruit production.

Editor: Baochuan Lin, Naval Research Laboratory, United States of America

Funding: This work was supported by a grant received from the Florida Citrus Advanced Technology Program. The funders had no role in study design, data collection and analysis, decision to publish, or preparation of the manuscript.

Competing Interests: The authors have declared that no competing interests exist.

* E-mail: amdandekar@ucdavis.edu

Introduction

Huanglongbing (HLB) or "citrus greening" is the most destructive citrus disease today [1,2]. HLB is caused by phloem-limited bacteria of the genus *Candidatus* Liberibacter. Three species of this pathogen are currently associated with the disease in citrus. The Asian form, *Candidatus* Liberibacter *asiaticus* (CaLas), is found in all HLB-affected countries outside Africa. The African form, *Ca. L. africanus* (CaLaf), and the American form, *Ca. L. americanus* (CaLam), are currently found only in Africa and Brazil, respectively [1]. CaLas, present in Asia, Brazil, Florida, and Louisiana, has recently been cultured [3]. The genome sequence of CaLas was obtained using a metagenomic approach from vascular tissues [4] and an infected psyllid [5]. CaLas is transmitted by *Diaphorina citri* Kuwayama, the Asian citrus psyllid (ACP). Typical symptoms of citrus greening disease include blotchy, mottled, and variegated leaf chlorosis, followed by tree decline. Infected leaves become upright, followed by leaf drop and twig dieback at later stages. Early flowering is also observed in CaLas-infected sweet orange. Initial symptoms are very similar to those caused by mineral nutrient stresses such as zinc, magnesium, or iron deficiency, hampering accurate diagnosis.

Understanding host responses to pathogen infection is essential to clarify the mechanisms of plant-microbe interactions and to develop novel strategies for therapy. Studies have been conducted to identify key genes and proteins induced by HLB in leaf tissues [6,7,8]. These studies showed that key pathways and processes such as cell defense, transport, photosynthesis, carbohydrate metabolism, and hormone metabolism are affected by the disease [6]. Clarifying host responses and disease development in fruit peel and leaves is critical for disease detection at the earliest stages. CaLas is typically found in leaves, but can be present in bark, root, flower (petal and pistil), and fruit (peduncle, columella, and seed coat) of infected trees [9]. Although pathogen detection is typically conducted on leaf tissues, fruit peel can also be employed for the analysis of host responses. The analysis of host responses in both leaves and fruits is important to understand the mechanisms of the fruit disorder and tree decline induced by the disease.

CaLas-infected peel tissues often show a characteristic color inversion as the fruit changes from green to yellow/orange [1]. HLB also results in fruits that are small, asymmetric, and lopsided, with a bent fruit axis, small or aborted seeds, and a strong yellow to brown stain in vascular bundles within the axis at the peduncular end [2].

Microarray technology has been used in numerous studies of host response to infection by pathogens including bacteria, viruses, and fungi [10,11,12]. However, this technology can reveal the expression of only those genes represented on the array. Possible misleading interpretations of microarray results can occur due to non-specific hybridization. Next-generation DNA sequencing technology can reveal very rare and unknown transcripts, offering a more precise and accurate picture of the transcriptome. These tools, already applied to plants [12,13,14,15], assume extensive prior knowledge of the organism under investigation. For plant species that lack whole-genome sequence information, an extensive EST database is required. Misleading results due to multiple mapping locations for the same sequence might occur. Indeed, data obtained are usually confirmed with qRT-PCR analysis or integrated with proteomic and metabolomic analyses. In addition, analysis of the deep transcriptome profile using biological network theory can help define gene regulatory networks. Protein networks are increasingly used to describe the molecular basis of disease-related subnetworks [16] and to define protein-protein interaction networks (PPI) that regulate disease resistance in plants and plant-pathogen interactions [17]. Bioinformatic tools are now available for visualizing and characterizing statistical properties of these networks (e.g. the BNArray, GeneReg and igraph packages of R, and other software such as Centibin, Graphviz, Pajek, and Cytoscape).

At present, no therapeutic treatments are available for HLB, and removal of infected trees and insect control are the main management strategies to limit or prevent its spread. Traditional diagnostic approaches rely on symptom recognition in the field, confirmed by PCR based on primers developed for individual *Candidatus* Liberibacter species. [18,19,20,21]. These practices were very useful to speed up pathogen detection and accelerate management procedures, although the pathogen may elude detection at asymptomatic stages [21,22,23]. This is probably due to the fact that the pathogen is phloem-limited and not uniformly distributed within the tissues of infected trees [9].

This study examines global changes in host gene expression due to CaLas infection in fruit peel. It aims to elucidate metabolic changes induced by the disease in the fruit. Using next generation sequencing technology, mRNA transcripts from fruit peel sample types representing various stages of disease were compared. Fruits displaying symptoms were compared to asymptomatic fruits from the same tree and to apparently healthy fruits taken from trees free of HLB symptoms in the same orchard. Fruits were categorized into one of the three stages based on qRT-PCR in addition to commonly observed symptoms.

Results

Transcriptome profiling using RNA-Seq

Between 24 and 41 million 85-nt paired-end reads were obtained from each of four cDNA libraries derived from mature fruit peel at different disease stages: fruit peel from uninfected trees in an orchard with no HLB present; fruit peel from apparently healthy trees (PCR-positive at the time of sample collection, healthy without any symptoms) in an orchard with HLB; symptomatic and asymptomatic fruit peel from the same trees infected with CaLas (PCR-positive with HLB symptoms). These

reads were aligned to the NCBI citrus unigene set, with 42 to 46% of reads per sample mapping to a unigene. Six pairwise comparisons were made between symptomatic, asymptomatic, apparently healthy, and healthy control fruit peel to calculate changes in expression (log fold ratio) of individual genes (Tables S1, S2, S3, S4, S5, S6). The comparison between asymptomatic and symptomatic stages of the disease identifies genes related to the appearance of symptoms. Interestingly, the overall expression profiles of apparently healthy and asymptomatic fruit peel were very similar, indicated by fewer differentially expressed transcripts in this comparison.

The four fruit types were also tested for the presence of Citrus Tristeza Virus (CTV) using CTV CP reference sequence T36 (M76485) to show cross-responses to multiple pathogens. CTV was detected in all four sample types. No significant differences were observed among the three fruit categories from the infected orchard (Figure S1).

Functional analysis of RNA-Seq data

was performed to determine how the four fruit types can be separated based on their overall transcriptome profile (Figure S2). Linear combination of transcriptomic data generated vectors or groups to best explain overall variance in the data set without prior assumptions about whether and how clusters might form. It was clearly evident that apparently healthy and asymptomatic fruits showed close similarities and were separated from the other two fruit types. Healthy fruits and symptomatic fruits were clearly distinguished from each other. The 21 target genes most specific to each category are listed in Table S7. Three complementary methods were used for functional analysis of the transcriptomic data: Fisher's Exact Test and Gene Ontology (GO) descriptions, PageMan gene set enrichment analysis, and pathway enrichment analysis using Pathexpress.

The Fischer Exact Test as provided in Blast2GO [24] is useful to determine the specific GO terms affected by the disease. Some GO terms were significantly over-represented among differentially expressed genes obtained from pairwise comparisons (Table S8). Among these, several GO terms associated with cell wall biogenesis, modification, and organization and related metabolic processes were over-represented among more abundant transcripts in healthy control fruit than in apparently healthy, asymptomatic, or symptomatic fruit. Interestingly, GO terms related to photosynthetic reactions (Photosystem I and II, rubisco activity, thylakoid localization) were over-represented in fruits showing the typical symptoms of HLB infection. Over-represented GO terms in four pairwise comparisons (symptomatic vs. apparently healthy, symptomatic vs. asymptomatic, symptomatic vs. apparently healthy and asymptomatic, and asymptomatic vs. symptomatic) were analyzed to determine which GO terms correlated significantly with HLB disease (Figure 1). It is interesting to note that when symptoms are clearly evident, GO terms of small carbohydrate metabolism (monosaccharide, sucrose, and galactose catabolic and metabolic processes) and other important pathways of primary metabolism such as the pentose-phosphate cycle are over-represented. As expected from visual analysis of fruit symptoms, many gene functions related to photosynthesis were over-represented at the symptomatic stage (e.g., electron carrier activity, chloroplast, photosystems, ribulose bisphosphate carboxylase complex and activity, thylakoid, photosynthetic dark reaction) compared with all other conditions. GO terms for ion transport were also over-represented at the symptomatic stage. The high number of GO terms for oxidoreductase activity is consistent with the hypothesis that the disease induces oxidative stress. Several categories of GO terms involved in vesicle transport

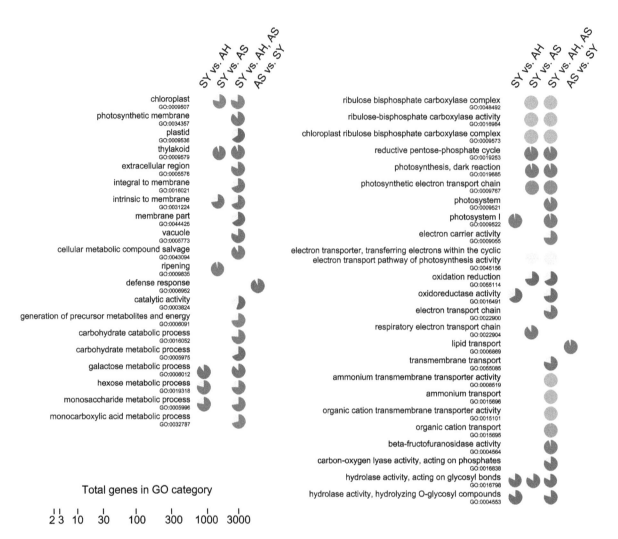

Total genes in GO category

Figure 1. Fisher's Exact Test analysis. Over-represented GO terms in four pairwise comparisons: symptomatic (SY) vs. apparently healthy (AH), symptomatic vs. asymptomatic (AS), symptomatic vs. apparently healthy and asymptomatic, asymptomatic vs. symptomatic. The size of the colored sector of each pie chart corresponds to the proportion of that GO term in the test set compared to all the genes to which that GO term has been assigned, where 100% colored means all of the genes in the reference set are in the test set. Colors correspond to the number of genes in the GO category on a log scale, as indicated by the scale bar.

and cell wall biogenesis, metabolism, and organization were over-represented among genes expressed at the symptomatic stage. Interesting defense-related and lipid transport GO terms were over-represented at the asymptomatic stage.

PageMan software [25] was used to visualize functional classes that were significantly affected by HLB disease (Figure S3). This method pinpoints which subcategory of genes were upregulated and downregulated in each main gene category based not only on metabolic pathways but also on cell functions. Increased expression in diseased samples is seen in photosynthesis, N-metabolism, amino acid synthesis, isoprenoids, jasmonate and salicylic acid, and several transcription factors. Functional classes with decreased expression include sucrose and starch biosynthesis, glycolysis, gibberellins, DNA and protein synthesis, and flavonoids metabolism. Additional changes in expression can be seen in the comparison between samples from the disease-free location and apparently healthy fruits (asymptomatic but from the infected orchard). These changes, eventually affected by environmental variability, probably were induced in early stages of HLB disease. Differential expression of transcripts comparing different stages of HLB infection and their functions were visualized using MapMan

software [26]. This provided more specific information on pathways and functions identified by Fisher's Exact Test and PageMan. For MapMan data analyses, a mapping file composed of NCBI *Citrus sinensis* unigenes was used.

The third enrichment method was conducted using the Pathexpress web-tool [27] to determine which metabolic pathways were significantly affected by the disease by comparing symptomatic and asymptomatic fruits (Table 1). The MapMan graphical metabolic overview identifies transcripts that are differentially expressed in symptomatic and apparently healthy fruit, with each colored square representing a single annotated gene in a particular pathway (Figure 2). Several genes involved in light reactions of photosynthesis, mitochondrial electron transport, sucrose metabolism, glycolysis, and fermentation were up-regulated. In contrast, gene transcripts for cell wall modification and degradation, pectin esterase activity, and cellulose synthesis were mostly down-regulated. We also identified several differentially expressed genes involved in secondary metabolic pathways, including terpenes, flavonoids, and phenylpropanoids.

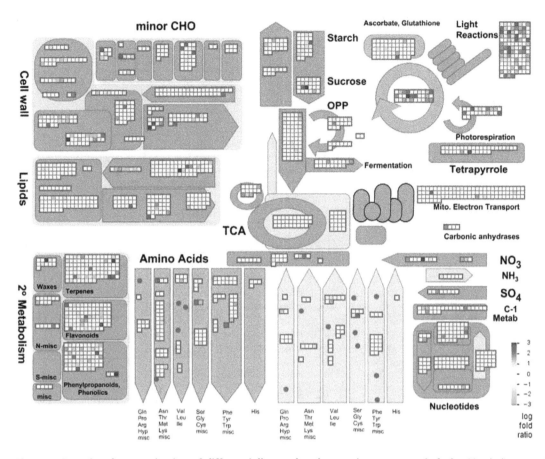

Figure 2. Functional categorization of differentially regulated genes in symptomatic fruits. Metabolism overview in MapMan depicting differential gene expression in symptomatic and apparently healthy fruits from the infected orchard. Log fold ratios are indicated as a gradient between red (up-regulated) and green (down-regulated).

Photosynthesis and carbohydrate metabolism

The expression of several genes involved in photosynthesis and carbohydrate metabolism increased significantly in symptomatic fruits, when compared to asymptomatic fruit from the same tree or different trees. Transcripts for oxygen-evolving enhancer 3 (PsbQ), photosystem II subunit Q-2, photosystem II reaction center protein J, and other genes encoding different subunits of photosystem II were highly abundant in symptomatic fruit. In addition, genes encoding subunits of cytochrome b6/f (complex subunit 8, subunit IV) and genes encoding ATP synthase subunits (ATPase subunit III, alpha, beta, F, and complex CF0) were induced (Figure 3A). Several transcripts encoding subunits of photosystem I increased, including chlorophyll A apoprotein subunit G, photosystem I reaction center subunit (PSI-N), and D1 subunit.

Several genes for enzymes involved in the first steps of glycolysis were differentially expressed. Up-regulation was observed for genes encoding protein serine/threonine kinase and fructose-2,6-biphosphatase (Figure 3B). There were significant changes in transcripts related to carbohydrate metabolism in symptomatic and asymptomatic fruit. The citrus orthologs of two *Arabidopsis* genes encoding different isoforms of invertase (AB276108 and EY662586) were induced in symptomatic fruit. In starch metabolism, transcription of glucose-1-phosphate adenylyltransferase was diminished, while several genes involved in starch degradation were significantly differentially expressed (Figure 3C). Expression of genes involved in raffinose synthesis were differentially regulated, while transcripts involved in galactinol metabolism were abundant in symptomatic fruits (Figure 3D).

Table 1. Differentially regulated pathways (up- and down-regulated) involved in HLB response in the symptomatic fruit as determined using Pathexpress.

Regulated Pathways	Between Trees	Within Tree
Starch and sucrose	$5*10^{-4}$	0.03
Carbon fixation	$9*10^{-4}$	0.02
Ascorbate and aldarate	0.03	n.s.
Phenylpropanoid	0.03	0.02
Alpha-linolenic acid	0.04	$2*10^{-3}$
Pentose and glucoronate	0.06	n.s.
Pentose phosphate	0.07	n.s.
Fructose and mannose	0.08	n.s.
Cyanoamino acid	n.s.	0.06
Glycerophospholipids	n.s.	0.06
Flavonoids	n.s.	0.06

Analysis was performed on the list of genes differentially expressed at a significance level of p<0.1 comparing symptomatic to asymptomatic fruits within tree and between different trees "N.s." means not significant.

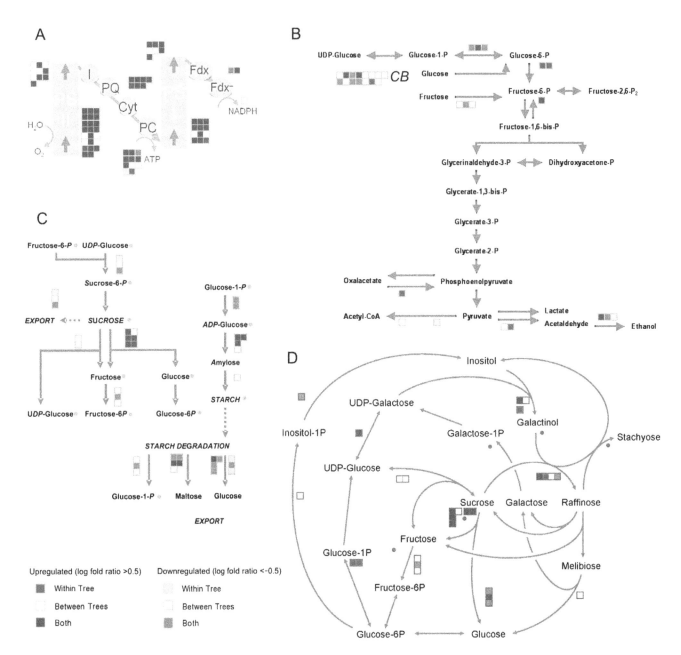

Figure 3. MapMan views of differentially regulated primary metabolism genes. Symptomatic fruit was compared to asymptomatic fruit within the same tree (AS), different tree (AH) and both trees. A) Light reactions of photosynthesis. B) Glycolysis. "CB" block shows gene expression in cytosolic branch genes of small carbohydrate metabolism. C) Sucrose and starch metabolism D) Raffinose metabolism. Colored data points indicate up or down expression with log fold ratios >0.5 and <−0.5, respectively.

Hormone-related pathways

Significant transcriptional changes in response to CaLas infection were observed for a group of genes involved in hormone biosynthesis, mobilization, and signal transduction. Log fold ratios for differentially expressed genes in symptomatic fruits compared to apparently healthy and asymptomatic fruits are shown (Figure 4). Transcripts related to GH3-like proteins involved in auxin synthesis (IAA-amino acid conjugate hydrolase and GH3.1 and GH3.4) were more abundant in symptomatic fruit. Transcripts for GRAM-domain containing protein, involved in the abscisic acid pathway, were more abundant. Several genes involved in ethylene biosynthesis and signal transduction were upregulated in HLB-affected fruit including ACO4, ethylene

receptor 1 (ETR1), ethylene response element binding factors (ERF1 and ERF2), ethylene forming enzyme (EFE), while ACO1 and an ethylene-responsive protein were downregulated. Interestingly, genes for ATHK1 and Snakin-1, involved in the cytokinin and gibberellin pathways, were repressed in infected fruit (Figure 4A).

Transcription factors

Several transcription factors were differentially expressed in HLB-infected fruit, including those belonging to the AP2-EREBP, helix-loop-helix, bZIP, C2H2 zinc finger, C2C2-CO-like, MYB-related, WRKY, and AS2 gene families (Figure 4B). Among genes for bZIP, haloacid dehalogenase-like hydrolase, GBox binding

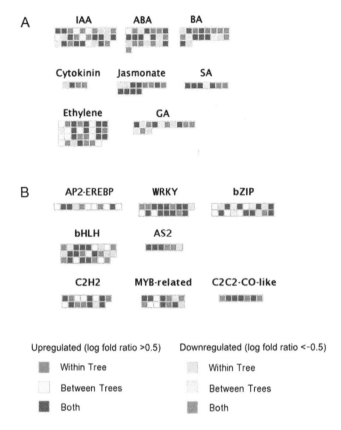

Figure 4. HLB-altered expression of hormone and transcription factors. Differential expression of (A) hormone-related transcripts and (B) transcription factors. Color scale indicates the comparisons between symptomatic asymptomatic and apparently healthy fruit, same as used in Figure 3.

factor 3 and 6 and GBF1 were upregulated at asymptomatic and symptomatic stage, while elongated hyocotyl 5 was downregulated. Some, but not all, bHLH proteins were upregulated in HLB disease. Two AS2 transcription factors (lateral organ boundaries gene family), LBD4 and LBD40, were upregulated while LBD41, LBD21, and LOB were downregulated. Interstingly most of the C2H2 transcription factors were upregulated in response to HLB as were C2C2-CO-like proteins (B-Box types). Several WRKY genes, known to be involved in biotic stress responses, showed increased expression as the disease progressed, including WRKY54, WRKY65, WRKY70, WRKY53, WRKY54, WRKY18, and WRKY50. Fewer WRKY transcripts were less in abundant: WRKY32, WRKY45, and WRKY47.

Volatiles and defense response

Up-regulation of genes involved in jasmonic acid (JA) biosynthesis such as allene oxide synthase, 12-oxophytodienoate reductase, and jasmonic acid-carboxylmethyltransferase was observed in infected fruit, showing that defense responses were induced (Figure 5A). It is interesting to note that while Lipoxygenase2 (EY727780) expression was down, expression increased for Lipoxygenase1/3 (EY663264). Transcriptional changes were also observed in the non-mevalonate or 1-deoxy-D-xylulose/2-C-methyl-D-erythritol-4-phosphate (DOXP/MEP) pathway for isoprenoid biosynthesis located in the plastids. Transcript abundance of 4-hydroxy-3-methylbut-2-en-1-yl diphosphate reductase (ISPH) and the geranylgeranylpyrophosphate synthase 1 genes involved in

carotenoid synthesis was elevated in CaLas-infected fruit. Additionally, genes encoding several types of terpene synthases involved in mono- and diterpene biosynthesis, which affect aroma composition and nutritional properties of citrus fruit, were differentially expressed in HLB-infected fruit (Figure 5B). The salicylic acid-mediated response was also activated as shown by the higher abundance of transcripts for genes encoding salicylic acid methyltransferases (Figure 5C).

Protein degradation

Differentially expressed genes involved in protein degradation were identified at the symptomatic stage (Figure 6A). Upregulated genes included the ubiquitin-ring family of proteins: zinc finger C3HC4-type ring finger, zinc ion binding ring-H2 protein, F-box family protein, UBQ10, and UBQ3, while transcripts for other C3HC4 ring finger proteins were downregulated. At the same time, transcripts for heat shock protein 82 were downregulated at both asymptomatic and symptomatic stages (Figure 6B). This gene is a major hub in the protein-protein interaction network deduced from the *Arabidopsis* protein-protein knowledgebase (Figure 6C; Table S1).

Real time PCR validation

Real-time PCR (RT-PCR) analyses were conducted using all four fruit types to validate the expression patterns of a subset of differentially expressed genes identified by next generation sequencing (Figure 7; Table 2). Among hormone-related transcripts, indole-3-acetic acid amido synthetase (GH3.4) was expressed more at all disease stages. Ethylene responsive factor 1 (ERF-1) expression was higher at the symptomatic stage. Transcripts for ent-kaurenoic acid hydroxylase 2 (KAO2) and gibberellin-responsive protein GASA1 were less abundant in infected fruits. The jasmonic acid and salicylic acid pathways were affected by HLB disease. Transcript abundance of lipoxigenase2 (Lox2), which is involved in jasmonate biosynthesis, was lower in fruit from infected trees at asymptomatic and symptomatic stages than in healthy trees from the disease-free location. Salicylic acid methyltransferase (SAM) was strongly up-regulated in infected fruits. Transcription factors are key players in transducing signals generated in response to pathogen infection. Among them, the WRKY family was highly involved in HLB response. WRKY70 was highly up-regulated in symptomatic fruits and its transcript abundance was also significantly higher in asymptomatic, apparently healthy fruit. Transcripts of RD26, NAC-1, and Myb-related transcription factors (MYB TF) were more abundant at asymptomatic stages. Heat shock protein 82 (HSP82), the hub with the highest degree in the protein-protein interaction network, was down-regulated in all infected fruits, with likely consequences for overall fruit metabolism.

Glucose-1-phosphate adenylyltransferase (APL4) was highly down-regulated in symptomatic fruits while invertase (β-fructofuranosidase) was clearly up-regulated in the peel of apparently healthy and symptomatic fruits from the infected orchard. Sucrose symporter (SucSymport) was up-regulated in infected, symptomatic fruits.

The terpenoid pathway was affected by HLB disease, as shown by the down-regulation of terpene synthase cyclase (TerpeneSC). Induction of lipid transfer protein (NNLTP) was observed in the absence of symptoms. The expression of other genes such as acidic cellulase and methyltransferase2 (MTransf2) was diminished in infected fruits.

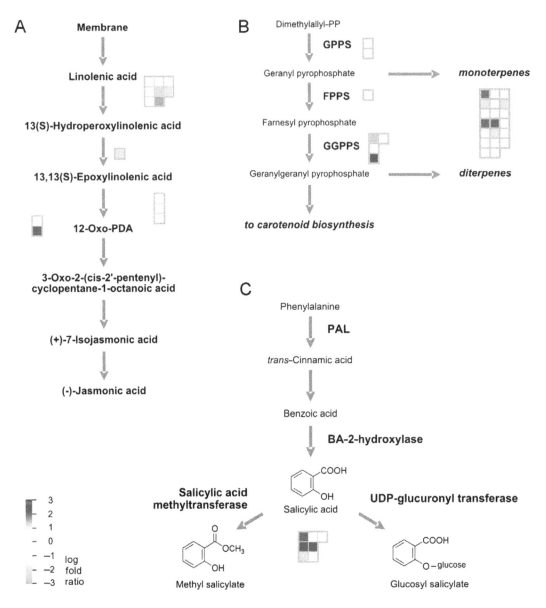

Figure 5. HLB-altered expression in symptomatic fruit in volatile and defense response pathways. Log fold ratios are indicated by a gradient from red (up-regulated) to green (down-regulated). Comparison was made between symptomatic and apparently healthy fruits from the infected orchard. A) Jasmonate biosynthesis, B) Terpenoid and carotenoid biosynthesis, C) Salicylic acid pathway.

Discussion

Huanglongbing, a highly destructive disease of citrus, threatens citrus-producing areas worldwide [1]. Previous studies monitored transcriptional changes in leaves on a large scale using microarrays to investigate host responses and reveal the mechanisms underlying disease development [6,7]. The present work focuses on transcriptional regulation in fruit peel to analyze the response to CaLas infection at different disease stages. Next-generation sequencing technology (NGS), as used here, can find differential expression of an increased number of transcripts, many of which may not be present in EST databases or represented in microarrays. NGS data can be used for specific transcriptome assemblies that become resource datasets for annotation of genomes and annotation of differentially regulated genes and proteins analyzed using any "omic" technique. However, the absence of a completed genome sequence limited the advantages of RNA-Seq technologies.

The aim of this work was to provide data using NGS technology for a comprehensive analysis of metabolic changes in fruit induced by HLB disease. These findings will uncover the fruit disorder mechanisms and facilitate development of short-term therapeutic strategies for already-infected trees. Toward this end, the experimental design included four types of fruit: healthy control fruit from an HLB-free orchard, apparently healthy fruit from non-symptomatic trees in an orchard affected by HLB, and asymptomatic and symptomatic fruit from infected symptomatic trees in the affected orchard. This design made it possible to identify differentially expressed genes at different disease stages. A comparison between apparently healthy and asymptomatic fruit revealed genes induced early in disease development. Comparing asymptomatic and symptomatic fruit identified genes involved in the host response during disease progression. Comparing symp-

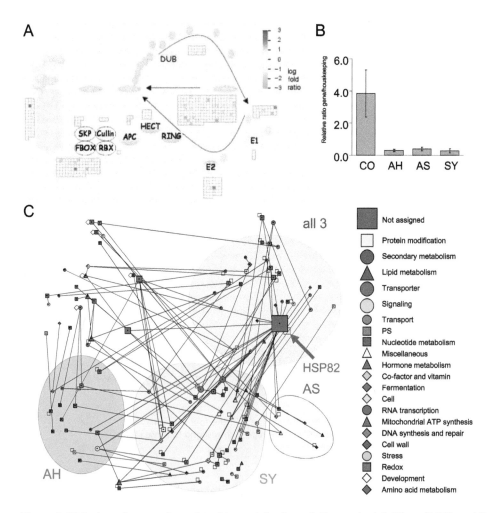

Figure 6. HLB-altered expression related to protein degradation and misfolding. A) Differentially expressed genes involved in ubiquitin-dependent degradation, showing comparison between symptomatic and apparently healthy fruits. B) Pattern of expression of heat shock protein 82 (HSP82) in HLB-free location (CO), apparently healthy (AH), asymptomatic (AS), and symptomatic (SY) fruits. C) Protein-protein interaction network predicted in *Citrus* based on the *Arabidopsis* knowledgebase. Networks between proteins encoded by HLB-regulated genes were divided into four different clusters. Node legend indicates functional classes of transcripts.

tomatic and apparently healthy fruit revealed host response genes related to the presence of HLB symptoms. Healthy fruit from the HLB-free location were also compared to the three sample types from the infected location. Differences in gene expression in these comparisons are likely to result from environmental and agronomic variability due to the difference in location, in addition to the effects of HLB.

A range of 1154 to 1762 differentially regulated genes (log fold ratio <-1.5 and >1.5) were found using RNA-Seq comparing the three categories of fruits from the infected orchard with those taken from a location free of HLB (Tables S1, S2, S3). In qRT-PCR analysis, 31 of 33 differentially regulated genes confirmed the pattern of expression found by RNA-Seq. Comparisons among the three fruit categories in the infected orchard resulted in fewer differentially regulated genes (301–780; Tables S4, S5, S6). This was expected, since samples within the same orchard grew under similar environmental and agronomic conditions.

A similar contrast between samples from the infected and uninfected orchards is seen in principal component analysis. The transcripts that most strongly characterize the asymptomatic and apparently healthy samples (MF-AS, MF-AH) are common between the two samples. The symptomatic (MF-SY) and

uninfected orchard samples (MF-CO) are both distinct groups and widely separated in the biplot (Figure S2). A visual summary outlines the most important transcriptional changes in the networks among genes, pathways, and cell functions in the fruit peel (Figure 8). Gene set enrichment analysis identified several pathways significantly affected by HLB as symptoms appear, considering both within-tree and between-tree comparisons, such as those for phenylpropanoids, starch and sucrose metabolism, carbon fixation, ascorbate, and alpha-linolenic acid (Figure 1, Figure S3).

Other pathways were also affected by HLB in the two comparisons within the same orchard. Transcripts encoding different subunits of the photosystem II reaction center and cytochrome b6-f complex subunit were more abundant in symptomatic HLB-infected fruit than in healthy fruit. Photosynthesis is central to all aspects of plant biology, since it provides energy for growth and reproduction, but its regulation by biotic and abiotic stresses is still unclear. The induction of photosynthetic light reactions in the fruit is consistent with the observation that symptomatic HLB-infected fruit often remains green. The retention of green color and increase of photosynthesis reactions is probably linked to the lower amount of ethylene detected in

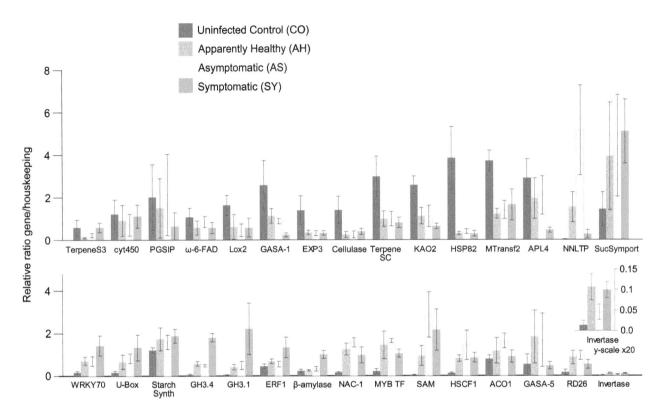

Figure 7. qRT-PCR analysis. Pattern of expression of genes belonging to different pathways in the four categories of fruits: symptomatic (SY), asymptomatic (AS), apparently healthy (AH) and healthy from HLB-free location (CO). Gene names correspond to those listed in Table 2. Bar heights for expression of Invertase also shown at 20-fold magnification (lower-right inset).

symptomatic fruits [28]. Photosynthesis is usually downregulated by pathogen attacks. The upregulation of photosynthetic reactions in the fruit does not contradict this common consideration because CaLas infections typically occur in young leaves. The transcriptomic changes observed in the fruit are probably linked to the source-sink disruption caused by leaf infections.

Protein degradation and modification pathways were significantly changed by CaLas infection, as shown by the upregulation of genes such as C3HC4-type ring finger proteins involved in ubiquitin-mediated degradation. Interestingly, heat shock proteins HSP82 and HSP70, highly interactive proteins in the PPI network inferred in citrus, were down-regulated at different stages of the disease. Heat shock proteins are highly conserved proteins induced in cells subjected to elevated temperatures or other environmental stresses [29,30]. These proteins act as molecular chaperones to stabilize, reduce misfolding, or facilitate refolding of proteins that have been denatured during stress events. In plant cells, HSP70 and HSP90 are involved in signal transduction leading to plant defense responses. Both proteins interact with a salicylic acid-induced protein kinase and their silencing affected the hypersensitive response in *Nicotiana benthamiana* while inducing non-host resistance to *Phytophthora infestans* [31]. HSP90 also modulates the innate immune responses involving gene-for-gene specific interactions, acting as a scaffold protein in a complex that mediates signal transduction [32]. Based on these findings, we speculate that down-regulation of heat shock proteins observed at different stages of HLB disease might increase protein misfolding in the fruit peel.

Genes encoding ATP synthase gamma and delta chains were also induced in symptomatic fruit, supporting the idea that CaLas

may act as an energy parasite by scavenging ATP from its host with a pathogen-specific ATP/ADP translocase. Indeed, the recently sequenced genome of CaLas revealed the presence of an ATP/ADP translocase in addition to ATP synthase [33]. ATP scavenging may be a possible mechanism of pathogenicity, affecting the fruit peduncle, columella, and seed coat.

Photosynthesis regulation in infected fruit peel may also reduce transport of sugars and ions such as nitrate, sulfate, and potassium. In this study, several differentially expressed genes were associated with transport of ions, including ammonium, sulfate, and phosphate (Table S1). Inorganic ions can modify sugar metabolism and photosynthesis [34,35]. Polarized vesicle trafficking, transport, and secretion of plant materials are associated primarily with non-specific resistance during host-pathogen interactions. ABC transporters play important roles in this process, since they are also involved in virulence, host range, and symptom elicitation [36].

Citrus proteins related to sugar and PDR/ABC transporters implicated in secretion of antimicrobial terpenoids [36] were specifically induced by HLB at the symptomatic stage (Table S1). Interestingly, among the genes identified in the CaLas genome were those for phosphate and zinc uptake into the cell [33]. This implies that mineral uptake by the pathogen may be enhanced due to induction of endogenous genes as well as host genes. However, phloem necrosis induced by CaLas contributes to the impaired nutritional transport functions and source-sink communication observed in this study.

Starch accumulation in leaf chloroplasts of sweet orange trees infected with CaLas has previously been demonstrated and is characteristic for HLB [1]. Genes encoding the large subunit of ADP-glucose pyrophosphorylase, the key enzyme catalyzing the

Table 2. List of genes analyzed by quantitative RT-PCR, including P-values for each of three pairwise comparisons from ANOVA of the expression data.

| Accession | Name | P-value | | |
		AH vs. CO	AS vs. CO	SY vs. CO
EY685091	Acidic cellulase	0.038	0.039	0.056
AF321533	ACC oxidase (ACO-1)	0.352	0.02	0.615
EU861194	APL4	0.276	0.323	0.009
EY710632	β-Amylase	0.856	0.276	0.003
EY710254	Cytochrome P450 (Cyt450)	0.632	0.368	0.841
EY660393	ERF-1	0.06	0.419	0.037
EY747590	Expansin 3 (Exp3)	0.059	0.055	0.054
CK935746	GASA-1	0.110	0.068	0.026
EY721422	GASA-5	0.171	0.219	0.795
EY693577	GH3.1	0.009	0.017	0.037
CF837667	GH3.4	0.001	0.0003	0.0001
EY745561	Heat Stress TF C-1 (HSCF1)	0.002	0.031	0.005
DY306049	HSP82	0.014	0.015	0.014
AB276107	Invertase	0.009	0.062	0.003
CV714426	Ent-kaurenoic acid oxidase (KAO2)	0.012	0.019	0.002
EY677115	Lipoxgenase2 (Lox2)	0.081	0.023	0.048
EY719680	Methyltransferase2 (MTransf2)	0.001	0.003	0.015
EY719020	Myb related TF (MYB TF)	0.031	0.00007	0.003
EY670400	NAC-1	0.002	0.0004	0.019
EY746283	NNLTP	0.021	0.014	0.087
EY683840	Omega-6-FAD	0.157	0.454	0.136
CX078423	Glycogenin-2/PGSIP	0.691	0.936	0.231
DY306001	RD26	0.024	0.006	0.054
EY747349	SA-methyltransferase (SAM)	0.029	0.01	0.02
EY710657	Starch synthase	0.178	0.137	0.03
AY098894	Sucrose symporter (SucSymport)	0.178	0.11	0.021
EY709847	Terpene synthase 3 (TerpeneS3)	0.089	0.171	0.997
CV886175	Terpene synth. cyclase (TerpeneSC)	0.028	0.027	0.019
EY655547	U-Box	0.067	0.008	0.029
EY675876	WRKY70	0.008	0.019	0.009

Names of genes (or abbreviated gene names in parentheses) correspond to bar clusters in Figure 7. Symptomatic (SY), asymptomatic (AS), apparently healthy (AH), healthy control (CO). P-values<0.05 are considered significant.

first and limiting step in starch biosynthesis, were up-regulated in infected leaves [6,37]. Interestingly, a gene encoding the large subunit of glucose-1-phosphate adenylyltransferase was down-regulated in infected and symptomatic fruit in the present study.

Also of interest are contrasting patterns of expression for different isoforms of starch synthase and starch cleavage genes, leaving the dynamics of starch accumulation in the fruit unclear (Figure 9). This is in contrast to experimental observation of starch accumulation in infected leaf tissues. In leaves, this accumulation was linked to up-regulation of genes involved in starch biosynthesis

and down-regulation of its conversion to maltose (DPE2 and MEX1) [37]. It has been hypothesized that starch accumulation is an effect of phloem plugging/necrosis during HLB infection, although it usually occurs before these symptoms are visible [38]. Differences in starch accumulation are probably linked to the different types of organs: mature leaves are "source" and fruit are "sink". However, further analysis of fruit starch accumulation must be performed to validate this hypothesis.

An increased abundance of transcripts for genes involved in the first steps of glycolysis and sucrose metabolism was observed in fruits from infected trees. In the cytosol, an invertase gene was up-regulated in asymptomatic and symptomatic fruit, affecting the sugar balance and communication between sink and source tissues. We speculate that this leads to increased glucose and fructose and decreased sucrose in fruit cells, which should be further validated with carbohydrate analysis. It is important to note that elevated glucose and fructose has been demonstrated in citrus leaves infected with CaLas [39]. Interestingly, different types of invertases were up-regulated in leaves (cell wall) [37] and fruits (vacuolar). Over-expression of yeast invertase in the cell wall of transgenic tobacco (*Nicotiana tabacum*) disrupted sucrose export, allowing soluble sugars and starch to accumulate, which consequently inhibited photosynthesis and resulted in stunted growth and bleached or necrotic leaf areas [40].

As previously suggested for leaves [38], it is possible that the differential expression of key genes involved in sucrose and starch metabolism, as observed in CaLas-infected citrus fruit, might affect the osmotic potential and induce plasmolysis, thus altering the ripening process and producing typical HLB symptoms. However, further analysis of sugar concentrations will be necessary to clarify the causes and effects of disease symptoms in the fruit. Other studies on CaLas-infected leaves have shown increased sucrose and glucose, but not fructose [37]. In fruit, it is possible that altered fructose and glucose concentrations might be responsible for physiological disorders and affect source-sink relationships with leaves.

The gene set enrichment analysis confirmed that sucrose and starch metabolism were highly affected by the disease. Integrated analysis of leaf and fruit data indicates that sugar and starch metabolism play a key role in the metabolic dysfunction induced by HLB disease. However, few studies have been conducted to address the effects of altering sugar metabolism on resistance to pathogen infections. Invertase plays a key role in the activation of stress responses and may function as an extracellular indicator for pathogen infection [41,42]. Indeed, transgenic plants overexpressing sugar metabolism enzymes such as a heterologous invertase from yeast have helped clarify source-sink relationships [4]. The expression of different viral movement proteins in transgenic plants and the resulting effects on photosynthesis, carbohydrate accumulation, and partitioning emphasize the importance of sugars in activating defense responses against biotic attacks [43,44,45].

Sink metabolism may be essential to satisfy the energy requirements of activating the cascade of defense responses. Interestingly, sucrose synthase was also more abundant in symptomatic fruit while sucrose transporter genes were downregulated by HLB. The Genevestigator database [46], indicated that these proteins may be down-regulated by hormones such as ethylene, methyl-jasmonates, and indol-3-acetic acid. The concept of sucrose metabolism regulators as a potential target for HLB therapeutics is intriguing.

Increasing evidence indicates an extensive cross-talk between sugar, hormone, and light signal transduction networks in plants [47]. Hormone pathways were significantly altered in fruit peel in

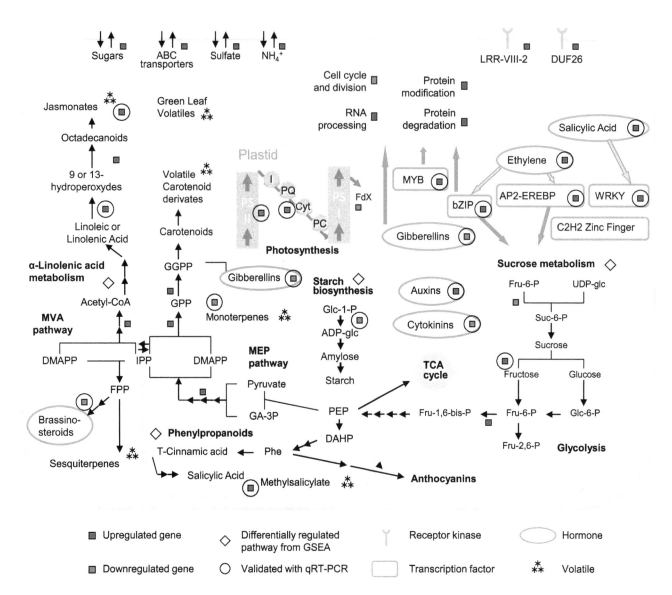

Figure 8. Fruit metabolism and regulatory pathways in CaLas-infected fruits. Genes, pathways, and cell functions that were differentially expressed are indicated with a square (red for up-regulated, green for down-regulated). Significantly differentially regulated pathways in gene set enrichment analysis are indicated in yellow.

response to CaLas infection. Two genes involved in auxin synthesis, GH3.1 and GH3.4, were induced in affected fruit. Gibberellin and cytokinin-related genes were mostly down-regulated in symptomatic fruits. Gibberellin regulation has been observed in other fruit disorders such as albedo breakdown disorder (unpublished data) and applications of GA₃ before fruit color break can reduce the occurrence of some fruit disorders. It is possible that sugar metabolism changes observed in the fruit might be linked with the down-regulation of gibberellins that regulate energy and carbohydrate metabolism. Previous studies have demonstrated that regulation of gibberellic acid-induced gene expression is affected by sugar and hormone signaling [48]. That cytokinins play a role in sugar regulation has been demonstrated [49]. Therapeutic approaches using small-molecule hormones such as cytokinins and gibberellins may allow modification of fruit metabolism to mitigate the negative impact of HLB on fruit quality and productivity.

Ethylene regulates a variety of developmental processes and stress responses in plants, including seed germination, cell elongation, senescence, fruit ripening, and defense. Nonetheless, ethylene can promote either disease resistance or susceptibility, depending on the host–pathogen interaction [50,51]. In our study, considerable changes were observed in the transcriptional profiles of genes related to ethylene biosynthesis and signal transduction. ACC synthase and ACC oxidase play pivotal roles in ethylene biosynthesis and their expression is often affected by pathogen attack [50]. It was unclear how ethylene concentration changes in fruit in response to HLB. Different isoforms of ACC synthase and ACC oxidase were up- or down-regulated. In addition, several transcripts involved in ethylene signaling and response were more abundant in symptomatic fruits, suggesting that HLB may have a stronger effect on ethylene signaling than ethylene biosynthesis. The general up-regulation of ethylene-related genes does not agree with the lower ethylene concentration previously reported in HLB symptomatic fruits [28]. This may be due to differences in fruit

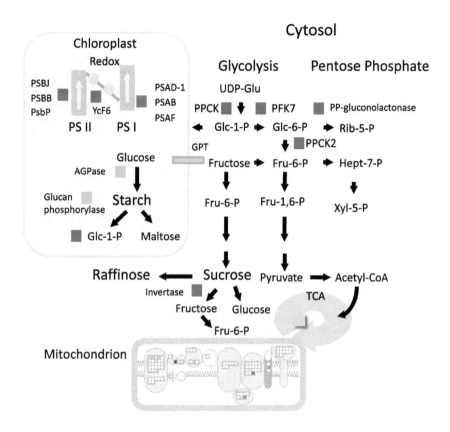

Figure 9. HLB-regulation of primary metabolism in symptomatic fruits. Differentially regulated genes and pathways involved in primary metabolism are indicated with a square (red for up-regulated, green for down-regulated).

developmental stages and health status. Furthermore, gene networks responsible for ethylene biosynthesis and perception are still not fully elucidated. A large number of genes annotated as ethylene-related occur as parologs playing tissue-specific roles, and many play additional roles in biotic stress response. Their expression may be drastically affected by the complex gene regulatory network involved in immune responses without directly affecting ethylene levels. The increased number of ERF- and AP2/EREBP-related genes modulated by HLB supports this notion. These factors control expression of many PR proteins and defense response effectors [50]. In addition, the induction of ethylene biosynthesis and signal transduction could profoundly modify fruit metabolism by accelerating senescence linked to the typical fruit malformations caused by HLB.

Salicylic acid and jasmonates are hormones involved in activating defense responses to pests and pathogens. Pathways for both hormones were activated in CaLas-infected citrus. It is known that different hormone-regulated defense pathways are activated depending on whether the pathogen is a necrotroph or biotroph [4]. The mechanisms of CaLas pathogenesis are poorly understood. The putative bacterial pathogen is closely related to bacterial families with symbiotic properties (i.e., *Rhizobiaceae*) [1]. However, the gene expression in challenged fruit showed an up-regulation of jasmonate-induced defense responses, typical of a host response to localized necrotroph invasion. This pathway is also stimulated by long distance signaling involving volatile compounds. Salicylic acid methyl-transferase was up-regulated in early, asymptomatic disease stages, potentially leading to production of volatile methylsalicylate. Nonexpressor of pathogenesis-

related genes1 (NPR1) was up-regulated in symptomatic fruits (Table S1). In *Arabidopsis*, NPR1 is required for SA-mediated suppression of JA-dependent defenses [52]. Also, ethylene modulates the NPR1 dependency of SA-JA antagonism, compensating for enhanced allocation of NPR1 to functions in SA-dependent activation of PR genes [53]. In plants, IRE1 (inositol-requiring 1) gene is considered to be involved in unfolded protein response (UPR) mediated by SAR with the engagement of BiP [54]. HLB upregulation was observed for IRE1a (EY679744), a gene closely involved in the UPR in plants [55]. This gene expression change suggests that UPR may be activated in the infected fruit.

Plants employ a network of intertwined mechanisms to counter infection by pathogens and parasites. One line of defense is based on dominant disease resistance (R) genes encoding nucleotide-binding, leucine-rich repeat (NB-LRR) proteins that mediate resistance to pathogens possessing corresponding avirulence (Avr) genes [56]. Several leucine-rich repeat (LRR) receptor kinases were up-regulated in symptomatic fruit, implying that they may be receptors triggering a futile defense response against CaLas.

Innate immunity can be regulated by transcription factors. Several genes belonging to the WRKY family of transcription factors such as WRKY70, WRKY13, WRKY30, WRKY40, WRKY65, and WRKY31 were up-regulated at both asymptomatic and symptomatic stages of HLB. WRKY70 acts as a convergence point, determining the balance between SA- and JA-dependent defense pathways in addition to being required for R gene-mediated resistance. Its role in JA and SA signaling, however, has recently been questioned [57]. Similarly,

AtWRKY53 positively modulates SAR. Members of the WRKY family were implicated in regulating the transcriptional reprogramming associated with plant immune responses and they may act as positive and negative regulators of disease resistance [58]. Other genes encoding bZIP and C2H2 zinc finger transcription factors were up-regulated in CaLas-infected fruits. The regulation of bZIPs in HLB-affected citrus may be associated with reported modifications observed in energy metabolism [59].

In the cytosol, up-regulation of 3-hydroxy-3-methyl-glutaryl CoA reductase (HMG1), 12-oxophytodienoate reductase 2, and allene oxide synthase may induce an increase in jasmonate-derived volatiles. Several genes involved in terpenoid metabolism, such as terpene synthase 3 and terpene synthase cyclase, were down-regulated in symptomatic fruits. These enzymes are involved in the synthesis and transport of a variety of terpenes, gibberellins, brassinosteroids, alkaloids, and plant volatiles that play diverse roles in plant development and defense [60]. The regulation of genes involved in these pathways has important implications for volatile emissions and is associated with a variety of responses. Further studies on the volatile emission profiles of CaLas-infected fruits will clarify the role of these enzymes.

Next generation sequencing enabled us to identify genes differentially expressed in citrus fruit peel in response to CaLas infection, leading to a better understanding of the processes involved in HLB disease development. This study identified several genes that were differentially expressed at the asymptomatic stage that may aid disease detection at primary stages of infection, before the pathogen can be detected by PCR. However, their usefulness as HLB-specific induced genes cannot be determined until a similar expression analysis is conducted on the same tissues infected by other citrus pathogens such as Citrus tristeza virus, *Xylella fastidiosa*, and *Xanthomonas axonopodis*. In the fruit peel, HLB induced transcriptional changes in important pathways such as sucrose and starch metabolism, hormone signaling, and isoprenoid synthesis. WRKY transcription factors appear to help regulate defense responses to CaLas in the fruit. The induction of genes involved in the light reactions of photosynthesis might increase the occurrence of reactive oxygen species, leading to protein degradation and misfolding. The application of small-molecule hormones is another promising short-term strategy to mitigate the devastating negative physiological effects of HLB.

Materials and Methods

Plant material and experimental design

Four types of mature fruit peel were analyzed based on origin, phenotype, and presence of the pathogen (Figure 10). Three categories of fruit were collected from "Valencia" sweet orange (C. sinensis L. Osb.) trees located at the USHRL-USDA Farm in Fort Pierce (St. Lucie County, FL). Trees were analyzed by PCR for the presence of CaLas using leaf petioles from four to six leaves collected from different areas in the canopy. All trees at this location were found to be PCR-positive at the time of collection. The first two categories of fruit peel were collected from trees with typical HLB disease symptoms on leaves (blotchy mottle and chlorosis) and fruit (small, green, and irregular in shape). Fruit peel was collected from asymptomatic and symptomatic fruit of the same tree. The third category was fruit peel from apparently healthy trees at the same location. These trees were HLB-positive using leaf petioles at the time of sampling. The fourth category was healthy fruit from 'Valencia' PCR-negative trees at a disease-free location, the Citrus Research and Education Center (Lake Alfred, FL). Fruit

peduncles, stored at $-20°C$, from all collected fruits were analyzed by PCR for the presence of CaLas. Peduncles from apparently healthy trees were PCR-negative while those from all other fruits in the infected orchard were PCR-positive, except for one asymptomatic sample. Peduncles of healthy fruits from disease-free location were PCR-negative. Five to ten fruit were collected from each of five different trees per treatment group, representing five biological replicates. Fruit peel segments were cut and mixed, immediately frozen in liquid nitrogen, and stored at $-80°C$. Juice sacs were removed before extraction.

PCR detection of *Ca*. L. asiaticus

Petioles and peduncles were ground in liquid nitrogen with a mortar and pestle and 100 mg ground tissue was used for DNA extraction. DNA was extracted using the Plant DNeasy® Mini Kit (Qiagen, Valencia, CA) according to manufacturer's instructions, yielding 20 to 30 ng DNA per extraction. Real-time PCR assays were performed using primers HLBas (5′- TCGAGCGCG-TATGCAATACG -3′) and HLBr (5′- GCGTTATCCCGTA-GAAAAAGGTAG -3′) and probe HLBp (5′- AGACGGGT-GAGTAACGCG -3′) [61]. Amplifications were performed using an ABI 7500 real-time PCR system (Applied Biosystems, Foster City, CA) and the QuantiTect Probe PCR Kit (Qiagen) according to manufacturer's instructions. All reactions were carried out in duplicate in a 20 µL reaction volume using 5 µL DNA per reaction. Plants or fruits were considered PCR-positive when C_T (cycle threshold) values were below 32.

RNA extraction

One gram of peel tissue was ground in liquid nitrogen with a mortar and pestle and resuspended in 10 mL guanidinium isothiocyanate buffer [62]. Total RNA was extracted according to Strommer et al. [63] with slight modifications. Phenol/

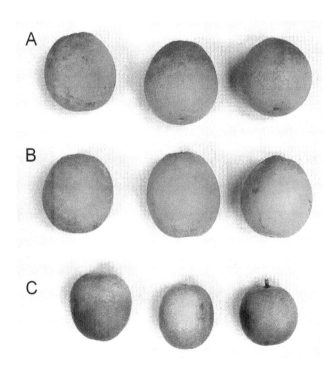

Figure 10. Stages of infection in the *Citrus* fruits studied. (A) Fruits of apparently healthy plants; (B) Asymtomatic fruits of infected plants; (C) Symtomatic fruits of infected plants.

chloroform/isoamylalcohol (25:24:1) extraction was followed by two extractions with chloroform/isoamylalcohol and precipitation of RNA with isopropanol at $-20°C$ overnight. RNA was pelleted by centrifugation at 10,000 g and 4°C for 1 h, resuspended in 5 mL water and precipitated overnight at 0°C with an equal volume of 8 M LiCl. After centrifugation at 10,000 g and 4°C for 1 h, RNA was washed twice with 70% ethanol, air-dried and dissolved in 500 µL of water. RNA was further purified using the RNeasy® MinElute Cleanup kit (Qiagen, Valencia, CA, USA) according to the manufacturer's instructions. RNA quality and purity were assessed using an Agilent Bionalyzer (Folsom, CA).

cDNA library construction and high throughput sequencing

RNA from the five biological replicates was equally pooled to 10 µg and then used to construct a cDNA library for each of the four fruit types. The cDNA libraries were constructed following the Illumina mRNA-sequencing sample preparation protocol (Illumina Inc., San Diego, CA). Final elution was performed with 16 µL RNase-free water. The quality of each library was determined using a BioRad Experion (BioRad, Hercules, CA). Each library was run as an independent lane on a Genome Analyzer II (Illumina, San Diego, CA) to obtain read lengths of up to 85 bp per paired end.

Sequence data processing and analysis

The raw Illumina reads were trimmed to remove low quality reads using custom scripts. Individual reads from each sample were aligned to the *Citrus sinensis* unigene set (15,808 sequences; NCBI Unigene Build #11, 4/20/09) with Burrows-Wheeler Transform (BWA) [64]. Read counts were generated with SAMtools [65] and custom scripts.

Six pairwise comparisons were made between the read counts for the four samples. For each pairwise comparison, the raw count data was normalized to control for different sequencing depths across samples with the DESeq Bioconductor package [66]. DESeq was developed for the statistical analysis of experiments with few or no replicates, accommodating this by treating samples from differing treatments or phenotypes as replicates in the estimation process. The results of this pooling are more conservative than in an experiment with more replicates.

Principal component analysis

Data analysis was performed to alleviate possible bias caused by the amount of collected material for each class or other confounding factors. A within-sample normalization process was applied to each sample to calculate the ratio of each target gene. This way, the sum of the ratios of all target genes was 1 while each target gene had its own count ratio within (0 to 1). For each sample, all target genes were first sorted in order from high to low. Then the top target genes with their accumulated ratio counting for 25% of the total were retained for further analysis, to make the differentially regulated gene selection more robust and focused on a relatively small number of strong target genes. By integrating all selected target genes for the four classes, we generated a complete list of potential target genes. In this list, some genes are shared among all four classes, while some are only observed in one class. Principal component analysis was then applied to the ratio matrix of this gene list to examine the contribution of each target gene to the separation of the classes. A biplot was constructed based on the first two principal components. The length of the loading vector for each target gene indicates the contribution of that gene to the separation based on the first two principal components. To

examine which genes contribute most to each class, two criteria for screening were employed. First, the mean value of the loading vector lengths (strengths) of all potential target genes was calculated and 80% of that value was set as the threshold for strength screening. Target genes with loading vector lengths larger than this threshold were considered strong. Secondly, the directional similarity was calculated between each target gene and each class and then 0.98 (i.e, the cosine of 10°) was set as the threshold value for similarity screening. Larger similarities indicate bigger contributions of a gene to a class.

Functional categorization and protein-protein network analysis

The *Citrus sinensis* unigene set was annotated using Blast2GO [67] to assign Gene Ontology (GO) terms to each gene. Lists of transcripts that were differentially expressed at a significant level ($p<0.01$, absolute value of Log Fold Change >1.5) in the pairwise comparisons were used as input for one-tailed Fisher's Exact Test in Blast2GO to identify GO terms that were significantly over-represented.

In addition, the differentially expressed genes were also functionally analyzed using the MapMan knowledgebase blasting to the TAIR database [25]. Gene set analysis was also performed using Pathexpress [26] for the highest *Arabidopsis* hit per *Citrus sinensis* unigene set (considering as a cutoff Log Fold Ratio $>1/ -1$). A protein-protein interaction network (PPI) was deduced for *Citrus* based on PPIs in *Arabidopsis* [68] using blast analysis. Networks were identified and visualized using Graphviz software.

Real time TaqMan PCR system

Real Time Taqman PCR analysis was conducted to validate RNA-seq data. Three biological replicates (a pool of 5 to 10 fruits from different plants) were used for each of the four types of fruit peel (healthy, apparently healthy, asymptomatic, and symptomatic). For each target gene, PCR primers and a TaqMan® probe were purchased as an assay mix from Applied Biosystems (Foster City, CA). DNase treatment and cDNA synthesis were performed in a combined protocol following instructions of the Quantitect Reverse Transcription Kit (Qiagen). A standard curve to determine the linearity of amplicon quantity vs. initial cDNA quantity was generated for each gene. Amplifications used 25 ng cDNA in a 20 µL final volume with TaqMan Universal PCR Master Mix and Taqman Assay ABI mixes (Applied Biosystems). Amplifications were performed on a StepOne Real Time PCR system (Applied Biosystems) using standard amplification conditions: 1 cycle of 2 min at 50°C, 10 min at 95°C, 40 cycles of 15 s at 95°C, and 60 s at 60°C. All PCR reactions were performed in duplicate. Fluorescent signals were collected during the annealing temperature and C_T values extracted with a threshold of 0.04 and baseline values of 3 to 10. *Citrus sinensis* elongation factor 1 alpha (EF-1α, accession AY498567) was used as an endogenous reference and $\Delta\Delta C_T$ was calculated by subtracting the average EF-1α C_T from the average C_T of the gene of interest.

Real Time Taqman PCR analysis was also conducted for CTV (Citrus Tristeza Virus) detection using the same RNA to determine the presence of the virus in the samples analyzed for HLB response. Primers were designed based on CTV CP reference sequence T36 (M76485) and the same protocol was followed as previously described.

ANOVA was performed considering the three biological replicates for each of the four fruit categories. P-values were determined for each of the six pairwise comparisons and values lower than 0.05 were considered significant.

Supporting Information

Table S1   Differentially expressed genes in symptomatic fruit in comparison to control (healthy in disease-free location), annotations and number of protein-protein interactions deduced from Arabidopsis knowledgebase.

Table S2   Differentially expressed genes in asymptomatic fruit in comparison to control (healthy in disease-free location), annotations and number of protein-protein interactions deduced from Arabidopsis knowledgebase.

Table S3   Differentially expressed genes in apparently healthy fruit in comparison to control (healthy in disease-free location), annotations and number of protein-protein interactions deduced from Arabidopsis knowledgebase.

Table S4   Differentially expressed genes in apparently healthy fruits in comparison to asymptomatic, annotations and number of protein-protein interactions deduced from Arabidopsis knowledgebase.

Table S5   Differentially expressed genes in symptomatic fruits in comparison to asymptomatic, annotations and number of protein-protein interactions deduced from Arabidopsis knowledgebase.

Table S6   Differentially expressed genes in symptomatic fruit in comparison to apparently healthy, annotations and number of protein-protein interactions deduced from Arabidopsis knowledgebase.

Table S7   21 genes making the strongest contribution to each of four classes. Five genes with count and log2foldchange

values also appear in two or more of Tables S1, S2, S3 and S6 as indicated.

Table S8   GO categories of transcripts found in significantly higher levels in pairwise comparisons (indicated as "up" and "down" groups), using Fisher's Exact Test.

Figure S1   CTV detection in Citrus using qRT-PCR. Analysis were conducted for the four categories of fruits: symptomatic (SY), asymptomatic (AS), apparently healthy (AH) and healthy from HLB-free location (CO).

Figure S2   Principal component analysis of differential gene expression in four HLB sample types. The length of the loading vector of each target gene indicates that gene's contribution to the separation of four categories of mature fruit peel by the first two principal components (C1, C2). MF-SY, symptomatic; MF-AS, asymptomatic; MF-AH, apparently healthy; MF-CO, HLB-free orchard.

Figure S3   Gene set enrichment analysis of Citrus fruits. PageMan analysis using Wilcoxon Rank Sum test, without multiple testing correction and cutoff = 1.0 Comparisons are shown between AS (asymptomatic) and SY (symptomatic), AH (apparently healthy) and SY (symptomatic), CO (HLB-free orchard) and AH (apparently healthy).

Author Contributions

Conceived and designed the experiments: AMD FM KDB. Performed the experiments: FM SLU UA MLP. Analyzed the data: FM RLR MB VB JF EL WZ DL RD CED AMD. Contributed reagents/materials/analysis tools: FM SLU UA MLP KDB AMD. Wrote the paper: FM AMD. Conceived and designed the experiments: AMD FM KDB. Performed the experiments: FM SLU UA MLP. Analyzed the data: FM RLR MB VB JF EL WZ DL RD CED AMD. Contributed reagents/materials/analysis tools: FM SLU UA MLP KDB AMD. Wrote the paper: FM AMD.

References

1. Bove JM (2006) Huanglongbing: A destructive, newly-emerging, century-old disease of citrus. Journal of Plant Pathology 88: 7–37.
2. Bove JM, Ayres AJ (2007) Etiology of three recent diseases of citrus in Sao Paulo State: Sudden death, variegated chlorosis and huanglongbing. Iubmb Life 59: 346–354.
3. Sechler A, Schuenzel EL, Cooke P, Donnua S, Thaveechai N, et al. (2009) Cultivation of 'Candidatus Liberibacter asiaticus', 'Ca. L. africanus', and 'Ca. L. americanus' Associated with Huanglongbing. Phytopathology 99: 480–486.
4. Tyler HL, Roesch LFW, Gowda S, Dawson WO, Triplett EW (2009) Confirmation of the Sequence of 'Candidatus Liberibacter asiaticus' and Assessment of Microbial Diversity in Huanglongbing-Infected Citrus Phloem Using a Metagenomic Approach. Molecular Plant-Microbe Interactions 22: 1624–1634.
5. Duan YP, Zhou LJ, Hall DG, Li WB, Doddapaneni H, et al. (2009) Complete Genome Sequence of Citrus Huanglongbing Bacterium, 'Candidatus Liberibacter asiaticus' Obtained Through Metagenomics. Molecular Plant-Microbe Interactions 22: 1011–1020.
6. Albrecht U, Bowman KD (2008) Gene expression in Citrus sinensis (L.) Osbeck following infection with the bacterial pathogen Candidatus Liberibacter asiaticus causing Huanglongbing in Florida. Plant Science 175: 291–306.
7. Kim JS, Sagaram US, Burns JK, Li JL, Wang N (2009) Response of Sweet Orange (Citrus sinensis) to 'Candidatus Liberibacter asiaticus' Infection: Microscopy and Microarray Analyses. Phytopathology 99: 50–57.
8. Fan J, Chen C, Yu Q, Brlansky RH, Li Z-G, et al. (2011) Comparative iTRAQ proteome and transcriptome analyses of sweet orange infected by Candidatus liberibacter asiaticus. Physiologia Plantarum DOI: 10.1111/j.1399-3054.2011.01502.x.
9. Tatineni S, Sagaram US, Gowda S, Robertson CJ, Dawson WO, et al. (2008) In planta distribution of 'Candidatus Liberibacter asiaticus' as revealed by

polymerase chain reaction (PCR) and real-time PCR. Phytopathology 98: 592–599.
10. Gibly A, Bonshtien A, Balaji V, Debbie P, Martin GB, et al. (2004) Identification and expression profiling of tomato genes differentially regulated during a resistance response to Xanthomonas campestris pv. vesicatoria. Molecular Plant-Microbe Interactions 17: 1212–1222.
11. Espinoza C, Medina C, Somerville S, Arce-Johnson P (2007) Senescence-associated genes induced during compatible viral interactions with grapevine and Arabidopsis. Journal of Experimental Botany 58: 3197–3212.
12. Panthee DR, Yuan JS, Wright DL, Marois JJ, Mailhot D, et al. (2007) Gene expression analysis in soybean in response to the causal agent of Asian soybean rust (Phakopsora pachyrhizi Sydow) in an early growth stage. Functional & Integrative Genomics 7: 291–301.
13. Boccara M, Sarazin A, Billoud B, Jolly V, Martienssen R, et al. (2007) New approaches for the analysis of Arabidopsis thaliana small RNAs. Biochimie 89: 1252–1256.
14. Navarro B, Pantaleo V, Gisel A, Moxon S, Dalmay T, et al. (2009) Deep Sequencing of Viroid-Derived Small RNAs from Grapevine Provides New Insights on the Role of RNA Silencing in Plant-Viroid Interaction. Plos One 4.
15. Donaire L, Wang Y, Gonzalez-Ibeas D, Mayer KF, Aranda MA, et al. (2009) Deep-sequencing of plant viral small RNAs reveals effective and widespread targeting of viral genomes. Virology 392: 203–214.
16. Ideker T, Sharan R (2008) Protein networks in disease. Genome Research 18: 644–652.
17. Holzmuller P, Grebaut P, Brizard JP, Berthier D, Chantal I, et al. (2008) "Pathogeno-Proteomics" Toward a New Approach of Host-Vector-Pathogen Interactions. In: Sparagano OAE, Maillard JC, Figueroa JV, eds. Animal Biodiversity and Emerging Diseases: Prediction and Prevention. Oxford: Blackwell Publishing. pp 66–70.

18. Jagoueix S, Bove JM, Garnier M (1996) PCR detection of the two 'Candidatus' liberobacter species associated with greening disease of citrus. Molecular and Cellular Probes 10: 43–50.

19. Hocquellet A, Toorawa P, Bove JM, Garnier M (1999) Detection and identification of the two Candidatus Liberobacter species associated with citrus huanglongbing by PCR amplification of ribosomal protein genes of the beta operon. Molecular and Cellular Probes 13: 373–379.

20. Teixeira DD, Danet JL, Eveillard S, Martins EC, Junior WCJ, et al. (2005) Citrus huanglongbing in Sao Paulo State, Brazil: PCR detection of the 'Candidatus' Liberibacter species associated with the disease. Molecular and Cellular Probes 19: 173–179.

21. Teixeira DC, Saillard C, Couture C, Martins EC, Wulff NA, et al. (2008) Distribution and quantification of Candidatus Liberibacter americanus, agent of huanglongbing disease of citrus in Sao Paulo State, Brasil, in leaves of an affected sweet orange tree as determined by PCR. Molecular and Cellular Probes 22: 139–150.

22. Lin H, Doddapaneni H, Bai XJ, Yao JQ, Zhao XL, et al. (2008) Acquisition of uncharacterized sequences from Candidatus Liberibacter, an unculturable bacterium, using an improved genomic walking method. Molecular and Cellular Probes 22: 30–37.

23. Wang Z, Yin Y, Hu H, Yuan Q, Peng G, et al. (2006) Development and application of molecular-based diagnosis for 'Candidatus Liberibacter asiaticus', the causal pathogen of citrus huanglongbing. Plant Pathology 55: 630–638.

24. Conesa A, Gotz S (2008) Blast2GO: A comprehensive suite for functional analysis in plant genomics. Int J Plant Genomics 619832.

25. Usadel B, Nagel A, Steinhauser D, Gibon Y, Bläsing OE, et al. (2006) PageMan: An interactive ontology tool to generate, display, and annotate overview graphs for profiling experiments. BMC Bioinformatics 7: 535.

26. Thimm O, Blasing O, Gibon Y, Nagel A, Meyer S, et al. (2004) MAPMAN: a user-driven tool to display genomics data sets onto diagrams of metabolic pathways and other biological processes. Plant Journal 37: 914–939.

27. Goffard N, Weiller G (2007) PathExpress: a web-based tool to identify relevant pathways in gene expression data. Nucleic Acids Research 35: W176–W181.

28. Rosales R, Burns JK (2011) Phytohormone Changes and Carbohydrate Status in Sweet Orange Fruit from Huanglongbing-infected Trees. Journal of Plant Growth Regulation 30;3): 312–321.

29. Almoguera C, Coca MA, Jordano J (1995) Differential accumulation of sunflower tetraubiquitin messenger-RNAs during zygotic embryogenesis and developmental regulation of their heat-shock response. Plant Physiology 107: 765–773.

30. Sabehat A, Lurie S, Weiss D (1998) Expression of small heat-shock proteins at low temperatures - A possible role in protecting against chilling injuries. Plant Physiology 117: 651–658.

31. Kanzaki H, Saitoh H, Takahashi Y, Berberich T, Ito A, et al. (2008) NbLRK1, a lectin-like receptor kinase protein of Nicotiana benthamiana, interacts with Phytophthora infestans INF1 elicitin and mediates INF1-induced cell death. Planta 228: 977–987.

32. Schulze-Lefert P (2004) Plant immunity: The origami of receptor activation. Current Biology 14: R22–R24.

33. Duan Y, Zhou L, Gottwald T (2009) Genome sequencing of "Ca. Liberibacter asiaticus". Phytopathology 99: S157–S157.

34. Sadka A, Dewald DB, May GD, Park WD, Mullet JE (1994) Phosphate modulates transcription of soybean vspb and other sugar-inducible genes. Plant Cell 6: 737–749.

35. Nielsen KK, Bojsen K, Roepstorff P, Mikkelsen JD (1994) A hydroxyproline-containing class-IV chitinase of sugar-beet is glycosylated with xylose. Plant Molecular Biology 25: 241–257.

36. van den Brule S, Smart CC (2002) The plant PDR family of ABC transporters. Planta 216: 95–106.

37. Fan J, Chen C, Brlansky RH, Gmitter Jr. FG, Li Z-G (2010) Changes in carbohydrate metabolism in Citrus sinensis infected with 'Candidatus Liberibacter asiaticus'. Plant Pathol 59: 1037–1043.

38. Etxeberria E, Gonzalez P, Achor D, Albrigo G (2009) Anatomical distribution of abnormally high levels of starch in HLB-affected Valencia orange trees. Physiological and Mol Plant Pathol 74: 76–83.

39. Albrecht U, Bowman KM (2008) Transcriptional response and carbohydrate metabolism of citrus infected with Candidatus Liberibacter asiaticus, the causal agent of huanglongbing in Florida. 11th Int. Citrus Congress. ICC, Wuhan, China.

40. von Schaewen A, Stitt M, Schmidt R, Sonnewald U, Willmitzer L (1990) Expression of a yeast-derived invertase in the cell wall of tobacco and Arabidopsis plants leads to accumulation of carbohydrate and inhibition of photosynthesis and strongly influences growth and phenotype of transgenic tobacco plants. EMBO Journal 9: 3033–44.

41. Herbers K, Takahata Y, Melzer M, Mock HP, Hajirezaei M, et al. (2000) Regulation of carbohydrate partitioning during the interaction of potato virus Y with tobacco. Mol Plant Pathol 1: 51–59.

42. Salzman RA, Tikhonova I, Bordelon BP, Hasegawa PM, Bressan RA (1998) Coordinate accumulation of antifungal proteins and hexoses constitutes a developmentally controlled defense response during fruit ripening in grape. Plant Physiology 117: 465–472.

43. Stitt M, Sonnewald U (1995) Regulation of metabolism in transgenic plants. Annual Review of Plant Physiology and Plant Molecular Biology 46: 341–368.

44. Roitsch T (1999) Source-sink regulation by sugar and stress. Current Opinion in Plant Biology 2: 198–206.

45. Lucas WJ, Wolf S (1999) Connections between virus movement, macromolecular signaling and assimilate allocation. Current Opinion in Plant Biology 2: 192–197.

46. Zimmermann P, Hirsch-Hoffmann M, Hennig L, Gruissem W (2004) GENEVESTIGATOR. Arabidopsis microarray database and analysis toolbox. Plant Physiology 136: 2621–2632.

47. Gibson SI (2005) Control of plant development and gene expression by sugar signaling. Current Opinion in Plant Biology 8: 93–102.

48. Perata P, Matsukura C, Vernieri P, Yamaguchi J (1997) Sugar repression of a gibberellin-dependent signaling pathway in barley embryos. Plant Cell 9: 2197–2208.

49. Sakamoto T, Sakakibara H, Kojima M, Yamamoto Y, Nagasaki H, et al. (2006) Ectopic expression of KNOTTED1-like homeobox protein induces expression of cytokinin biosynthesis genes in rice. Plant Physiology 142: 54–62.

50. Broekaert WF, Delaure SL, De Bolle MFC, Cammuel BPA (2006) The role of ethylene in host-pathoven interactions. Annual Review of Phytopathology 44: 393–416.

51. Pieterse CMJ, Leon-Reyes A, Van der Ent S, Van Wees SCM (2009) Networking by small-molecule hormones in plant immunity. Nature Chemical Biology 5: 308–316.

52. Spoel SH, Koornneef A, Claessens SMC, Korzelius JP, Van Pelt JA, et al. (2003) NPR1 modulates cross-talk between salicylate- and jasmonate-dependent defense pathways through a novel function in the cytosol. Plant Cell 15: 760–770.

53. Leon-Reyes A, Spoel SH, De Lange ES, Abe H, Kobayashi M, et al. (2009) Ethylene Modulates the Role of Nonexpressor of pathogenesis-related genes1 in Cross Talk between Salicylate and Jasmonate Signaling. Plant Physiology 149: 1797–1809.

54. Wang D, Weaver ND, Kesarwani M, Dong X (2005) Induction of protein secretory pathway is required for systemic acquired resistance. Science 308: 1036–1040.

55. Moreno AA, Mukhtar MS, Blanco F, Boatwright JL, Moreno I, et al. (2012) IRE1/bZIP60-mediated unfolded protein response plays distinct roles in plant immunity and abiotic stress responses. Plos One 7: e31944.

56. Postel S, Kemmerling B (2009) Plant systems for recognition of pathogen-associated molecular patterns. Semin Cell Dev Biol 20: 1025–1031.

57. Ren DT, Liu YD, Yang KY, Han L, Mao GH, et al. (2008) A fungal-responsive MAPK cascade regulates phytoalexin biosynthesis in Arabidopsis. Proceedings of the National Academy of Sciences of the United States of America 105: 5638–5643.

58. Eulgem T, Somssich IE (2007) Networks of WRKY transcription factors in defense signaling. Current Opinion in Plant Biology 10: 366–371.

59. Baena-Gonzalez E, Rolland F, Thevelein JM, Sheen J (2007) A central integrator of transcription networks in plant stress and energy signalling. Nature 448: 938–U910.

60. Mercke P, Kappers IF, Verstappen FWA, Vorst O, Dicke M, et al. (2004) Combined transcript and metabolite analysis reveals genes involved in spider mite induced volatile formation in cucumber plants. Plant Physiology 135: 2012–2024.

61. Li WB, Hartung JS, Levy L (2006) Quantitative real-time PCR for detection and identification of Candidatus Liberibacter species associated with citrus huanglongbing. Journal of Microbiological Methods 66: 104–115.

62. Chomczynski P, Sacci N (1987) Single-step method of RNA isolation by acid guanidinium hiocyanate-phenol-chloroform extraction, Analytical Biochemistry 162: 156–159.

63. Strommer J, Gregerson R, Vayda M (1993) Isolation and characterization of plant mRNA. In: Greenberg BM, Glick BR, eds. Methods in Plant Molecular Biology and Biotechnology. Boca Raton: CRC Press Inc. pp 49–65.

64. Li H, Durbin R (2009) Fast and accurate short read alignment with Burrows-Wheeler Transform. Bioinformatics 25: 1754–1760.

65. Li H, Handsaker B, Wysoker A, Fennell T, Ruan J, et al. (2009) The Sequence Alignment/Map format and SAMtools. Bioinformatics 25: 2078–2079.

66. Anders S, Huber W (2010) Differential expression analysis for sequence count data. Genome Biology 11: R106.

67. Gotz S, Garcia-Gomez J, Terol J, Williams T, Nueda M, et al. (2008) High-throughput functional annotation and data mining with the Blast2GO suite. Nucleic Acids Res 36: 3420–3435.

68. Geisler-Lee J, O'Toole N, Ammar R, Provart NJ, Millar AH, et al. (2007) A predicted interactome for Arabidopsis. Plant Physiology 145: 317–329.

8

Effect of Streptomycin Treatment on Bacterial Community Structure in the Apple Phyllosphere

Erika Yashiro[¤], **Patricia S. McManus***

Department of Plant Pathology, University of Wisconsin-Madison, Madison, Wisconsin, United States of America

Abstract

We studied the effect of many years of streptomycin use in apple orchards on the proportion of phyllosphere bacteria resistant to streptomycin and bacterial community structure. Leaf samples were collected during early July through early September from four orchards that had been sprayed with streptomycin during spring of most years for at least 10 years and four orchards that had not been sprayed. The percentage of cultured phyllosphere bacteria resistant to streptomycin at non-sprayed orchards (mean of 65%) was greater than at sprayed orchards (mean of 50%) ($P = 0.0271$). For each orchard, a 16S rRNA gene clone library was constructed from leaf samples. Proteobacteria dominated the bacterial communities at all orchards, accounting for 71 of 104 OTUs (determined at 97% sequence similarity) and 93% of all sequences. The genera *Massilia*, *Methylobacterium*, *Pantoea*, *Pseudomonas*, and *Sphingomonas* were shared across all sites. Shannon and Simpson's diversity indices and Pielou's evenness index were similar among orchards regardless of streptomycin use. Analysis of Similarity (ANOSIM) indicated that long-term streptomycin treatment did not account for the observed variability in community structure among orchards ($R = -0.104$, $P = 0.655$). Other variables, including time of summer, temperature and time at sampling, and relative distance of the orchards from each other, also had no significant effect on bacterial community structure. We conclude that factors other than streptomycin exposure drive both the proportion of streptomycin-resistant bacteria and phylogenetic makeup of bacterial communities in the apple phyllosphere in middle to late summer.

Editor: Jack Anthony Gilbert, Argonne National Laboratory, United States of America

Funding: This work was supported by United States Department of Agriculture Cooperative State Research, Education and Extension Service projects WIS04828 and WIS01425 and the Vaughan-Bascom fund of the Department of Plant Pathology, University of Wisconsin-Madison. The funders had no role in study design, data collection and analysis, decision to publish, or preparation of the manuscript.

Competing Interests: The authors have declared that no competing interests exist.

* E-mail: psm@plantpath.wisc.edu

¤ Current address: Department of Fundamental Microbiology, University of Lausanne, Lausanne, Switzerland

Introduction

The use of antibiotics in agriculture is controversial because of the possibility of selection for antibiotic resistant bacteria on farms and subsequent horizontal transfer of resistance genes to clinically important bacteria [20,44]. Most attention has focused on the practice of adding antibiotics to animal feed, because many of the antibiotics used as growth supplements for animals are also important in the treatment of human disease [16]. However, the use of antibiotics for the control of plant diseases has drawn scrutiny ever since streptomycin was first used for this purpose in the late 1950s [7]. In the United States, the application of antibiotics to plants accounts for less than 0.5% percent of total antibiotic use, with application of streptomycin and oxytetracycline to tree fruits and nursery plants for the control of fire blight accounting for the majority of antibiotic use in plant agriculture [18]. Nevertheless, streptomycin-resistant plant pathogens, including *Erwinia amylovora*, the fire blight pathogen, have emerged. In some cases *strA* or the gene pair *strA-strB*, which encode aminoglycoside phosphotransferases, are carried on transposon Tn*5393* [5,18]. Variants of Tn*5393* have been described in bacteria inhabiting diverse ecological niches, including soil, plants, animals and humans [23,35], suggesting the possibility for further horizontal transfer of resistance genes in the environment or on food. Because of the

potential environmental and human health risks associated with antibiotic use, in 2004 the European Union implemented tighter restrictions on the use of streptomycin on plants [15]. Similar reasoning led the National Organic Standards Board in the United States to recommend phasing out use of streptomycin and oxytetracycline in organic tree fruit production by 2014 (http://www.ams.usda.gov/AMSv1.0/getfile?dDocName=STELPRDC5091714).

The heightened regulations and general concerns surrounding the use of streptomycin and other antibiotics in plant agriculture have been an impetus behind studies addressing the impact of antibiotics on bacterial communities in cropping systems. Rodríguez et al. [30] found that bacterial community diversity, the proportion of bacteria resistant to streptomycin or gentamicin, and the presence of tetracycline resistance genes in bacteria on lettuce leaves did not differ among eight farms where those antibiotics were used and one non-exposed organic farm. Likewise, Rodríguez-Sánchez et al. [31] reported that use of oxytetracycline and gentamicin on coriander did not affect the abundance of bacteria resistant to those antibiotics, the presence of tetracycline- or gentamicin-resistance genes, or the presence of broad-host-range plasmids from the IncP-1 and IncQ incompatibility groups, which have been implicated in horizontal transfer of antibiotic resistance genes [26,38]. Studies on the effects of streptomycin on bacterial communities in agricultural settings are contradictory. For example, Tolba et al. [40] compared soil from a streptomycin-treated apple orchard in Germany with non-exposed soil from the same research station and found that although the treated and non-treated soils did not differ in the proportion of bacteria resistant to streptomycin, the incidence of *strA* and *strB* was greater in treated soil. However, in a survey of soils from diverse habitats in Europe [42], the abundance of *strA*, *strB*, and genes encoding aminoglycoside nucleotidyltransferases was not greater in streptomycin-treated soil from an apple orchard in Germany compared to non-treated soil from the same location or soils from other polluted or pristine sources. Likewise, Duffy et al. [6] concluded that the use of streptomycin in orchards had no effect on indigenous bacterial communities in soil or in the phyllosphere. By contrast, the incidence of streptomycin-resistant bacteria was greater in the phyllosphere, but not the soil, of pear trees from a nursery that had received 15 sprays of streptomycin over a 2-year period compared to samples from nurseries where streptomycin had not been used [39]. These previous studies have shed light on community membership and antibiotic resistance mechanisms, but they have lacked replicated treatment and control sites, making it impossible to account for site-to-site variability in bacterial communities unrelated to antibiotic exposure.

In the current study we test the hypothesis that the use of streptomycin in commercial apple orchards alters bacterial community structure on apple leaves. We compared phyllosphere bacterial communities in four commercial apple orchards where streptomycin had been used for several years with communities from four orchards where streptomycin had not been used. Our goal was to assess the long-term effects of streptomycin use, rather than the transient disruption of communities that might be expected in the spring immediately after spraying. To this end, we sampled leaves in middle to late summer, constructed 16S rRNA clone libraries from phyllosphere bacteria, identified operational taxonomic units (OTUs) from the cloned sequences, and then used diversity indices to describe communities. Because certain bacterial taxa have previously been identified as likely reservoirs of antibiotic-resistance genes in the phyllosphere [17,19,33,34], it was relevant to identify bacterial taxa in orchards, an objective served better by obtaining 16S rRNA gene sequences than fingerprinting methods. We used UniFrac and analysis of similarity (ANOSIM) to further test the effects of streptomycin treatment and other variables on the phylogenetic makeup of bacterial communities.

Materials and Methods

Ethics Statement

No specific permits were required for the described field studies. At each study site, the landowner granted us permission to collect apple leaves. The studies did not involve endangered or protected species.

Sample Collection

Apple leaves were collected in southern Wisconsin, USA, at eight commercial orchards during early July through early September, 2009 (Table 1). At all sites, fungicides, insecticides, and fertilizers were applied as needed to maintain tree health. At four orchards (Ep, SR, BFF, and LP) the sampled trees were sprayed with streptomycin sulfate at a concentration of 50 to 100 ppm up to three times annually during the bloom period (early to mid May) for at least the past 10 years. However, orchard BFF was not sprayed in 2008, and orchards BFF and LP were not

sprayed in 2009, the year of sampling, because weather conditions were not conducive to fire blight. At three orchards (DC, EL, and GPS) streptomycin had never been used, while at one orchard (BW) streptomycin was sprayed once prior to 1995. Despite this one spray, orchard BW was classified as "not sprayed" in this study. At the time of streptomycin application, leaves were expanding. None of the orchards was treated with oxetracycline, another antibiotic registered for use on apple trees.

At each orchard, 50 fully-expanded, healthy-appearing leaves from each of five or six arbitrarily selected trees (i.e., a total of 250 or 300 leaves per orchard) of the cultivar 'Gala' were collected at least 2 days after the last rainfall to minimize the effect of rain on bacterial community structure [10]. Leaves were collected from throughout the tree canopy by their petioles with gloved hands, packed individually in sterile plastic bags, stored on ice during transport to the laboratory, and then stored for up to 2 days at 4°C until processed. In the laboratory, using sterile tweezers and scissors, petioles were removed and discarded. Leaf blades were weighed individually and then sonicated in 10- to 13-leaf batches in 400 ml of sterile deionized water containing approximately two to three drops Tween 20 per liter in a tabletop ultrasonic cleaning bath (Branson model 3510, Branson Ultrasonic Corp., Danbury, CN, USA) for 5 min. The extracts from all the leaves from a single tree were bulked.

Bacterial Enumeration

An aliquot of the combined leaf extracts from each tree was serially diluted and then plated onto 0.1× tryptic soy agar (TSA) amended with cycloheximide (100 μg ml^{-1}) to inhibit fungal growth, as well as 0.1× TSA amended with cycloheximide (100 μg ml^{-1}) and streptomycin sulfate (50 μg ml^{-1}) to select for streptomycin resistant bacteria. After incubation for 7 days at room temperature and with ambient light, bacterial colonies were counted and the proportion resistant to streptomycin was calculated. A two-tailed Student's t-test was calculated on the mean percent streptomycin resistance between sprayed and non-sprayed orchards to determine the long-term effect of streptomycin use on bacterial populations. To examine the shorter term effect of streptomycin on the leaf bacterial community, colony counts from the 10 individual trees from the orchards that were sprayed in the sampling year (i.e., Ep and SR) were combined into a single dataset, while colony counts from the 10 trees from orchards that were not sprayed in the sampling year but had a history of streptomycin exposure (i.e., BFF and LP) were combined. A two-tailed Student's t-test was then calculated, with N = 10 for sprayed in sampling year and N = 10 for not sprayed in sampling year.

DNA Extraction and 16S rRNA Gene Clone Library Construction

Bacterial cells in leaf extracts from each tree were pelleted by centrifugation at 12,857×g at 4°C for 15 min. Pellets were then resuspended in residual supernatant and centrifuged at 16,870×g for 5 min in a microcentrifuge to further concentrate the bacteria. Genomic DNA was extracted from each individual cell pellet using the FastDNA spin kit (MP Biomedicals, Solon, OH, USA). PCR amplification of 16S rRNA genes was performed in triplicate using primers 27f and 1492r as described in Bräuer et al. [4] on two (orchard BFF) or four (all other orchards) extracts for a total of six or 12 independent reactions per orchard. Amplification products were combined for each orchard, purified using the Wizard SV Gel and PCR Clean-up System (Promega, Madison, WI, USA), and then cloned into the pGEM-T Easy vector according to the manufacturer's instructions (Promega). *Escherichia coli* DH5α cells

Table 1. Sampling sites and conditions.

Orchard	Streptomycin treatment[a]	Date of last streptomycin spray	GPS coordinates		Sampling date	Time of day[b]	Air temp (°C)
			Latitude	Longitude			
Ep	sprayed	14 May–09	42.974	−89.475	6 Jul–09	1100	27
SR	sprayed	15 May–09	43.317	−90.833	29 Jul–09	0955	29
BFF	sprayed	14 May–07	43.242	−88.026	14 Aug–09	1024	28
LP	sprayed	May–08[c]	43.348	−89.287	1 Sep–09	1100	20
DC	not sprayed	–	43.025	−89.221	8 Jul–09	0900	16
EL	not sprayed	–	42.745	−88.231	21 Jul–09	1049	25
GPS	not sprayed	–	42.997	−89.209	5 Aug–09	0950	24
BW[d]	not sprayed	Before 1995	42.644	−88.162	12 Aug–09	1035	NA[c]

[a]"Sprayed" orchards were sprayed with streptomycin sulfate at a concentration of 50 to 100 ppm up to three times annually during the bloom period (early to mid May) for at least the past 10 years. "Not sprayed" orchards were not sprayed with streptomycin.
[b]Central Daylight Time.
[c]Day of last spray at orchard LP is unknown.
[d]Temperature data not available at orchard BW.

were transformed with the ligation products, and a clone library was constructed for each orchard. The 16S rRNA gene insert of each clone was amplified using the M13f(-20) and M13r(-27) primers according to Bräuer et al. [4] and sequenced from the 5′ end using the 27f primer on an Applied Biosystems 3730xl automated DNA sequencing instruments (Applied Biosystems, Foster City, CA, USA) at the University of Wisconsin-Madison Biotechnology Center.

Taxonomic and Statistical Analyses of Bacterial Communities

The sequences were screened against chloroplast and other non-ribosomal DNA using BLASTn and the Ribosomal Database Project Classifier [1,45]. Potential chimeras were identified using Mallard and omitted from further study after manual verification [2]. Short sequences that did not cover the reference *E. coli* base positions 103 to 920 were also omitted from further study. The remaining sequences were aligned using the SILVA aligner (25) and imported into ARB [14]. Using a filter to exclude the nonoverlapping ends beyond positions 103 to 920, a neighbor-joining matrix was generated in ARB and exported. The fasta file from these same sequences was also exported and used for taxonomic classification of the bacterial communities using the RDP Classifier.

Mothur [32] was used to assign operational taxonomic units (OTUs) at the 97% sequence similarity level and to calculate Shannon and Simpson's diversity indices, Pielou's index of evenness, and rarefaction curves. For beta diversity analyses, PhyML was used to generate a maximum likelihood tree of all the sequences [8]. The settings used were: the general time reversible (GTR) correction model; optimized equilibrium frequencies; estimated proportion of invariable sites, four substitution rate categories; estimated gamma shape parameter; and SPR tree topology search operations. The validity of using the GTR correction model was also confirmed by applying jModelTest2 on smaller subsets of the 16S rRNA gene dataset [24]. Due to the computer intensive nature of the software, the full dataset was not processed. The tree generated by PhyML was then imported into mothur, and the weighted UniFrac metric that accounts for phylogenetic relationships among taxa [13] was calculated and visualized in a Principle Coordinate Analysis (PCoA) plot.

Hypothesis testing to determine which variables influenced bacterial community structure at the different orchards was performed on weighted UniFrac using the analysis of similarity (ANOSIM) with 10,000 iterations. A Mantel test with the Pearson correlation coefficient and 1000 iterations was used to determine the effect of relative distances between orchard sites on the bacterial communities at each sampling site. The hypothesis tests were repeated using Bray-Curtis distance (non-phylogenetic) defining OTUs at the 97%, 99%, and 100% (i.e., unique sequences) similarity thresholds. The Bonferroni correction was applied to correct for multiple comparisons.

Nucleotide Sequence Accession Numbers

The 16S rRNA gene sequences obtained in this study were deposited into the NCBI GenBank database under accession numbers JQ046416 to JQ048217.

Results

Proportion of Bacterial Communities Resistant to Streptomycin

At sprayed orchards 43% to 59% (mean of 50%) of the total culturable bacteria were resistant to streptomycin, whereas at non-sprayed orchards 57% to 72% (mean of 65%) of the total culturable bacteria were resistant to streptomycin (Table 2). A two-tailed Student's t-test indicated that the proportion of streptomycin resistant bacteria differed between the sprayed and non-sprayed orchards ($P = 0.0271$). Among sprayed orchards, the observation of higher CFU counts at orchards sprayed in 2009, the year of sampling (Ep and SR), versus those not sprayed in 2009 (BFF and LP), prompted an analysis of shorter-term effects of streptomycin on total CFU. The mean log_{10} CFU/g leaf tissue was 5.58 and 6.29 for sprayed and non-sprayed orchards, respectively. A two-tailed Student's t-test indicated that the difference in total CFU was highly significant ($P < 0.001$).

Phylogenetic Characterization of Phyllosphere Bacterial Communities

A total of 3199 16S rRNA gene clone sequences were obtained for the eight-orchard study. After excluding non-ribosomal DNA

Table 2. Bacterial CFU recovered from the apple phyllosphere and percentage that were resistant to streptomycin (strR).

Orchard	Streptomycin treatment	Log₁₀ CFU/g leaf (SEM)[a]	Log₁₀ strR CFU/g leaf (SEM)[b]	% strR (SEM)[c]
Ep	sprayed	5.59 (0.05)	5.36 (0.04)	59 (3.4)
SR	sprayed	5.58 (0.07)	5.21 (0.08)	43 (2.4)
BFF	sprayed	6.43 (0.13)	6.13 (0.15)	50 (2.7)
LP	sprayed	6.14 (0.13)	5.82 (0.14)	49 (4.7)
Mean				50 (3.3)
DC	not sprayed	5.65 (0.03)	5.49 (0.08)	72 (10.7)
EL	not sprayed	6.58 (0.10)	6.42 (0.09)	69 (5.3)
GPS	not sprayed	6.43 (0.14)	6.22 (0.14)	61 (3.8)
BW	not sprayed	5.63 (0.09)	5.38 (0.11)	57 (3.3)
Mean				65 (3.6)
P[d]				0.0271

[a]CFU on TSA amended with cycloheximide. Values in rows with an orchard designation are the mean and SEM of five or six trees per orchard.
[b]CFU on TSA amended with cycloheximide and streptomycin. Values in rows with an orchard designation are the mean and SEM of five or six trees per orchard.
[c]The mean % strR and SEM for the individual orchards was calculated across trees at each orchard (N = 5 or 6). The mean and SEM for the mean % strR was calculated separately for sprayed (N = 4) and not sprayed orchards (N = 4).
[d]P-values for the two-tailed t-test comparing sprayed (N = 4) and not sprayed (N = 4) orchards.

(notably chloroplast DNA), low quality or short sequences, and chimeras, 1802 sequences remained for analyses, with a range of 169 to 335 sequences per orchard (Table 3; Fig. 1). Across all orchards, 104 OTUs were identified, with a range of 19 to 35 OTUs per orchard (Table 3; Fig. 1). The phylum Proteobacteria comprised 93% of the total bacterial community, followed by Bacteroidetes, Actinobacteria, Firmicutes, Acidobacteria, Tenericutes, and Chlorflexi (Table 4). Sequences for five genera, *Massilia*, *Methylobacterium*, *Pantoea*, *Pseudomonas*, and *Sphingomonas*, were shared across all sites (Table 5). At individual orchards, the predominant genus was *Sphingomonas* (orchards Ep, BFF, LP, EL, and GPS), *Pseudomonas* (orchards SR and DC), or *Massilia* (BW) (Table 5). Because the two most abundant genera, *Sphingomonas* and *Pseudomonas*, frequently possess either intrinsic or acquired streptomycin resistance [11,19,34,39,41], we tested whether the abundance of these genera varied between sprayed and not sprayed orchards. The percentage of sequences representing *Sphingomonas* in sprayed and non-sprayed orchards was 31% (range 16 to 39%) and 37% (range 6 to 64%), respectively (Table S1). The percentage of sequences representing *Pseudomonas* in sprayed and non-sprayed orchards was 23% (range 10 to 37%) and 17% (range 5 to 38%), respectively (Table S1). Two-tailed Student t-tests revealed no difference in the proportions of *Sphingomonas* (P = 0.7470) or *Pseudomonas* (P = 0.5165) at orchards sprayed with streptomycin compared to non-sprayed orchards. Table S1 shows the phylogenetic characterization and abundance of all bacterial 16S rRNA gene sequences in this study.

Rarefaction curves indicated that more OTUs would have been obtained with additional sequencing (Fig. 1). Shannon and Simpson's indices indicated that bacterial diversity was generally similar across sites, although the community from streptomycin-sprayed orchard LP was more diverse than other communities (Table 3). Pielou's evenness index indicated that the bacterial communities were comprised of a few highly abundant taxa and many relatively rare taxa (Table 3). Two-tailed Student's t-tests indicated no differences in Shannon or Pielou's indices for communities from sprayed orchards compared to non-sprayed orchards (P = 0.293 for Shannon; P = 0.225 for Pielou's), whereas the Simpson's index indicated somewhat greater diversity

(P = 0.083) for communities from sprayed orchards compared to non-sprayed orchards.

Beta Diversity

The weighted Unifrac metric was used to calculate the pairwise distances between the phyllosphere bacterial communities of the eight orchards. The first two axes of the principal coordinates analysis (PCoA) represented 40.8% and 22.7%, respectively, of the total variation (Fig. 2). ANOSIM tests showed that none of the variables we tested accounted for the variability in community structure among orchards. There was no significant effect of streptomycin treatment (R = −0.104, P = 0.655), nor an effect of not spraying streptomycin during the sampling year regardless of long-term treatment (R = −0.177, P = 0.246). Although sampling of the eight orchards extended from early July to early September, the effect of sampling time did not explain the variability in community structure (ANOSIM groupings by month R = 0.198, P = 0.204; groupings by 2-week period R = 0.043, P = 0.657). The time of day of sampling, which ranged from 0900 to 1100 hr, did not account for community structure (R = 0.041, P = 0.398), nor did the ambient temperature at the time of sampling, which ranged from 16°C to 29°C (R = 0.094, P = 0.606). For total log₁₀ CFU, R = 0.625 and P = 0.030 (Bonferroni-corrected alpha value = 0.0056). For log₁₀ CFU on streptomycin-amended TSA, R = 0.056 and P = 0.594. Geographic proximity among the orchards also did not influence the community structure (Mantel r = 0.000, P = 0.639). The same conclusions (i.e., no significance for any of the variables) were obtained using Bray-Curtis distance, a non-phylogenetic metric of community similarity (Table S2).

Recalculating the weighted Unifrac after removing *Sphingomonas* from the dataset did not reveal any effects of streptomycin. The bacterial communities at orchards Ep, DC, and SR clustered together, while orchards EL and GPS became more distant from LP and BFF on a PCoA plot (Fig. S1). Closer examination revealed that *Sphingomonas* comprised 60% and 64% of the communities at EL and GPS, respectively, compared to 6% to 39% at the other sites (Table S1). Recalculating the weighted Unifrac values after removing both *Sphingomonas* and *Pseudomonas* from the dataset also did not reveal any effect of streptomycin on

Table 3. Diversity and evenness estimates for 16S rRNA gene libraries derived from phyllosphere bacteria of apple orchards differing in exposure to streptomycin.

Orchard	Streptomycin treatment	No. sequences	No. OTUs[a]	Shannon[b]	Simpson's 1-D[b]	Pielou's evenness
Ep	sprayed	204	26	2.12 (0.17)	0.83 (0.02)	0.40
SR	sprayed	205	21	1.84 (0.16)	0.78 (0.03)	0.34
BFF	sprayed	199	19	2.10 (0.15)	0.83 (0.03)	0.40
LP	sprayed	169	35	2.92 (0.17)	0.92 (0.02)	0.57
Mean for sprayed		194	25	2.25	0.84	0.43
DC	not sprayed	209	30	2.07 (0.19)	0.79 (0.04)	0.39
EL	not sprayed	264	29	2.02 (0.17)	0.76 (0.04)	0.36
GPS	not sprayed	335	25	1.66 (0.17)	0.62 (0.06)	0.28
BW	not sprayed	217	26	2.06 (0.18)	0.78 (0.04)	0.38
Mean for not sprayed		257	28	1.95	0.74	0.35
p [c]		0.109	0.582	0.293	0.083	0.225

[a]Number of unique OTUs determined at the 97% sequence similarity threshold.
[b]95% confidence intervals for the Shannon and Simpson's indices are indicated between parentheses. Simpson's index (D) is presented as 1-D; higher values indicate greater diversity.
[c]P-values for the two-tailed t-test comparing sprayed (N = 4) and not sprayed (N = 4) orchards.

the remaining bacterial community (Fig. S2; Table S3). Orchard BW remained an outlier on the PCoA plot, indicating that *Sphingomonas* was not responsible for the divergence in community structure for that site. Unique to orchard BW were 25 sequences belonging to a single OTU in the phylum Firmicutes, genus *Lactobacillus* (Table 3; Table S1). Also, sequences representing the genus *Massilia* were more abundant for orchard BW than other orchards (Table 5). If orchard BW was considered an outlier, recalculating ANOSIM tests resulted in R = 0.3904 and P = 0.046 (Bonferroni-corrected alpha value = 0.005) for the effect of spraying during the spring of the sampling year on bacterial communities in mid to late summer.

Discussion

The results indicate that the routine springtime use of streptomycin in apple orchards for the control of fire blight disease does not have long-term effects on the diversity or

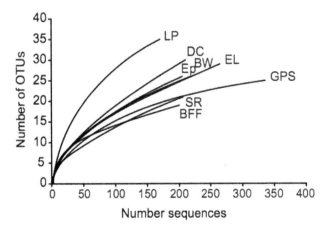

Figure 1. Rarefaction curves of the sequenced clones for each orchard. The sequences were binned into OTUs at the 97% similarity threshold.

phylogenetic composition of bacterial communities in the phyllosphere in middle to late summer. However, the proportion of cultured bacteria resistant to streptomycin was lower in orchards with a history of streptomycin exposure than in non-exposed orchards. We hypothesized that this unexpected result might be explained by a chance higher abundance of *Sphingomonas* and *Pseudomonas* in non-exposed orchards, since these genera are known to have a high level of stable, intrinsic or acquired resistance to streptomycin regardless of exposure [11,19,34,39,41]. In fact, we previously observed that the order Sphingomonadales accounted for 100% of cultured streptomycin-resistant bacteria at a research orchard where streptomycin had never been used [46]. Although we did not identify cultured bacteria in the current study, we estimated the abundance of *Sphingomonas* and *Pseudomonas* based on their abundance in clone libraries. The validity of this approach is supported by the finding that the relative abundance of each of these genera was similar among cultured isolates and 16S rRNA gene clones derived from the same field samples [47]. In the present study, the abundance of *Sphingomonas* and *Pseudomonas* sequences in clone libraries was similar between the sprayed and non-sprayed orchards, making it unlikely that those genera accounted for the difference in the percentage of streptomycin-resistant colonies. We previously found a greater abundance of Actinomycetales, specifically the genera *Curtobacterium* and *Frigoribacterium*, in the apple phyllosphere by culturing on TSA than by cloning 16S rRNA genes [47]. These genera, or other members of the Actinomycetales that are known producers of streptomycin [12,43] and therefore would also be resistant to streptomycin, might have been more abundant among the cultured bacteria than is suggested by clone libraries. However, there is no obvious reason to expect that they would be more abundant in non-sprayed than sprayed orchards. Thus, while we cannot explain the higher level of resistance in non-sprayed orchards, it is clear that springtime use of streptomycin over many years does not lead to elevated levels of streptomycin resistance in the apple phyllosphere later in the summer.

Despite a lack of long-term effects of streptomycin on bacterial phyllosphere communities, where the antibiotic was applied

Table 4. Taxonomy and abundance of bacterial 16S rRNA gene sequences in libraries constructed from apple leaves from orchards differing in streptomycin treatment.

Phylogenetic group	No. OTUs[a]	No. sequences (% of total)	Sprayed				Not sprayed			
			Ep	SR	BFF	LP	DC	EL	GPS	BW
Protobacteria	71	1677 (93)	193	203	188	153	193	246	319	182
Alpha	28	801	71	37	127	94	21	175	228	48
Beta	14	346	34	58	10	24	28	55	37	100
Delta	6	10	3	1	2	1	3	0	0	0
Gamma	23	520	85	107	49	34	141	16	54	34
Bacteroidetes	11	57 (3)	3	0	6	12	9	15	8	4
Actinomycetes	12	31 (2)	5	2	5	4	1	3	8	3
Firmicutes	4	28 (2)	1	0	0	0	2	0	0	25
Acidobacteria	3	4 (<1)	1	0	0	0	1	0	0	2
Tenericutes	1	3 (<1)	0	0	0	0	3	0	0	0
Chlorflexi	2	2 (<1)	1	0	0	0	0	0	0	1
Total	104	1802 (100)	204	205	199	169	209	264	335	217

Header above groups: **Orchard streptomycin treatment and no. sequences**

[a]Number of unique OTUs determined at the 97% sequence similarity threshold.

during spring of the year of sampling, the total bacterial population, but not the proportion of bacteria resistant to streptomycin, was reduced in mid-summer. When site BW, an outlier on PCoA plots due to observable differences in its bacterial community composition, was treated as an outlier in ANOSIM, we found weak evidence that exposure to streptomycin during the year of sampling affected the phylogenetic composition of communities.

The enterobacterium *Pantoea agglomerans* (formerly *Erwinia herbicola*) has been cited as a likely reservoir of mobile streptomycin resistance genes in apple orchards [18]. From a public health perspective, the presence of enterobacteria in orchards and on other food crops has received attention [3,21,30] because this group includes several food-borne pathogens as well as commensal bacteria that might serve as reservoirs for resistance genes in the human gut. In the current study, *Pantoea* was abundant in one non-sprayed and two sprayed orchards, and therefore was not obviously affected by long-term streptomycin use. We had

expected a greater abundance of *Pantoea* and the related enterobacterial genera *Erwinia* and *Enterobacter* across all orchards, since these genera are commonly isolated from apple tissues during spring [17,33]. On leaves of cottonwood, the abundance of members of the Enterobacteriacae was highly variable throughout the summer, with *Pantoea* dominating in early summer and diminishing by fall [29]. Thus, differences in the succession of bacterial communities, and the impacts of immigration, emigration, growth and death might explain the imbalance in the distribution of *Pantoea* among orchards.

Previous studies concerning the use of antibiotics in crop production report the proportion of antibiotic resistant bacteria in leaf, flower, or soil samples relative to antibiotic exposure, and some provide information on the presence of resistance genes and/or potential resistance plasmids [6,17,19,30,31,33,34,36,37,39,40,42]. Our study is unique, however, in that comparable treatment and control sites were replicated, making it possible to perform statistical comparisons between sprayed and non-sprayed sites and thereby

Table 5. Genera shared across all orchards.

Phylogenetic group	No. OTUs[a]	No. sequences (% of total)[b]	Sprayed				Not sprayed			
			Ep	SR	BFF	LP	DC	EL	GPS	BW
Sphingomonas	13	666 (37)	64	32	77	65	13	158	216	41
Pseudomonas	9	345 (19)	45	75	48	16	80	14	43	24
Massilia	2	312 (17)	28	53	9	21	26	52	34	89
Pantoea	2	146 (8)	39	32	1	5	57	1	2	9
Methylobacterium	3	83 (5)	7	4	24	22	3	8	9	6

Header above groups: **Orchard streptomycin treatment and no. sequences**

[a]Number of unique OTUs determined at the 97% sequence similarity threshold.
[b]Total number of sequences, including sequences for OTUs that were not shared across all orchards, was 1802.

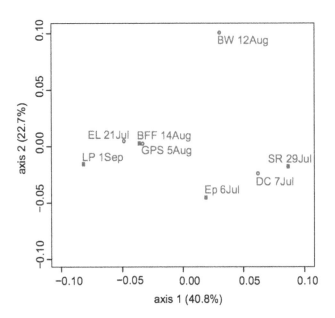

Figure 2. PCoA plot of weighted Unifrac generated from a maximum likelihood tree. Red squares indicate orchards that were sprayed with streptomycin; blue circles indicate orchards that were not sprayed.

determine the effects of many years of streptomycin exposure under conditions typical in commercial apple production. Moreover, past studies aimed at identifying potential reservoirs for antibiotic resistance genes in the phyllosphere have relied on culturing [5,17,19,33,34,39]. It is well established that culturing generally reveals only a small fraction of bacterial diversity [9,22,27], and recently we demonstrated greater richness of bacteria in the apple phyllosphere determined by 16S rRNA clone libraries than by culturing [47]. The current study provides the first culture-independent assessment of bacterial communities in apple orchards varying in exposure to streptomycin.

Our analyses were based on a modest number of sequences (1802 across eight orchards) relative to the thousands of sequences per sample that are routinely obtained through modern high-throughput technology. Indeed, rarefaction curves suggested that further sequencing would have revealed greater richness in our samples. While we cannot dismiss the potential ecological importance of rare taxa that might have been detected through deeper sequencing, the dominance in our dataset of a few taxa common to all sites makes it unlikely that deeper sequencing would have revealed significant differences in bacterial communities at sprayed and non-sprayed orchards, at least when analyzed by methods that take into account sequence abundance (e.g., weighted UniFrac).

We conclude that long-term streptomycin use does not increase the frequency of resistant bacteria or disrupt bacterial communities on apple leaves. Our findings contribute to a growing body of literature that indicates using streptomycin to control fire blight

has low environmental impact [6,28]. However, our conclusion does not absolve streptomycin of all risk associated with its use. For example, it is possible that streptomycin could select for novel resistance genes in apple orchards, even if the overall frequency of resistant bacteria is not increased. A greater diversity of mobile resistance genes in apple orchards could lead to horizontal transfer of resistance among a greater range of bacteria, which in turn could be consumed on fresh produce. However, for those concerned with regulating the use of streptomycin on crops, it will be critical to compare data from replicated treated and control sites, as was done in the current study, since resistance genes and mobile genetic elements are present even in environments with no known anthropogenic selection pressure [19,40,42].

Supporting Information

Figure S1 Weighted Unifrac PCoA plot of the orchard bacterial communities, excluding *Sphingomonas* species. The streptomycin treated sites were Ep, SR, BFF, and LP. The nontreated sites were DC, EL, GPS, and BW.

Figure S2 Weighted Unifrac PCoA plot of the orchard bacterial communities, excluding *Sphingomonas* and *Pseudomonas* species. The streptomycin treated sites were Ep, SR, BFF, and LP. The nontreated sites were DC, EL, GPS, and BW.

Table S1 Phylogenetic characterization and abundance of bacterial 16S rRNA gene sequences in libraries constructed from apple leaves differing in streptomycin treatment.

Table S2 ANOSIM and Mantel values from Bray-Curtis matrix of the bacterial communities at the 100%, 99%, and 97% similarity threshold. The Bonferroni corrected alpha is 0.0056.

Table S3 ANOSIM and Mantel values from weighted Unifrac matrix for the bacterial communities excluding *Sphingomonas* and *Pseudomonas* species. The Bonferroni corrected alpha is 0.0056.

Acknowledgments

We thank A Ives, J Handelsman, P Schloss, A Shade, and R Spear for helpful discussions and insights; and G Maynaud, C Rasmussen, A Bindrim Morgan, M Linske, G Taycher, and PR Tentscher for technical assistance. We also thank apple growers in Wisconsin for allowing us to sample in their orchards.

Author Contributions

Conceived and designed the experiments: EY PSM. Performed the experiments: EY. Analyzed the data: EY PSM. Wrote the paper: EY PSM.

References

1. Altschul SF, Gish W, Miller W, Myers EW, Lipman DJ (1990) Basic local alignment search tool. J Mol Biol 215: 403–410.
2. Ashelford KE, Chuzhanova NA, Fry JC, Jones AJ, Weightman, AJ (2006) New screening software shows that most recent large 16S rRNA gene clone libraries contain chimeras. Appl Environ Microbiol 72: 5734–5741.
3. Boehme S, Werner G, Klare I, Reissbrodt R, Witte, W (2004) Occurrence of antibiotic-resistant enterobacteria in agricultural foodstuffs. Mol Nutr Food Res 48: 533–531.

4. Bräuer SL, Yashiro E, Ueno NG, Yavitt JB, Zinder SH (2006) Characterization of acid-tolerant H-2/CO2-utilizing methanogenic enrichment cultures from an acidic peat bog in New York State. FEMS Microbiol Ecol 57: 206–216.
5. Chiou, C-S, Jones AL (1993) Nucleotide sequence analysis of a transposon (Tn*5393*) carrying streptomycin resistance genes in *Erwinia amylovora* and other Gram-negative bacteria. J Bacteriol 175: 732–740.

6. Duffy B, Walsh F, Pelludat C, Holliger E, Oulevet C, et al. (2011) Environmental monitoring of antibiotic resistance and impact of streptomycin use on orchard bacterial communities. Acta Hort 896: 483–488.

7. Goodman, RN (1961) Chemical residues and additives in foods of plant origin. Am J Clin Nutr 9: 269–276.

8. Guindon S, Gascuel O (2003) A simple, fast, and accurate algorithm to estimate large phylogenies by maximum likelihood. Syst Biol 52: 696–704.

9. Handelsman J, Smalla K (2003) Conversations with the silent majority. Curr Opin Microbiol 6: 271–273.

10. Hirano SS, Upper CD (1990) Population biology and epidemiology of *Pseudomonas syringae*. Annu Rev Phytopathol 28: 155–177.

11. Huang T-C, Burr TJ (1999) Characterization of plasmids that encode streptomycin-resistance in bacterial epiphytes of apple. J Appl Microbiol 86: 741–751.

12. Huddleston AS, Cresswell N, Neves MCP, Beringer JE, Baumberg S, et al. (1997) Molecular detection of streptomycin-producing streptomycetes in Brazilian soils. Appl Environ Microbiol 63: 1288–1297.

13. Lozupone CA, Hamady M, Kelley ST, Knight R (2007) Quantitative and qualitative beta diversity measures lead to different insights into factors that structure microbial communities. Appl Environ Microbiol 73: 1576–1585.

14. Ludwig W, et al. (2004) ARB: a software environment for sequence data. Nucleic Acids Res 32: 1363–1371.

15. Mayerhofer G, Schwaiger-Nemirova I, Kuhn T, Girsch L, Allerberger, F (2009) Detecting streptomycin in apple from orchards treated for fire blight. J Antimicrob Chemother 63: 1076–1077.

16. McEwen SA, Fedorka-Cray PJ (2002) Antimicrobial use and resistance in animals. Clin Infect Dis 34(Suppl. 3): S93–S106.

17. McGhee GC, Sundin, GW (2011) Evaluation of kasugamycin for fire blight management, effect on nontarget bacteria, and assessment of kasugamycin resistance potential in *Erwinia amylovora*. Phytopathology 101: 192–204.

18. McManus, PS, Stockwell VO, Sundin GW, Jones AL (2002) Antibiotic use in plant agriculture. Annu Rev Phytopathol 40: 443–465.

19. Norelli JL, Burr TJ, LoCicero AM, Gilbert MT, Katz BH (1991) Homologous streptomycin resistance gene present among diverse gram-negative bacteria in New York State apple orchards. Appl Environ Microbiol 57: 486–491.

20. O'Brien TF (2002) Emergence, spread, and environmental effect of antimicrobial resistance: How use of an antimicrobial anywhere can increase resistance to any antimicrobial anywhere else. Clin Infect Dis 34(Suppl. 3): S78–S84.

21. Ottesen AR, White JR, Skaltsas DN, Newell MJ, Walsh CS (2009) Impact of organic and conventional management on the phyllosphere microbial ecology of an apple crop. J Food Protect 72: 2321–2325.

22. Pace, NR (1997) A molecular view of microbial diversity and the biosphere. Science 276: 734–740.

23. Pazzella C, Ricci A, DiGiannatale E, Luzzi I, Carattoli A (2004) Tetracycline and streptomycin resistance genes, transposons, and plasmids in *Salmonella enterica* isolates from animals in Italy. Antimicrob Agents Chemother 48: 903–908.

24. Posada D (2008) jModelTest: Phylogenetic model averaging. Mol Biol Evol 25: 1253–1256.

25. Pruesse E, Quast C, Knittel K, Fuchs BM, Ludwig W, et al. (2007) SILVA: a comprehensive online resource for quality checked and aligned ribosomal RNA sequence data compatible with ARB. Nucleic Acids Res 35: 7188–7196.

26. Pukall R, Tschäpe H, Smalla K (1996) Monitoring the spread of broad host and narrow host range plasmids in soil microcosms. FEMS Microbiol Ecol 20: 53–66.

27. Rappé MS, Giovannoni S (2003) The uncultured microbial majority. Annu Rev Microbiol 57: 369–394.

28. Rezzonico F, Stockwell VO, Duffy B (2009) Plant agricultural streptomycin formulations do not carry antibiotic resistance genes. Antimicrob Agents Chemother 53: 3173–3177.

29. Redford AJ, Fierer, N (2009) Bacterial succession on the leaf surface: A novel system for studying successional dynamics. Microb Ecol 58: 189–198.

30. Rodriguez C, Lang L, Wang A, Altendorf K, Garcia F, et al. (2006) Lettuce for human consumption collected in Costa Rica contains complex communities of culturable oxytetracycline- and gentamicin-resistant bacteria. Appl Environ Microbiol 72: 5870–5876.

31. Rodríguez-Sánchez C, Altendorf K, Smalla K, Lipski A (2007) Spraying of oxytetracycline and gentamicin onto field-grown coriander did not affect the abundance of resistant bacteria, resistance genes, and broad host range plasmids detected in tropical soil bacteria. Biol Fertil Soils 44: 589–596.

32. Schloss PD, Westcott SL, Ryabin T, Hall JR, Hartmann, M, et al. (2009) Introducing mothur: open-source, platform-independent, community-supported software for describing and comparing microbial communities. Appl Environ Microbiol 75: 7537–7541.

33. Schnabel EL, Jones AL (1999) Distribution of tetracycline resistance genes and transposons among phylloplane bacteria in Michigan apple orchards. Appl Environ Microbiol 65: 4898–4907.

34. Sobiczewski P, Chiou C-S, Jones AL (1991) Streptomycin-resistant epiphytic bacteria and homologous DNA for streptomycin resistance in Michigan apple orchards. Plant Dis 75: 1110–13.

35. Sundin GW (2002) Distinct lineages of the *strA-strB* streptomycin-resistance genes in clinical and environmental bacteria. Curr Microbiol 45: 63–69.

36. Sundin GW, Bender CL (1993) Ecological and genetic analysis of copper and streptomycin resistance in *Pseudomonas syringae* pv. syringae. Appl Environ Microbiol 59: 1018–1024.

37. Sundin GW, Bender CL (1995) Expression of the *strA-strB* streptomycin resistance genes in *Pseudomonas syringae* and *Xanthomonas campestris* and characterization of IS6100 in *Xanthomonas campestris*. Appl Environ Microbiol 61: 2891–2897.

38. Sundin GW, Bender CL (1996) Dissemination of the *strA-strB* streptomycin-resistance genes among commensal and pathogenic bacteria from humans, animals, and plants. Mol Ecol 5: 133–143.

39. Sundin GW, Monks DE, Bender CL (1995) Distribution of the streptomycin resistance transposon Tn*5393* among phylloplane and soil bacteria from managed agricultural habitats. Can J Microbiol 41: 792–799.

40. Tolba S, Egan S, Kallifidas D, Wellington EMH (2002) Distribution of streptomycin resistance and biosynthesis genes in streptomycetes recovered from different soil sites. FEMS Microbiol Ecol 42: 269–276.

41. Vanbroekhoven K, Ryngaert A, Bastiaens L, Wattiau P, Vancanneyt M, et al. (2004) Streptomycin as a selective agent to facilitate recovery and isolation of introduced and indigenous *Sphingomonas* from environmental samples. Environ Microbiol 6: 123–1136.

42. van Overbeek LS, Wellington EMH, Egan S, Smalla K, Heuer H, et al. (2002) Prevalence of streptomycin-resistance genes in bacterial populations in European habitats. FEMS Microbiol Ecol 42: 277–288.

43. Waksman SA, Schatz A, Reynolds, DM (1946) Production of antibiotic substances by Actinomycetes. Annals of the New York Academy of Sciences 48: 73–86.

44. Wang HH, Schaffner DW (2011) Antibiotic resistance: How much do we know and where do we go from here? Appl Environ Microbiol 77: 7093–7095.

45. Wang Q, Garrity GM, Tiedje JM, Cole JR (2007) Naive Bayesian classifier for rapid assignment of rRNA sequences into the new bacterial taxonomy. Appl Environ Microbiol 73: 5261–5267.

46. Yashiro E, Spear R, Clinton-Cirocco K, McManus P (2008) Assessment of bacteria from apple leaves by culture-dependent and culture-independent methods. Phytopathology 98: S177.

47. Yashiro E, Spear RN, McManus PS (2011) Culture-dependent and culture-independent assessment of bacteria in the apple phyllosphere. J Appl Microbiol 110: 1284–1296.

9

Conservation Implications of Changes in Endemic Hawaiian Drosophilidae Diversity across Land Use Gradients

Luc Leblanc, Daniel Rubinoff*, Mark G. Wright

Department of Plant and Environmental Protection Sciences, University of Hawai'i, Honolulu, Hawai'i, United States of America

Abstract

Endemic Hawaiian Drosophilidae, a radiation of nearly 1000 species including 13 federally listed as endangered, occur mostly in intact native forest, 500–1500 m above sea level. But their persistence in disturbed forest and agricultural areas has not been documented. Thus, control efforts for agricultural pests may impact endemic species if previously undocumented refugia in agricultural areas may play a role in their conservation. To quantify whether invasive plants and agriculture habitats may harbor endemic Drosophilidae, we established standardized trapping arrays, with traps typically designed to control invasive fruit flies (Tephritidae), with 81 sites across native, disturbed and agricultural land use gradients on the islands of Hawai'i and Maui. We collected and identified, to species level, over 22,000 specimens. We found 121 of the possible 292 species expected to occur in the sampled areas, and the majority (91%) of the captured specimens belonged to 24 common species. Species diversity and numbers were greatest in the native forest, but 55% of the species occurred in the invasive strawberry guava belt and plantation forest, adjacent to and almost 500 m from native forest, and 22 species were collected in orchards and nonnative forest as far as 10 km from native habitats. Their persistence outside of native forest suggests that more careful management of disturbed forest and a reassessment of its conservation value are in order. Conservation efforts and assessments of native forest integrity should include the subset of species restricted to intact native forest, since these species are highly localized and particularly sensitive. Additionally, future efforts to control invasive pest fruit flies should consider the nontarget impacts of maintaining traps in and near native forest. This survey project demonstrates the utility of thorough biotic surveys and taxonomic expertise in developing both sensitive species lists and baseline diversity indices for future conservation and monitoring efforts.

Editor: Patrick O'Grady, University of California, Berkeley, United States of America

Funding: This study was made possible with funding from U.S. Department of Agriculture-Agricultural Research Service through a Specific Cooperative Agreement with University of Hawaii CTAHR under Project Number: 0500-00044-016-07, "Study of Attraction of Nontarget Organisms to Fruit Fly Female Attractants and Male Lures in Hawaii". Additional research funding was provided by USDA-National Institute of Food and Agriculture Agreement No. 58-5320-9-430, and Hatch projects HAW00942-H and HAW00956-H, administered by the College of Tropical Agriculture and Human Resources, University of Hawaii at Manoa. The funders had no role in study design, data collection and analysis, decision to publish, or preparation of the manuscript.

Competing Interests: The authors have declared that no competing interests exist.

* E-mail: rubinoff@hawaii.edu

Introduction

Conservation of threatened ecosystems presents a special challenge, distinct from species-centric efforts, since there may be large concentrations of endangered plants and animals vulnerable to extinction in relatively small areas. Further, adequate knowledge of the taxonomy for the more obscure components of biodiversity, like insects, is often lacking, particularly with regard to species-level vulnerability and management needs. Such challenges are global in nature, but perhaps most evident in Hawai'i, which has endured high rates of extinction across a broad range of the flora and fauna. Because most of the macrofauna is extinct [1,2], insects not only represent the most significant portion of remaining endemicity, but also the best guides to saving overall biodiversity in highly threatened systems. Yet, for the most part, insect biodiversity is poorly known and the varying sensitivity of particular species, even those that are known to be vulnerable, has rarely been studied. A first step towards assessing the conservation needs of an endemic entomofauna, and

therefore more broadly conserving their essential roles in native ecosystems [3,4], should include standardized surveys of insect diversity across gradients representing habitats in various states of degradation. In so doing we could generate data that not only reveals the impacts of land use on a large segment of biodiversity, but also identify those members of the native community which best indicate intact habitat for conservation prioritization. Such a methodology would be of broad utility across many threatened ecosystems, particularly those that have already suffered the extinctions of their most charismatic fauna, but may still harbor important components of the original biodiversity. However, such a methodology requires the use of a group that is adequately diverse so as to provide relative measures of sensitivity to habitat quality across multiple species. It is also necessary to have some level of taxonomic expertise to ensure accurate species level identifications, which dramatically increase the accuracy and value of such surveys.

Endemic Hawaiian Drosophilidae species are an exceptionally diverse assemblage, of approximately 1000 species [5] radiating from a single colonization event [6]. Five hundred and sixty five described endemic species (416 *Drosophila* and 149 *Scaptomyza*) and 32 introduced species are known to occur in Hawai'i [7,8,9]. The larvae of the endemic species are extremely specialized, occurring in decaying leaves, bark, fruit, flowers, or the sap flux of plants belonging to 36 angiosperm families, fern rachis, fungus, and even spider eggs or green algae in streams [10,11,12,13]. Most (74.7%) species breed on a single substrate and host plant family and 49% of the Angiosperm-breeding species are monophagous on a single plant genus [13]. Thus, the presence or absence of a suite of drosophilid species may be used to assess habitat integrity and monitor the impact of nonnative plant encroachment across gradients.

The majority of endemic drosophilids occur between 500–1500 m in elevation [14], in four of the Hawai'i forest ecosystems defined by Fosberg [15]: the wet ohi'a (*Metrosideros polymorpha* Gaudich.) forest, the cloud forest (above and contiguous with the ohi'a forest), the drier mixed mesophytic forests, where more unusual hosts including fungi are common, and the dryland sclerophyll forest [16]. Very few species are recorded at lower altitudes, in the leeward dry forest, or the high arid regions above 2100 m [16], and even fewer have ever been collected in nonnative forest [14].

Numerous threats related to human activity and invasive species contributed to the documented decline [17,18] of endemic drosophilids, to the point that 13 of the picture wings are on the United States Federal Endangered Species list and parts of their critical habitat protected [19]. Major threats include grazing and weed dispersal by feral ungulates, invasions of nonnative plants such as strawberry guava (*Psidium cattleianum* Sabine), conversion of endemic habitats into agricultural and pasture land, wildfires in mesic scrubland, and predation by invasive ants [16,20] and yellowjackets [18]. Adding to these threats may also be the unintended nontarget impact of practices to control or eradicate insect pests, such as fruit flies (Tephritidae) [21,22]. In places with very high levels of regional endemicity, like the Hawaiian Islands, such nontarget impacts could threaten species with distributions restricted to a few square kilometers on the slope of a single volcano.

We sought to use the presence of introduced and endemic Drosophilidae to understand the impacts of land use across a gradient on an extremely diverse endemic and introduced insect fauna. We asked how the diversity of both endemic and invasive insects might change through standardized trapping along a transect from endemic forest into adjacent agricultural lands. Specifically, would endemic insects be wholly restricted to intact native forest; which species are more resilient to habitat alteration; is the presence of invasive species indicative of poorer quality habitat; and are any species useful as indicators of not just intact native or agricultural lands, but transitional areas that may also be of some conservation value? To answer these questions we generated data on the occurrence and abundance of 121 endemic drosophilid species across a gradient from intact native forest to intensive agricultural use to help establish a standardized sampling baseline for endemic insect diversity across land use regimes. We present detailed data on the presence of endemic Drosophilidae in native forest, the persistence of populations in adjacent nonnative forest belts, and their presence in more distant agricultural environments on Hawai'i and Maui islands. Because we identified all individual flies to species, specific measures of how much disturbance different endemic species can tolerate is available, as well as the impacts of pest control practices in mixed and native forest. More broadly, this survey data is relevant to the species-specific impact of a land use gradient on a diverse endemic radiation of conservation concern.

Materials and Methods

Sites and Traps

We surveyed endemic and introduced drosophilids using a standard trapping protocol established to monitor nontarget insect attraction fruit fly male lures and food attractants [21,22]. Traps were maintained at 81 sites, in six broad locations on Hawai'i and Maui Islands, covering a diversity of ecosystems, from farmland to invasive forest, strawberry guava belts, mixed and endemic forest. While trapping procedures across sites were not always precisely replicated, they were consistent within each gradient, allowing for comparison across land use regimes. All sites, referred to by their numbers throughout this paper, are described in details and mapped on Figure 1, Figure2, Figure 3, and Figure 4.

Ethics

Permits for collecting insects in the State Forest Reserves were delivered by the Hawai'i State Department of Land and Natural Resources (Betsy Gagné) and access permits were granted by its Division of Forestry and Wildlife offices. Required additional access permits to the Waikamoi Preserve and the Ko'olau Forest Reserve were granted by the Nature Conservancy (Pat Bily) and the East Maui Irrigation Company, respectively. We also acknowledge the private landowners and growers in Waimea and Kula for their hospitality and permission for access and use of their farms.

Hawai'i Island

Nine sites were maintained in a 20 km long transect along the Stainback Highway, from the Pana'ewa Rainforest Zoo near Hilo (138 m above sea level) up to 1,045 m. The upper four sites (1,045–706 m) (sites 1 to 4) were in native wet montane ohi'a-dominated forest, and the lower sites 5 to 9 (522–138 m) were in invasive strawberry guava dominated forest (three sites), a citrus orchard and a mixed fruit orchard. Fifteen additional sites were in a 35 km transect on the Saddle Road, from the junction of Kaūmana and Saddle Roads (439 m) to Pu'u Huluhulu (2,012 m) (sites 10 to 24). The lowest site was in invasive strawberry guava forest, while the other sites were in native montane wet herbland bogs over recent lava flows (three sites), wet montane ohi'a forest (two sites), and wet (six sites) and dry (three sites) ohi'a-dominated kīpuka forests. Six sites were along the upper Hāmākua Ditch Trail (North Kohala Forest Reserve), from the far end of the flume (1,019 m) to the entrance of the Reserve, off the Waimea water reservoir (906 m) (sites 25 to 30). The forest entrance site was in the strawberry guava belt, adjacent to the mixed native wet montane ohi'a forest (where the other sites were located). Five trapping sites were in the agricultural community of Waimea (744–872 m), about 4 km southwest of the Kohala sites (sites 31 to 35). Two sites were in backyards with a diversity of fruit trees, one site in a citrus orchard, one site in a large feral stand of common guava (*Psidium guajava* L.), and the last site at the foot of the North Kohala Forest Range, in a forest dominated by invasive tropical ash [*Fraxinus uhdei* (Wenzig) Lingelsh].

Maui Island

Fourteen sets of traps were maintained in nine sites across the agricultural community of Kula (517–1,138 m) (sites 36 to 44). Sites covered a variety of common tree crops: persimmon (*Diospyros kaki* L. fil.) orchards (six sets of traps), coffee plantations (two sets in maintained and two sets in abandoned plots), two sets

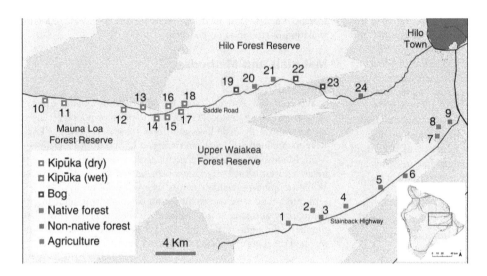

Figure 1. Trapping sites on Hawai'i Island along Stainback Highway and Saddle Road.

in nonnative forest adjacent to orchards (one next to persimmon and the other next to persimmon and coffee), and two sets in citrus and mixed fruit orchards. Forest ecosystems on Maui were covered with a 37-site transect, on the northern slope of the Haleakalā Volcano, with one site every 150 m, along two linear intersecting transects (sites 45 to 81). The first 2 km long transect was along the Maile Trail, from the Flume road in the Makawao Forest Reserve (1,294 m) (mixed native/invasive forest) into the Waikamoi Nature Conservancy Reserve (intact mesic forest dominated by ohi'a and koa) up to near 'Ukulele Camp at 1,583 m (sites 60 to 45). The second (4 km) transect was along the Flume Road, from the entrance of Makawao Forest Reserve (1,284 m) first along

nonnative plantation forest dominated by *Pinus* sp., *Eucalyptus* sp., or tropical ash (sites 81 to 78), then into mixed native mesic forest (sites 60 to 64), continuing into the Ko'olau Forest Reserve (wet native ohi'a-koa forest) to the junction of the Pipeline Road (1,285 m) (sites 65 to 71), and for 1 km along the Pipeline Road forest down to 1,184 m elevation (mixed native/invasive wet forest) (sites 72 to 77).

Traps and Lures

We surveyed using the bucket and MultiLure traps typically used for fruit fly monitoring, since these not only attracted endemic insects but also allowed us to evaluate the potential for

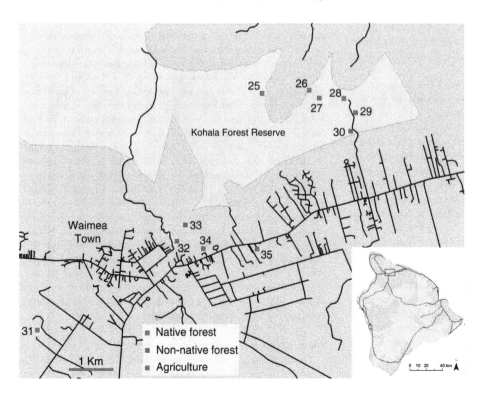

Figure 2. Trapping sites on Hawai'i Island in the Kohala Forest Reserve and Waimea agricultural area.

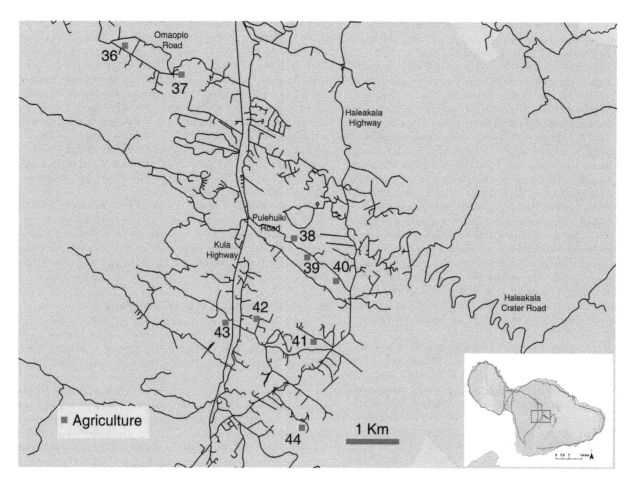

Figure 3. Trapping sites on Maui Island in the Kula agricultural area.

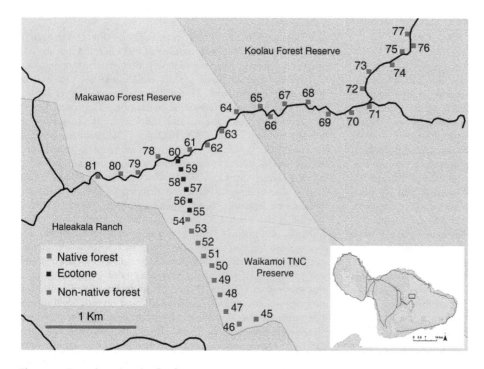

Figure 4. Trapping sites in the forest reserves on Maui Island.

nontarget impacts of pest fruit fly suppression. Additionally, at some sites, we used mushroom-bait and pan traps, which are known to be attractive to endemic insects [23,24].

Bucket traps [21] consisted of 1.3-liter white plastic cups with two lateral holes on opposite sides near the top to allow insect entry; they were covered with a plastic plate to prevent trap flooding by frequent rain. Aluminum tie wire was used to hang traps in trees. Each trap had one of four lure treatments. For male lure treatments, lure plugs with the fruit fly attractants cue-lure or methyl eugenol (Scentry Biologicals, Billings, Montana, USA) were suspended from the trap inner hook. For the third treatment, decaying fruit flies [*Bactrocera dorsalis* (Hendel)] were placed in pouches made of gauze at the bottom of the trap, in the liquid preservative, to emulate the accumulation of decaying fruit flies attracted to male lures, since many insects are attracted to this resource [21]. The last treatment was an unbaited control trap. Bucket traps baited with fermented mushroom bait in a saturated $4 \times 2 \times 2$ cm sponge hung below the trap ceiling [23] were used, instead of traps baited with decaying flies, at the Maui forest sites. A 25×45 mm strip containing 10% dichlorvos (Vaportape ® II, Hercon Environmental, Emingsville, Pennsylvania, USA) was attached to the inner hook of all traps, to rapidly kill captured arthropods.

To more completely assess the drosophilid diversity across sites, we also used MultiLure® traps (Better World Manufacturing Inc, Fresno, CA), baited with the fruit fly food attractant BioLure® (Suterra LLC, Bend, OR), consisting of three components (ammonium acetate, trimethylamine hydrochloride, and putrescine) in separate packets with slow-release membranes, attached to the inner surface of the trap cover.

Yellow pan traps, made of yellow Solo® 12oz plastic bowls (15 cm diameter, 4 cm deep) (Solo Cup Company, Lake Forest, Illinois, USA) were used in the Maui forest sites. They effectively attract a broad range of flying insects, including Drosophilidae [24].

Additionally, 200 ml of a 20% solution of propylene glycol (Sierra Antifreeze®, Old World Industries, Northbrook, Illinois, USA) was used in all the traps to retard decay of captured arthropods, and facilitate identification of specimens.

Trapping Procedure

At each site, traps were hung in trees 1.5–2.0 m above the ground, and at least 10 m apart to avoid interference between traps. Trap contents were cleared weekly, unless otherwise indicated. Traps on Hawai'i Island, four buckets and one BioLure per site, were maintained from June to August 2005, for 10 weeks on Stainback, 9 weeks on Saddle and in Waimea, and 8 weeks in Kohala. Bucket traps with the male lure traps and the lure-free controls were maintained and serviced continuously through the season. Traps with decaying flies as bait and the BioLure traps were maintained continuously at Waimea sites and for one week straight, every other week, in Stainback (five collections starting at week 1) and Kohala and Saddle (three collections starting at week 3). In Kula (Maui), four buckets and one BioLure trap were set at each site starting in May 2006, near the end of the persimmon flowering season, and left until the end of harvest season, in late November. Traps were serviced weekly for 13 weeks (until late August), and subsequently monthly for the last 3 months. Traps baited with decaying flies were used only during the first 13 weeks, because monthly servicing intervals would have caused a complete decay of the trapped drosophilids themselves. In the Maui forest transects, we maintained at each site three bucket traps (both male lures and the unbaited control) continuously for 12 weeks (June-August 2006). After 6 weeks, all trap sites were shifted downslope

by 75 m along the transects, to maximize habitat coverage. To avoid impacting endemic insect populations, BioLure traps were intermittently used for 14 days, as 3 days on week 3, 4 days on week 4, and again 3 days on week 9 and 4 days on week 10. Traps with decaying flies were not used in the Maui forest, because similar or closely related species present there were previously found to be attracted to BioLure on Hawai'i Island [21] and we did not want to overly impact populations of particular species by trapping them more intensively. Instead, we maintained mushroom baited traps, one at each site, intermittently for one week on weeks 4, 6, 10 and 12, as well as pan traps, three at each site, placed on the ground and maintained for 7 days, on weeks 4 and 10 to broadly assess drosophilid diversity.

Positions of traps at all sites on both islands were re-randomized every 3 weeks to minimize effect of trap position on catches. Pouches of decaying flies were replaced weekly, while male lure plugs, pesticide strips and BioLure membranes were used for the duration on Hawai'i Island and Maui forests, and replaced once after 13 weeks in Kula.

Sample Processing and Data Analysis

All Drosophilidae were counted, sexed and identified to species level whenever feasible. Reference collections of voucher specimens have been deposited at the University of Hawai'i Insect Museum (Mānoa), and the Bishop Museum, in Honolulu (HI).

Trapping data for endemic Drosophilidae are reported here, with the alien species reported in more details separately [7]. All counts from each sample were converted to number of flies per trap per day. Because drosophilids were not attracted to the male lures [21], capture data from bucket traps with the two male lures and the unbaited control traps were analyzed together, as bucket traps. Similarly, traps with decaying flies, BioLure and mushroom bait attracted comparatively large numbers of mostly the same species of drosophilids, and their data are treated together under the "food lure" category. Data from pan traps are treated separately. The EstimateS software [25] was used to generate the species accumulation curve, with 50 randomizations without replacement.

To analyze differences in fly community assemblages in each of the habitats sampled, mean number of flies collected in traps was calculated across sampling dates within each trapping site. Canonical correspondence analysis (CCA) was used to analyze fly community structure at the "species group" level in the different habitats, and to describe effects of environmental factors on community composition and distribution of fly species in different habitats, using CANOCO 4.5 [26]. Problems typically associated with unbalanced experiments are reduced by screening for collinearity among environmental variables (habitats) using the inflation factor in CANOCO, which reduces potential bias [26,27]. Environmental variables included in the analysis were habitat type (endemic forest, kīpuka and bogs, mixed endemic/non-indigenous forest, nonendemic forest, fruit and coffee orchards, and invasive strawberry guava or common guava). Fly counts were log n+1 transformed, and rare species were downweighted in the analyses [26]. Significance of correspondence axes was assessed using Monte Carlo tests (499 permutations), as implemented in CANOCO 4.5. These analyses were conducted on species groups of Drosophilidae [8], and on individual species in separate analyses. For the species group analysis, data from both islands were combined, as the members of the groups share similar ecological characteristics [13], and members of all groups were present on both islands. The species level analyses were separated by island, because very few species were collected on both islands. Preliminary examination of the data indicated that a linear model was more appropriate than a unimodal model, and therefore Redundancy Analysis (RDA) was used to analyze the species occurrence in different habitats (CANONO 4.5). Results of the ordinations were plotted (CCA or RDA axes 1 and 2) with environmental variables to visualize the changes in community assemblages in different habitats, and associations of species groups and species with specific habitats.

Results

Two hundred and ninety-one of the 328 endemic Drosophilidae species from Maui and Hawai'i Islands were known or likely to occur in the sampled areas, on the basis of literature and museum specimen label data. We collected over 22,000 specimens, representing 121 of these expected species, plus 32 undescribed new species, suggesting our sampling methods were relatively thorough (Table 1). Species richness was higher on Maui (75 species) than Hawai'i (50 species). Additional sampling would have revealed additional species, as reflected in the species accumulation curves (Fig. 5), but locating nearly half of the known species in the study sites, and such large numbers, allows for inference regarding the distribution of these endemic insects with respect to habitat use. The contribution of different trapping methods to detecting a more saturated sample of species richness in the different habitats is demonstrated in Fig. 5; if only a single method had been used, the species accumulation curves would have undersampled the drosophilid assemblage substantially. While food lures attracted large numbers of the spoon tarsus, *Antopocerus*, *Engiscaptomyza*, *haleakalae* and picture wing groups, bucket and pan traps collected a number of other species underrepresented in food lure samples, such as the *Elmomyza* on Hawai'i island and the two most numerous of the modified mouthpart species in our samples, *D. comatifemora* and *D. hirtitarsus*, caught almost exclusively in the pan and bucket traps, respectively. Similarly, the two bristle tarsus species were collected exclusively in pan traps on Maui. Other species may turn out to be widespread when non-traditional sampling methods are used, as for *S. undulata*, thought to be very rare until pan traps were found to collect them in large numbers [14]. The complete list of the species collected, their island-wide distributions, on the basis of our data, literature surveys and museum data, is presented in Table S1 (supporting information) to this paper and as part of a searchable database (Drosophilidae of Hawai'i. Available: www.herbarium.hawaii.edu/drosophila/ Accessed 2013 Mar 8).

The dominant groups, in number of specimens collected and the highest proportion of expected species actually captured, were the leaf breeder (*Antopocerus*, spoon tarsus, split tarsus) and fungus breeder (*haleakalae* group) groups of *Drosophila*, and the subgenera *Elmomyza* and *Engiscaptomyza* of *Scaptomyza* (Table 1, Fig. 6). *Antopocerus*, spoon tarsus, and *Elmomyza* accounted for 77% of all captures on Hawai'i Island, and *haleakalae*, spoon tarsus, *Elmomyza* and *Engiscaptomyza* were numerically dominant on Maui, with 78% of all captures. The majority (91%) of captures were of 24 common species (online support material). New island distribution records are also reported for fourteen species, further demonstrating the effectiveness of our survey methods: *D. brunneisetae* Hardy, *D. macrochaetae* Hardy, *S. chauliodon* (Hardy), *S. cryptoloba* Hardy, *S. mutica* Hardy are new to Hawai'i Island, and *D. paucitarsus* Hardy & Kaneshiro, *S. articulata* Hardy, *S. basiloba* Hardy, *S. brunnimaculata* Hardy, *S. diaphorocerca* Hardy, *S. levata* Hardy, *S. nigrosignata* Hardy, *S. setosiscutellum* (Hardy), and *S. xanthopleura* Hardy are new to Maui.

Host plant and breeding substrate are known for 71 of the 121 collected species, and most breed in the leaves and/or bark of either the Araliaceae (*Cheirodendron* and *Tetraplasandra* spp) (34

species) and/or the Campanulaceae (mostly *Clermontia* spp) (19 species), or on fungi (12 species), suggesting strong associations with the native flora.

Only four species of the charismatic picture wings were found on Hawai'i, all infrequently (except *D. ochracea* Grimshaw), consistent with their documented decline in the island's wet forests attributed to the destruction of *Clermontia* host trees by feral pigs and predation by invasive yellowjacket wasps [17,18,28]. In contrast, 14 species of picture wings were regularly trapped in the Maui native forest, especially in the Waikamoi reserve, where host *Clermontia* trees abound and efforts are underway to eradicate pigs from the fenced reserve.

At least a few individuals of 23 of the 29 expected introduced drosophilid species were collected [7], but three species dominated the captures and were common even in the endemic habitats (Table 2): *D. sulfurigaster bilimbata* Sturtevant (47% of introduced drosophilids), *D. immigrans* Bezzi (4%) and *D. suzukii* (Matsumura) (46%). The first two breed on decaying guava [14] and other introduced fruit, rather than the usual substrates of the endemic species, and *D. suzukii*, a severe pest of small fruits abundant at all trapping sites, was bred from endemic raspberry (*Rubus hawaiiensis* A. Gray) [13]. Assuming that *D. suzukii* can infest most or all of the nine species of *Rubus* of Hawai'i, and possibly the endemic ōhelo (*Vaccinium* spp) as well, then hosts are commonly available from sea level to at least 2000 m, in mesic to wet environments [29], potentially sustaining the large populations of this pest fly observed in Hawai'i. Its impact as a potential pest of endemic raspberry has not been documented. Since guava is invasive and not used as a host by endemic drosophilids, the potential impact of the most common invasive drosophilids in the Hawaiian forest is likely to be limited. Possible competition of *D. immigrans* with endemic species for breeding sites has been suggested [14], but not yet investigated.

Endemic species were also captured in nonnative forest, adjacent to endemic habitats (Table 2). In the Kohala Forest Reserve on Hawai'i, they were common in the strawberry guava forest belt (site 30), almost 100 m from endemic forest (Fig. 7). In the Maui forest transect, six of the 13 most common endemic species were regularly captured in mostly nonnative forest patches in the endemic forest reserves (sites 55–60), and in adjacent nonnative plantation forest (sites 78–81), up to almost 500 m distant from endemic forest (Fig. 8). These distant occurrences suggest that either they are breeding in the nonnative habitat or they are dispersing through it. In either case it has implications for nontarget impacts of pest tephritid fruit fly control. Similarly, three endemic species were trapped in alien mountain ash forest, almost 500 m distant from endemic forest in the Kohala Mountains (site 33). One of them, *S. lobifera* Hardy, occurred beyond the forest in feral guava stands (site 32) and a backyard (site 35), at least 1 Km from the native Forest Reserve.

Although the majority of endemic flies were collected in endemic forest sites and adjacent nonnative forest, a diversity of endemic species were captured in small numbers in farmland and nonnative forest and orchards, more than 10 Km from endemic forest (Table 3, Fig. 5c–d). On Maui, fourteen species were repeatedly trapped at four of the Kula farmland locations, in coffee and persimmon orchards and adjacent nonnative forest, more than 5 Km from any endemic forest. Larvae of several of these taxa breed in fungus, and may not be host specific, possibly rendering them less effective for habitat quality assessments, and simultaneously more vulnerable to nontarget impacts of fruit fly control. Endemic drosophilids in Kula were more common in the nonnative forest adjacent to orchards (0.11 per trap per day) than in the orchards themselves (0.004 per trap per day) [22]. Because pest fruit flies were uncommon in these forest patches, growers can

Table 1. Numbers of expected and captured species of Drosophilidae during studies of nontarget attraction to fruit fly (Tephritidae) lures on Hawai'i and Maui Islands, summarized by group.

Group	Total described in Hawai'i	Expected at sites	Collected at sites	New species	Hawai'i Is.	Maui	Hawai'i and Maui
		Number of species			**Proportion of total endemic species captured**		
Endemic Drosophila							
Antopocerus	15	8	7	0	35.46	5.69	16.33
Bristle tarsus	18	14	4	6	0.06	0.13	0.10
Ciliate tarsus	21	10	4	4	0.56	0.37	0.44
haleakalae	54	29	21	1	6.11	31.26	22.28
Modified mouthparts	106	54	7	5	5.42	7.39	6.69
Nudidrosophila	28	7	2	0	0.42	0.63	0.55
Picture wing	120	50	20	0	1.74	2.65	2.32
Split tarsus	24	15	8	1	5.18	3.77	4.27
Spoon tarsus	12	10	9	0	19.02	9.65	13.00
Misc and unplaced	18	6	1	0	0.00	0.05	0.03
Endemic Scaptomyza							
Alloscaptomyza	8	4	2	3	0.46	0.26	0.34
Bunostoma	8	5	4	0	0.01	0.50	0.32
Celidosoma	1	0	0	0	0.00	0.00	0.00
Elmomyza	86	55	22	4	22.81	24.72	24.04
Engiscaptomyza	6	4	2	0	0.00	12.04	7.74
Exalloscaptomyza	6	2	1	0	0.00	0.02	0.01
Grimshawomyia	3	3	1	0	1.25	0.00	0.44
Rosenwaldia	6	3	0	0	0.00	0.00	0.00
Tantalia	6	3	2	1	0.00	0.72	0.46
Titanochaeta	12	7	3	7	1.22	0.15	0.53
Unplaced Scaptomyza	7	3	1	0	0.27	0.00	0.10
Immigrant Drosophilidae	32	29	23	0	n.a.	n.a.	n.a.

avoid impacting fringe populations of endemic insects by restricting trapping and protein bait spraying to the orchards themselves [22]. Similarly, six endemic species were trapped at all five lower sites of the Stainback Highway, even in strawberry guava forest and fruit orchards. The picture wing *D. ochracea*, which breeds on bark of *Freycinetia arborea* Gaudich, was repeatedly trapped in four of the five Stainback sites, despite being a relatively uncommon species in Ola'a [18]. This suggests that sparsely distributed *Cheirodendron* and other host trees observed at low altitude along Stainback Highway may sustain small populations of endemic drosophilids in otherwise inhospitable disturbed environments. Such information is of great relevance for conservation planning and habitat restoration.

Canonical correspondence analysis (CCA) of the species groups data (Fig. 9) showed a distinct gradient from endemic to invasive plant habitats (CCA axis 1, $P = 0.0020$; species-environment relation variance explained by axis 1:64.8%; and cumulatively by axis 1 and 2:86.0%), with immigrant species consistently associated with strawberry guava and fruit orchards, and to a lesser extent, introduced forest species and coffee or bog habitats. While some endemic groups (e.g. modified mouthparts) were associated with mixed forest and primarily endemic forest, the majority of endemic groups were most closely associated with endemic forest, and were only captured in low numbers in mixed

and introduced forest habitats. The second axis on the CCA ordination for species groups was defined primarily by mixed forest, and kīpuka habitats (Fig. 9). Some groups (notably the ciliated tarsus clade), were closely associated with the Saddle Road Kīpuka habitats, and were completely unassociated with alien habitats.

Redundancy Analysis (RDA) results for individual species on each island are presented with the fly species grouped into clusters, circling delineated by eye, on the RDA ordinations for clarity of presentation. RDA on Hawai'i Island (Fig. 10) showed a marginally significant ($F = 12.46$; $P = 0.0640$, species-environment relation variance explained by axis 1:78.6%; and cumulatively by axis 1 and 2:89.5%) association of adventive species with strawberry guava (clusters A and B, right quadrants of RDA ordination). A further distinct group of species (cluster C, lower left quadrant, dominated by endemic species), was associated with Kīpuka habitat and endemic forest, with a minimal association of those species with introduced forest and common guava. The relatively wide spread in the ordination of the 45 species in cluster C along the second RDA axis indicates that this assemblage of species is associated with endemic forest, but do occur to some extent in adjacent introduced forest. A smaller set of four endemic species and seven introduced species (cluster D, upper left

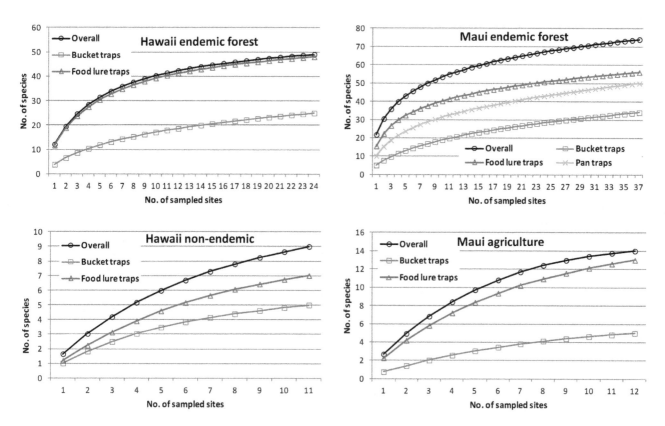

Figure 5a–d. Species accumulation curves for endemic Drosophilidae collected in Hawai'i. Data presented separately for different types of traps in endemic forest sites of Hawai'i (a) and Maui (b) and nonnative forest and agricultural sites of Hawai'i (c) and Maui (d).

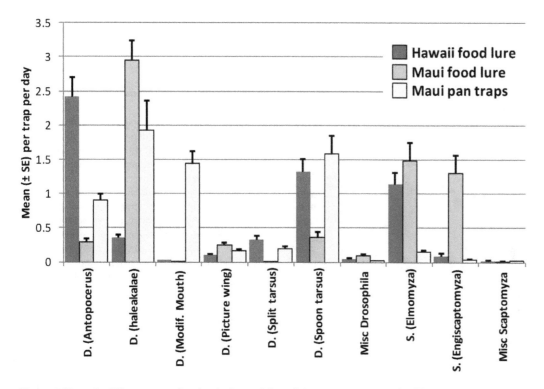

Figure 6. Mean (± SE) captures of endemic _Drosophila_ and _Scaptomyza_, summarized by taxonomic groupings. Data presented based on captures in food lure and yellow pan traps maintained in native forest sites on Hawai'i (2005) and Maui (2006) islands.

Table 2. Mean captures (mean ± SE per trap per day) of endemic and immigrant Drosophilidae in food lure and bucket traps in endemic and adjacent ecotone or nonendemic habitats.

Transect	Habitat	No. trapping sites	No. endemic species trapped	Endemic drosophilids in food lure traps	Endemic drosophilids in bucket traps	Immigrant drosophilids in food lure traps	Immigrant drosophilids in bucket traps
Stainback	Endemic	4	22	4.09±0.76	0.27±0.05	13.29±5.63	0.06±0.02
	Nonendemic	5	5	0.04±0.01	0.03±0.01	247.9±74.0	7.41±2.44
Kohala	Endemic	5	18	10.68±1.51	0.23±0.03	12.73±2.23	0.12±0.02
	Ecotone	1	10	4.89±0.80	0.24±0.06	35.8±24.4	0±0
Waimea	Nonendemic	5	5	0.19±0.05	0.02±0.01	11.18±1.79	1.90±0.56
Saddle	Endemic	14	42	5.13±0.71	0.27±0.02	3.64±0.52	0.03±0.01
	Nonendemic	1	1	0.02±0.02	0.02±0.01	206±185	1.50±0.66
Maui forest	Endemic	28	70	8.17±0.73	0.28±0.02	4.99±0.70	0.01±0.001
	Ecotone	5	33	3.04±0.95	0.05±0.01	5.56±1.60	0.004±0.002
	Nonendemic	4	24	2.25±0.56	0.07±0.02	9.93±2.00	0.002±0.001
Kula	Nonendemic	9	14	0.05±0.01	0.01±0.002	2.34±0.41	0.08±0.01

quadrant), were associated with fruit, bog and residential area habitats.

RDA for the Maui (Fig. 11) samples showed highly significant ($F = 7.64$; $P = 0.0080$; species-environment relation variance explained by axis 1:75.9%; and cumulatively by axis 1 and 2:97.7%) clusters of species associated with fruit and coffee plantations (lower right quadrant), an intermediate group associated, to some extent, with most habitats examined (clustered in upper left and right quadrants), albeit most strongly associated with endemic- and introduced forest habitats, and a third cluster associated primarily with endemic forest. Clusters A and B are distinctly separated from clusters C and D in the diagram. Clusters C and D are essentially not separable along the first RDA axis (Fig. 8), but the second RDA axis appears to distinguish the two clusters along a gradient defined by endemic (cluster C) and introduced (cluster D) forest. Cluster D includes 22 endemic species that are associated with introduced habitat to some extent. In summary, the RDA results showed that a substantial proportion of introduced fly species were associated with introduced vegetation, primarily fruit trees. Endemic fly species were either strongly or completely associated with endemic habitats, or possibly forest habitats that include native plant species, but some of them were able to persist in disturbed habitats, possibly on isolated islands of host plants.

Discussion

Our results demonstrate the importance of standardized surveys and intensive taxonomy with identifications to species level in contributing to the development of useful measures of habitat quality. Not all endemic species are dependent on high quality native habitat, and a few are largely independent of it. Thus, while most endemic species are more strongly associated with native habitat, relatively few are strongly enough associated to be of the highest value in conservation assessments and used for monitoring environmental health. Such species may be selected among those in clusters C of both Figures 10 and 11. In addition to being restricted to native habitat, ideal attributes of these species or clades should be that they are commonly encountered, easily attracted to bait, diverse in species, and with their taxonomy and host plant associations well documented. These most sensitive species, which could indicate highest quality native habitat and serve as indicators of environmental health should be selected among the *Antopocerus*, spoon tarsus and split tarsus clades. Some of the picture wing species may be considered, because they may be observed and identified in the field without killing them, but several of them do persist in areas distant from endemic forest. While umbrella species *per se* are controversial [30,31], the use of a broad suite of species, culled from habitat gradient surveys, as employed in this study, can help generate a viable and more nuanced measure of ecosystem integrity less vulnerable to the vagaries of more species-specific management practices.

Conversely, while most alien species surveyed were more strongly associated with disturbed habitats, many were also present or abundant in the highest quality native forest. Thus, a species-level diagnosis of trap composition was needed, not just a total of invasive species, or abundance to understand how invasive flies interact with native habitats. Our species-level approach reveals that some invasive species are useful in assessing habitat integrity, while others invade broadly and probably should be discounted from habitat quality assessments.

Many endemic drosophilids are still relatively common and fairly diverse in endemic habitats, due in part to significant efforts to conserve endemic plants and their associated insect faunas by excluding pigs and other ungulates in fenced forest reserves [18],

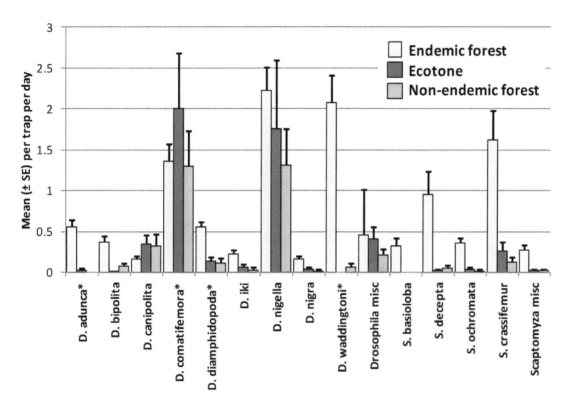

Figure 7. Mean (± SE) captures of endemic *Drosophila* and *Scaptomyza* in native forest and adjacent strawberry guava belt. Sites located in the Kohala Forest Reserve, Hawai'i Island (2005).

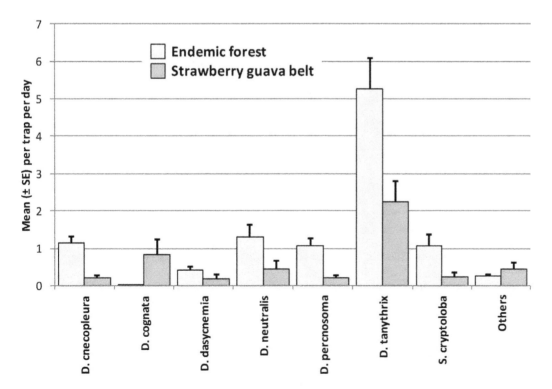

Figure 8. Mean (± SE) captures of endemic *Drosophila* and *Scaptomyza* in native and adjacent ecotone and nonnative forest. Sites located in the Makawao-Waikamoi-Ko'olau forest transect of Maui Island (2006), in endemic forest, nonendemic ecotone within endemic forest, and nonendemic forest adjacent (up to 400 m distant) to endemic forest. Data are from captures in food lure traps (mushroom bait and BioLure), except for species with asterisks (*), for which pan trap capture data are used, because of their higher captures than in food lure traps.

Table 3. Captures of endemic *Drosophila* and *Scaptomyza* in bucket and food lure traps maintained in nonendemic forest and agricultural ecosystems on Maui and Hawai'i Islands.

Location/species	Total captured	No. sites	Habitats	Known hosts[a]
Maui: Kula				
D. fuscifrons	10	2	Persimmon, forest next to orchard	Unknown (possibly fungus)
D. hirtitarsus[b]	118	1	Coffee, persimmon, forest next to orchard (mostly)	Sap flux of *Nestegis* (Oleaceae)
D. polita	23	2	Persimmon, forest next to orchard (mostly)	Fungal body
D. quinqueramosa	2	2	Forest next to orchard	Unknown (possibly fungus)
S. buccata	3	1	Persimmon	Unknown
S. confusa	47	3	Coffee (mostly), persimmon	Unknown
S. crassifemur	1	1	Coffee	Unknown
S. decepta	15	2	Coffee, persimmon, forest next to orchard	Unknown
S. diaphorocerca	2	2	Coffee, forest next to orchard	Unknown
S. hackmani	1	1	Forest next to orchard	Leaf, flower, fruit or bark of *Cheirodendron* (Araliaceae), *Clermontia* (Campanulaceae), *Rubus* (Rosaceae) and *Melicope* (Rutaceae)
S. levata	1	1	Persimmon	Unknown
S. mauiensis	2	2	Persimmon, coffee	Flowers of *Ipomoea* (Convolvulaceae)
S. nasalis	3	1	Coffee, forest next to orchard	Unknown
S. xanthopleura	21	3	Coffee, persimmon, forest next to orchard	Fungal body
Hawai'i: Waimea				
D. bipolita	5	1	*Fraxinus uhdei* forest	Unknown (possibly fungus)
D. cnecopleura	1	1	Backyard	Leaves of *Cheirodendron* (Araliaceae)
S. chauliodon	1	1	*Fraxinus uhdei* forest	Spider egg mass
S. lobifera	57	3	*Fraxinus uhdei* forest, common guava stand, backyard	Unknown
S. palmae	1	1	Backyard	Palm, flowers of *Hibiscadelphus* (Malvaceae)
Hawai'i: Stainback				
D. bipolita	3	2	Strawberry guava, invasive forest	Unknown (possibly fungus)
D. hirtitarsus (sp nr)[b]	19	2	Strawberry guava, invasive forest	Unknown
D. infuscata	2	2	Strawberry guava, invasive forest	Stem and bark of *Clermontia* (Campanulaceae), *Nestegis* (Oleaceae), and *Freycinetia* (Pandanaceae)
D. murphyi	1	1	Mixed fruit orchard	Bark of *Cheirodendron*, *Tetraplasandra* (Araliaceae) and *Clermontia* (Campanulaceae)
D. ochracea	19	4	Strawberry guava, invasive forest, fruit orchards	Stem and bark of *Freycinetia* (Pandanaceae)
S. cryptoloba	4	2	Strawberry guava, invasive forest	Leaf/frass of *Charpentiera* (Amaranthaceae) and *Clermontia* (Campanulaceae)

[a]Host record information from reference 13.
[b]*D. hirtitarsus* is complex of very similar species, with the true *D. hirtitarsus* restricted to Maui, and those on the others islands are yet unstudied sibling species (35).

(e.g. the Ola'a and the Waikamoi Forests). Although we collected 41.5% of the expected species, additional described species may persist, at least as small isolated populations. The species accumulation curves in native forest (Fig. 5a–b) might have approached asymptotes had our sampling been more targeted and less randomized, with active visual searches for adults attracted to bait applied to trees, sampling of larval substrates, and strategically placing our traps near host trees. However, our goal was not to locate particular species, but rather to examine the potential of standardized trapping gradients to identify informative species for conservation management. Future surveys specifically targeting

assessments of native habitat integrity should use these more directed techniques to detect the most sensitive species we identified in this study.

The presence of endemic species in mixed alien forest is attributable to dispersal from endemic forest, but probably also persistence on isolated endemic host trees and adaptation to introduced hosts. Heed [10] repeatedly reared nine species from leaf litter of a single isolated *Cheirodendron* tree in Kīpuka Puaulu (Volcanoes N.P.), and *D. reynoldsiae* Hardy and Kaneshiro was reared from the bark of four endemic *Reynoldsia* trees isolated in nonnative forest at Kulu'ī Gulch in the Ko'olau mountains [11].

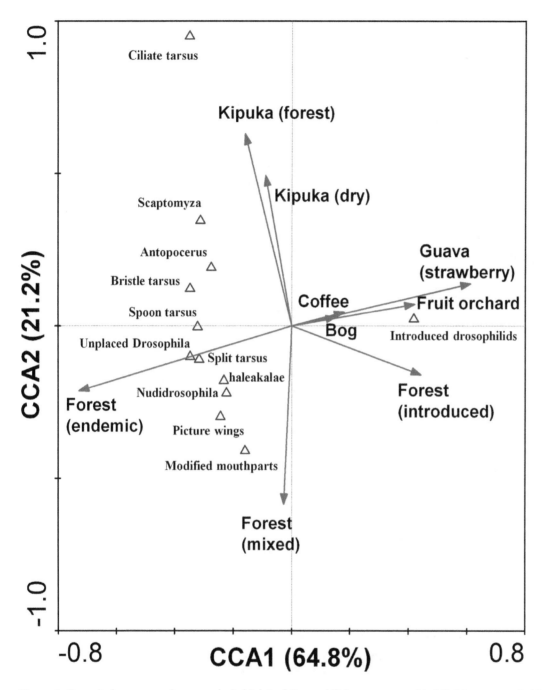

Figure 9. Canonical correspondence analysis biplot of Drosophilidae groups and habitat types sampled in Hawai'i. The length of arrows shows the degree of influence of each habitat variable.

Thus, the importance of preserving even remnant populations of native plants cannot be underestimated in maintaining endemic Hawaiian drosophilid diversity. Even single trees may serve as 'island' refugia from which restoration efforts can be initiated [10]. Thirteen endemic species were reared from plant genera that are not native to Hawai'i, and the majority of these are oligophagous or polyphagous [13]. With intensive sampling, even some of the apparently more sensitive picture wings may be collected at very low altitudes, such as *D. ochracea* in our study and the polyphagous *D. crucigera* Grimshaw, reared from nonnative *Erythrina* bark at 120 m altitude at the Lyon Arboretum (O'ahu) [11], and still present at that site in 2010 (L.L. unpublished data). The ubiquity

of some invasive Drosophilidae across even the most intact native forests in our study suggests that these alien species can't serve as indicators of lower habitat quality, emphasizing the need for species level identifications, which allow the use of more restricted, and informative, species.

The technology used to manage pest fruit flies, although with some undesirable nontarget effects, is a major improvement over the past practice of organophosphorous insecticide cover sprays, and has been widely adopted by Hawaiian growers in recent years [32]. Our study suggests that a few precautions can help conserve endemic drosophilids [21]. Food lure traps and protein bait spraying should be restricted to within orchards and avoided in the

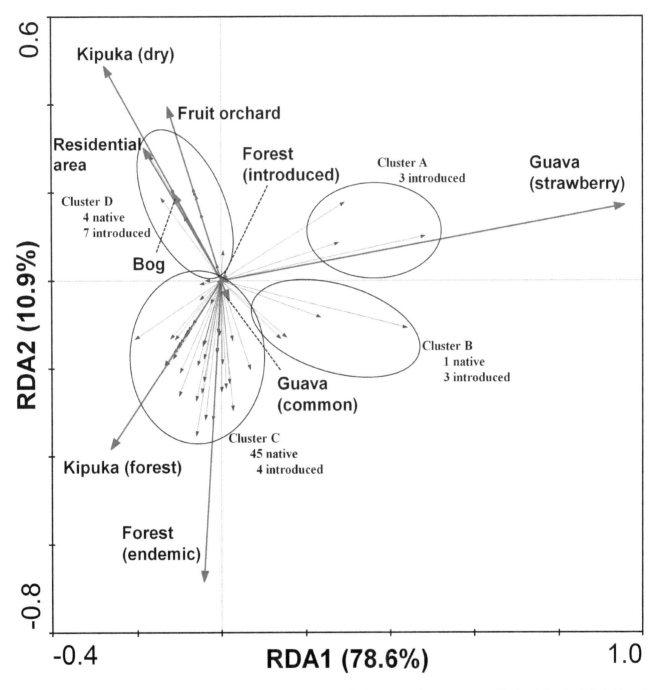

Figure 10. Redundancy analysis biplot of Drosophilidae species and habitat associations on Hawai'i Island. Species included in each group are listed in Table S1 (supporting information). Length of habitat vectors indicates influence of each on the ordination; species vectors indicate strength of association of each species with any habitat vector; cosine of the angle between vectors estimates correlation, smaller angles show higher correlations.

vicinity of isolated endemic host plants that may sustain small breeding populations of endemic drosophilids. The deployment of improved male lure formulations that do not require traps [32] should be encouraged for fruit fly control through male annihilation. If these relatively simple measures are taken, the potential nontarget impacts of fly control on native insects can be minimized.

Endemic Drosophilidae represent the most species diverse lineage in Hawai'i and a globally unique resource for understanding the process of evolution [33]. Our research demonstrates that most endemic fly species are restricted to native forest, and that a subset of species is highly sensitive and may be useful indicator species. Standardized gradient trapping schemes like those employed here are also important in generating baseline data for long term habitat monitoring. Over time, the retreat and extinction of particular species, even from forest reserves, may provide important indications of subtle changes in ecosystem integrity that are not otherwise obvious [34]. Interestingly, there is a subset of endemic species that are able to persist in invaded forest, presumably where remnant native host trees occur in very

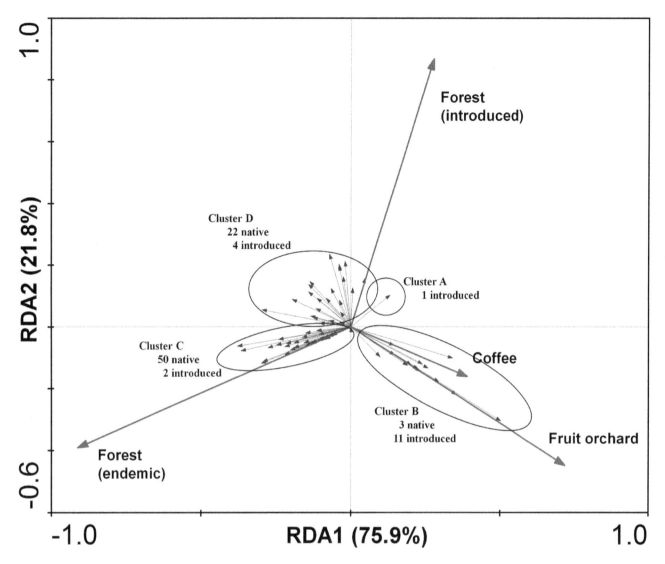

Figure 11. Redundancy analysis biplot of Drosophilidae species and habitat associations on Maui. Species included in each group are listed Table S1 (supporting information). Length of habitat vectors indicates influence of each on the ordination; species vectors indicate strength of association of each species with any habitat vector; cosine of the angle between vectors estimates correlation, smaller angles show higher correlations.

low densities. This is encouraging for conservation efforts since it suggests that even highly degraded indigenous forest areas may maintain, in the short term at least, some of their endemic plant and insect components. Thus conservation efforts in areas that appear to be almost completely overwhelmed by invasive plants may not be a lost cause.

Kealoha Kinney (University of Hawai'i, Hilo) for their invaluable help with fieldwork. Matt Shipley and Clinton Kanada (University of Hawai'i, Mānoa) assisted in processing samples. Roger Vargas (U.S. Department of Agriculture, Agriculture Research Servive, Hilo, Hawai'I, USA), Ken Kaneshiro and Andy Taylor (University of Hawai'i, Mānoa) provided advice on data analysis. We also thank Patrick O'Grady (University of California, Berkeley) and Karl Magnacca (University of Hawai'i, Hilo) for discussions on drosophilid taxonomy and verification of our species determinations.

Acknowledgments

We are grateful to Neil Miller, Nancy Chaney (U.S. Department of Agriculture, Agriculture Research Servive, Hilo, Hawai'I, USA) and

Author Contributions

Conceived and designed the experiments: LL. Performed the experiments: LL. Analyzed the data: LL MGW. Wrote the paper: LL DR MGW.

References

1. Pimm SL, Moulton MP, Justice IJ (1994) Bird extinctions in the Central pacific. Philos Trans R Soc, B 344: 27–33.

2. Paulay G, Starmer J (2011) Evolution, insular restriction, and extinction of oceanic land crabs, exemplified by the loss of an endemic *Geograpsus* in the Hawaiian Islands. PLoS ONE 6(5): e19916. doi:10.1371/journal.pone.0019916.

3. Howarth FG (1990) Hawaiian terrestrial arthropods: an overview. Bishop Museum Occasional Papers 30: 4–26.

4. Meyer III WM, Ostertag R, Cowie RH (2011) Macro-invertebrates accelerate litter decomposition and nutrient release in a Hawaiian rainforest. Soil Biol Biochem 43: 206–211.

5. Kaneshiro KY (1993) Introduction, colonization, and establishment of exotic insect populations: fruit flies in Hawai'i and California. American Entomologist 39(1): 23–29.

6. O'Grady PM, Desalle R (2008) Out of Hawai'i: The origin and biogeographic history of the genus *Scaptomyza* (Diptera, Drosophilidae). Biological Letters 4: 195–199.

7. Leblanc L, O'Grady PM, Rubinoff D, Montgomery SL (2009) New immigrant Drosophilidae in Hawai'i, and a checklist of the established immigrant species. Proc Hawai'i Entomol Soc 41: 121–127.

8. O'Grady PM, Magnacca KN, Lapoint RT (2010) Taxonomic relationships within the endemic Hawaiian Drosophilidae. Bishop Museum Occasional Papers 108: 1–34.

9. Magnacca KN, Price DK (2012) New species of Hawaiian picture wing *Drosophila* (Diptera: Drosophilidae), with a key to species. Zootaxa 3188: 1–30.

10. Heed WB (1968) Ecology of the Hawaiian Drosophilidae. University of Texas Publications 6818: 387–418.

11. Montgomery SL (1975) Comparative breeding site ecology and the adaptive radiation of picture-winged *Drosophila* (Diptera: Drosophilidae) in Hawai'i. Proc Hawai'i Entomol Soc 22: 65–103.

12. Kaneshiro KY (1997) R.C.L. Perkins' legacy to evolutionary research on Hawaiian Drosophilidae (Diptera). Pac Sci 51: 450–461.

13. Magnacca KN, Foote D, O'Grady PM (2008) A review of the endemic Hawaiian Drosophilidae and their host plants. Zootaxa 1728: 1–58.

14. Hardy DE (1965) Insects of Hawai'i. Vol. 12. Diptera: Cyclorrhapha. II, Series Schizophora, section Acalypterae I. Family Drosophilidae. Honolulu: University of Hawai'i Press.

15. Fosberg FT (1972) Guide to excursion III, revised edition. Tenth Pacific Science Congress. Honolulu: University of Hawai'i.

16. Carson HL, Hardy DE, Spieth HT, Stone WS (1970) The evolutionary biology of the Hawaiian Drosophilidae. In: Hecht MK, Steere WC, editors. Essays in evolution and genetics in honor of Theodosius Dobzhansky (A supplement to Evolutionary Biology). New York: Appleton-Century-Crofts. 437–543.

17. Carson HL (1986) *Drosophila* populations in the Ola'a tract, Hawai'i Volcanoes National Park, 1971–1986. In: Smith CW, editor. Proceedings of the third conference in natural sciences Hawai'i Volcanoes National Park. Cooperative National Park Resources Studies Unit. Honolulu: University of Hawai'i at Manoa, Department of Botany. 3–9.

18. Foote D, Carson HL (1995) *Drosophila* as monitors of change in Hawaiian ecosystems. In: Laroe ET, Farris GS, Puckett CE, Doran PD, Mac MJ, editors. Our living resources. A report to the nation on the distribution, abundance, and health of U.S. plants, animals, and ecosystems. Washington, D.C: United States Department of Interior, National Biological Service. 368–372.

19. USFWS (United States Fish and Wildlife Service) (2007) Endangered and threatened wildlife and plants: revised proposed designation of critical habitats for 12 species of picture-wing flies from the Hawaiian Islands: proposed rule. Fed Regist 72: 67427–67522.

20. Krushelnycky PD, Loope LL, Reimer NJ (2005) The ecology, policy, and management of ants in Hawai'i. Proc Hawai'i Entomol Soc 37: 1–25.

21. Leblanc L, Rubinoff D, Vargas RI (2009) Attraction of nontarget species to fruit fly (Diptera: Tephritidae) male lures and decaying fruit flies in Hawai'i. Environ Entomol 38: 1446–1461.

22. Leblanc L, Vargas RI, Rubinoff D (2010) Attraction of *Ceratitis capitata* (Diptera: Tephritidae) and endemic and introduced nontarget insects to BioLure bait and its individual components in Hawai'i. Environ Entomol 39: 989–998.

23. Kaneshiro KY, Ohta AT, Spieth HT (1977) Mushrooms as baits for *Drosophila*. Drosophila Information Service 52: 85.

24. Kitching RL, Bickel D, Creagh AC, Hurley K, Symonds C (2004) The biodiversity of Diptera in Old World rain forest surveys: a comparative faunistic analysis. Journal of Biogeography 31: 1185–1200.

25. Colwell RK (2009) EstimateS: Statistical estimation of species richness and shared species from samples. Version 8.2. User's Guide and application http://purl.oclc.org/estimates Accessed 2013 Jan 3.

26. ter Braak CJF, Šmilauer P (2002) CANOCO Reference manual and CanoDraw for Windows user's guide: Software for canonical community ordination (version 4.5). Ithaca, NY, USA (www.canoco.com): Microcomputer Power.

27. Bazelet CS, Samways MJ (2011) Relative importance of management vs. design for implementation of large scale ecological networks. Landscape Ecology 26: 341–353.

28. Muir C, Price DK (2008) Population structure and genetic diversity in two species of Hawaiian picture-winged *Drosophila*. Mol Phylogenet Evol 47: 1173–1180.

29. Gerrish G, Stemmerman L, Gardner DE (1992) The distribution of *Rubus* species in the State of Hawai'i. Cooperative National Park Resources Studies Unit, Technical Report 85. Honolulu: University of Hawai'i in Manoa.

30. Rubinoff D (2001) Evaluating the California gnatcatcher as an umbrella species for conservation of Southern California coastal sage scrub. Conserv Biol 15: 1374–1383.

31. Roberge J, Angelstam P (2004) Usefulness of the umbrella species concept as a conservation tool. Conserv Biol 18: 76–85.

32. Vargas RI, Mau RFL, Jang EB, Faust RM, Wong L (2008) The Hawai'i fruit fly areawide pest management programme. In: Koul O, Cuperus GW, Elliott N, editors. Areawide Pest management: Theory and Implementation. Oxfordshire, UK: CAB International. 300–325.

33. Kaneshiro KY (1983) Sexual selection and direction of evolution in the biosystematics of Hawaiian Drosophilidae. Annu Rev of Entomol 28: 161–178.

34. Vorsino AE, King CB, Haines WP, Rubinoff D (2013) Modeling the habitat retreat of the rediscovered endemic Hawaiian moth *Omiodes continuatalis* Wallengren (Lepidoptera: Crambidae). PLoS ONE 8(1): e51885. doi:10.1371/journal.pone.0051885.

35. Magnacca KN, O'Grady PM (2009) Revision of the *modified mouthparts* species group of Hawaiian *Drosophila* (Diptera: Drosophilidae): The *ceratostoma*, *freycinetiae*, *semifuscata*, and *setiger* subgroups and unplaced species. Univ Calif Publ Entomol 130: viii+94pp.

10

Carbon Sequestration by Fruit Trees - Chinese Apple Orchards as an Example

Ting Wu[1], Yi Wang[1], Changjiang Yu[1], Rawee Chiarawipa[1], Xinzhong Zhang[1], Zhenhai Han[1]*, Lianhai Wu[2]*

1 College of Agronomy and Biotechnology, China Agricultural University, Beijing, China, **2** Rothamsted Research, North Wyke, Okehampton, United Kingdom

Abstract

Apple production systems are an important component in the Chinese agricultural sector with 1.99 million ha plantation. The orchards in China could play an important role in the carbon (C) cycle of terrestrial ecosystems and contribute to C sequestration. The carbon sequestration capability in apple orchards was analyzed through identifying a set of potential assessment factors and their weighting factors determined by a field model study and literature. The dynamics of the net C sink in apple orchards in China was estimated based on the apple orchard inventory data from 1990s and the capability analysis. The field study showed that the trees reached the peak of C sequestration capability when they were 18 years old, and then the capability began to decline with age. Carbon emission derived from management practices would not be compensated through C storage in apple trees before reaching the mature stage. The net C sink in apple orchards in China ranged from 14 to 32 Tg C, and C storage in biomass from 230 to 475 Tg C between 1990 and 2010. The estimated net C sequestration in Chinese apple orchards from 1990 to 2010 was equal to 4.5% of the total net C sink in the terrestrial ecosystems in China. Therefore, apple production systems can be potentially considered as C sinks excluding the energy associated with fruit production in addition to provide fruits.

Editor: Carl J. Bernacchi, University of Illinois, United States of America

Funding: The study was funded by the State Science and Technology Support Program of China (NC2010BF0079), and both the Key Laboratory of Beijing Municipality of Stress Physiology and Molecular Biology for Fruit Trees and the Key Laboratory of Biology and Genetic Improvement of Horticultural Crops (Nutrition and Physiology) of Ministry of Agriculture. The funders had no role in study design, data collection and analysis, decision to publish, or preparation of the manuscript.

Competing Interests: The authors have declared that no competing interests exist.

* E-mail: rschan@cau.edu.cn (ZH); lianhai.wu@rothamsted.ac.uk (LW)

Introduction

The contribution of orchards to carbon (C) cycling ranging from C storage [1,2], root respiration [3,4,5,6,7] and net CO_2 flux [8] has been published. Measured and simulated components of the C balance in apple trees have also been reported [9,10]. However, there are very few studies that consider the impacts of orchard management practices on the environment and C cycle using a systems approach. The potential for C credits based on standing biomass for orchards is limited compared to forest stands growing in the same climatic zone. It is roughly estimated that New Zealand orchards could sequester less than 70 t C ha^{-1} within their lifespan of at least 25 years, while forest stands can sequester much more, e.g. *Pinus radiata*, reaching 300–500 t C ha^{-1} under New Zealand conditions [11]. However, orchards and forests may sequester similar amounts of C in the first few years after their establishment [11]. Although the comparison of C sequestration between forests and orchards has been made, the results were not produced based on the same criteria, e.g. living biomass was only considered without including indirect C emissions associated with orchard management practices. The content of soil organic matter depends largely on the periodic input of organic materials and the decomposition rate of soil organic matter [12]. Any increase of the soil C pool is the result of biotic C inputs that comes from CO_2 fixation directly or indirectly by plants in agroecosystems. The fixed C is partitioned to different organs of the fruit trees, which

depends on number of factors, e.g. genotype [13], tree age [14], orchard density [15], fruit production [16], training systems [17] and orchard management [18]. Meanwhile, it would be very difficult to quantify some components of C balance in an orchard system, e.g. the natural fall of flowers and fruits, microbial respiration and rhizodeposition in the overall C balance of an orchard [8]. Some of the C allocated to roots will return to the air as root respiration and enter the soils as rhizodeposits. The amount of C losses through these channels varies. It was reported that one quarter to one third of respiration occurring in a soil is from roots of higher plants, and rhizodeposition accounts for 2–30% of total dry matter production in young plants [19]. In contrast, C immobilised in short-life organs, like fruits and leaves, has other pathways [8]. Normally, C from fruits is removed from the orchard system through harvesting while that from leaves is translocated into the woody perennial parts of the apple trees or converted to soil organic C through decomposition after the leaves become litterfall.

China is the largest apple producer in the world with c.a. 35×10^6 t in the 2011/2 season (USDA). Fruit trees are an important component in the Chinese agricultural sector with 8.67 million ha of orchards, of which apple orchards cover 1.99 million ha [20]. Considering its storage capacity of fixed C and the size of planting area, the orchards in China could play an important role in the C cycle of terrestrial ecosystems and contribute to C sequestration. However, there is little knowledge on how to improve the C sequestration potential in Chinese orchards.

To accurately estimate C stocks and fluxes in orchards located in different regions, it is desirable to set a baseline that can eliminate the uncertainties caused by the variations in fruit tree structure, stocks and fluxes among and within geographical sites combined with direct field measurements from local sample plots. In natural and semi-natural forests, the C carrying capacity can be used as an indicator for the baseline [21,22]. The indicator is made up of a matrix of factors to quantify forest degradation and sequestration in terms of C losses and gains due to land-use changes [21,23,24,25]. Because orchards are managed systems, the estimation of C sequestration capacity is different from the natural forest. Therefore, it is necessary to build a new matrix for orchard ecosystems to accommodate management factors. The aim of this paper is 1) to set up an assessment matrix of C sequestration capability based on measured data to evaluate the C sequestration capability of apple trees of a range of ages; and 2) to assess the status of C sequestration by apple orchards in China.

Materials and Methods

Study Site

A field-modelling study was conducted in the apple orchards located in Changping District, Beijing (116°28′55″E, 40°20′34″N) between 2009 and 2010. The climate of this region is temperate continental monsoon with an annual average temperature of 12.1°C. The soil of the orchards is deep sandy loam or loam with approximately 1.2% of soil organic matter content in the 20 cm topsoil. Three orchards of Fuji/Makino apple trees, 5, 18 and 22 years old, were chosen (i.e. apple trees were planted in 2005, 1992 and 1988 separately), to represent juvenile, mature and over-mature phases in the life cycle of apple trees. Information on fertilizer use and irrigation application in the orchards was obtained via the Beijing Agricultural Technical Extension Station. General information about the orchards is shown in Table 1. Four frames [26] in each orchard were randomly set up but avoiding the influence of diseases and insects on the trees within the frames to determine the value of stem diameter. The value with the highest probability within the orchard using the normal distribution of the stem diameters was taken to identify an apple tree to be harvested for biomass estimation. Three frames from the four quadrats in the orchard were chosen for fine-root observations with minirhizotron, soil core samplings for fine-roots and soil respiration measurements. The area of the quadrat was 2×4 m^2 for the 5-year-old orchard and 3×5 m^2 for the 18- and 22-year-old orchards, considering the planting densities of the orchards with different ages.

Analysis of Carbon Sequestration Capability

Matrix analysis was used to investigate the capability of C sequestration. Three indirect sources for C emission: irrigation, and chemical fertilizer and manure applications were identified. A set of the potential assessment factors, U_i ($i = 1,2,3$, representing the 5-, 18- and 22-year-old orchards, respectively), was formed with the subset in the following order: long residence woody (stem + branch + coarse root), leaf, fruit, fine root, pruning, soil respiration, irrigation and fertilizer application. Because the pruned branches would normally leave the orchard systems immediately, this component is treated as a source of C emissions within the orchards in this study. There are two columns in U_i: the C capture elements (U_{i1}) and the C emission elements (U_{i2}), which were used to evaluate the contribution of the assessment factors to C sequestration in an orchard. The C capture by long residence of

Table 1. General information of the Fuji apple tree orchards at Changping, Beijing.

	5-yr-old	18-yr-old	22-yr-old
Orchard size (ha)	0.33	1.67	0.33
Nitrogen application (kg N ha^{-1} yr^{-1})	135	149	149
Phosphorus application(kg P ha^{-1} yr^{-1})	135	149	149
Potassium (kg P ha^{-1} yr^{-1})	135	149	149
Animal manure (kg ha^{-1} yr^{-1})*	87000	99000	97500
KNO$_3$ (kg ha^{-1} yr^{-1})	300	330	330
ZnSO$_4$ (kg ha^{-1} yr^{-1})	300	330	330
Bacterial fertilizer (kg ha^{-1} yr^{-1})	300	330	330
Irrigation amount (m^3 ha^{-1} yr^{-1})	3220	3300	3300

*OM content: about 25%; nutrient content: N - 1.6%, P - 1.5% and K - 0.8%.

woody, leaf or fruit was represented with the C content of increased biomass. The contribution to C emissions from fertilizer or irrigation is calculated by amount of fertilizer application or irrigation multiplied by a conversion coefficient which was adopted from *West and Marland* [27].

A vector (B_i) was used to assign weighting factors for each subset in U_i. The values in the vector reflect the contribution of the factors in the overall assessment for orchard i. The elements in B_i include: weighting factors of long residence woody, leaf, fruit, fine root, pruning, soil respiration, irrigation and fertilizer application in order. The matrix V_i derived from the product of the matrix and its corresponding vector was the outcome of the C capture and emission from the orchard i:

$$V_i = B_i U_i \tag{1}$$

The measurement of the potential assessment factors described were as below.

Biomass Estimation

Three representative apple trees from each orchard were identified in September, 2009 and the increments in stem diameter at 20 cm from the ground of the trees were measured in October, 2009 to October, 2010. Increment of stem diameter measurements one year apart were used to estimate the biomass increment for the experimental year. The trees were divided into leaves, branches, main stems and coarse roots, and all fresh weights were measured in the field. Then sub-samples from each fraction were dried to determine total dry matter of each fraction. The sub-samples were oven-dried at 65°C until constant mass was reached. The dried sub-samples were ground after weighting before total C analysis was made by EA 1108 elemental analyser (Italy, Carlo Erba) to determine C content. The data were used to fit allometric equations to quantify the relationship between the biomass of different parts (leaf, stem, branch, coarse root) and stem diameter. Because of the life cycle of an apple tree, fruit production would begin to decline in the over-mature phase year by year. Therefore, a parabola equation was used to quantify the relationship between fruit production and stem diameter. Statistical analysis was made using SPSS for Windows (Rel. 11.5.0, 2002. Tokyo: SPSS Inc.).

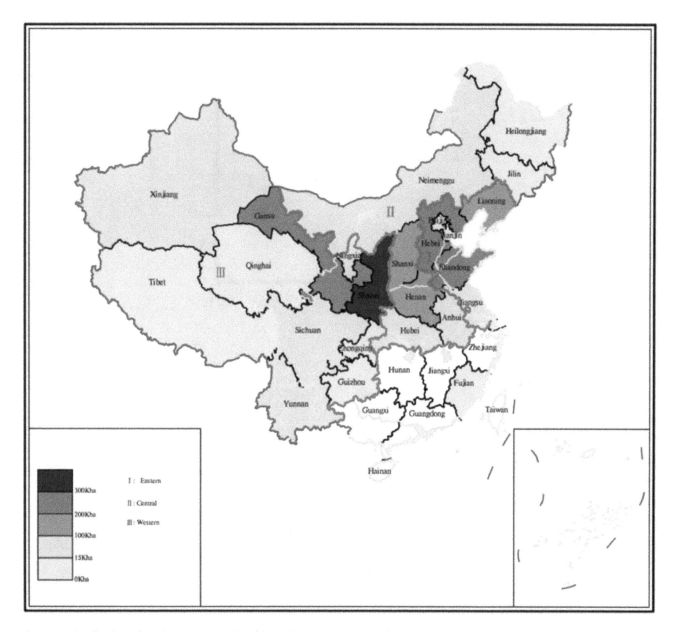

Figure 1. Distribution of apple grown area in China. The regions without colour indicate non-apple grown areas.

Fine Roots Observation

The turnover rate of fine roots is an important parameter to indicate the contribution of the fine roots to C sequestration. Minirhizotron and soil coring method were used to estimate the turnover rate.

In July, 2009 before the experiment started, nine 90 cm-long minirhizotron tubes of 5 cm diameter were inserted into the soil at 45° angle with the horizon in the second quadrat of each orchard to allow fine roots to settle in the soil surrounding the tubes. Minirhizotron images were collected every ten days using a BTC-10 minirhizotron microscope (Bartz Technology, USA) from October 2009 to October 2010. This generated 64 images from each minirhizotron tube. The length of fine roots was calculated for all the images using the I-CAP software. The collected information was used to calculate the length of fine roots (cm) per unit area. A fine root turnover index, defined as the ratio of fine root mortality in a year (cm cm-1) to initial fine root length (cm

cm^{-1}) within the minirhizotron window [26], was calculated using the data with the reported method. Within a quadrat from each orchard, ten soil cores of 4 cm in diameter and 80 cm in depth were sampled. The soil columns were separated into depths 0–20, 20–40, 40–60 and 60–80 cm. The soil samples were transferred in plastic bags. Roots were manually picked out from the samples, washed and sorted (<2 mm), then they were oven-dried at 80°C until constant mass were reached. Fine root dry matter density (mg DM cm^{-3}) was estimated based on the relationship between the weight and the surface area of sampled fine roots, which was used to estimate the biomass of fine roots.

Soil Respiration

Soil respiration (and temperature) at 10 cm soil depth in the three quadrats in each orchards were measured every ten days between June 2009 and June 2010 (excluding first three months in 2010 when the top soil was frozen) using developed closed gas-

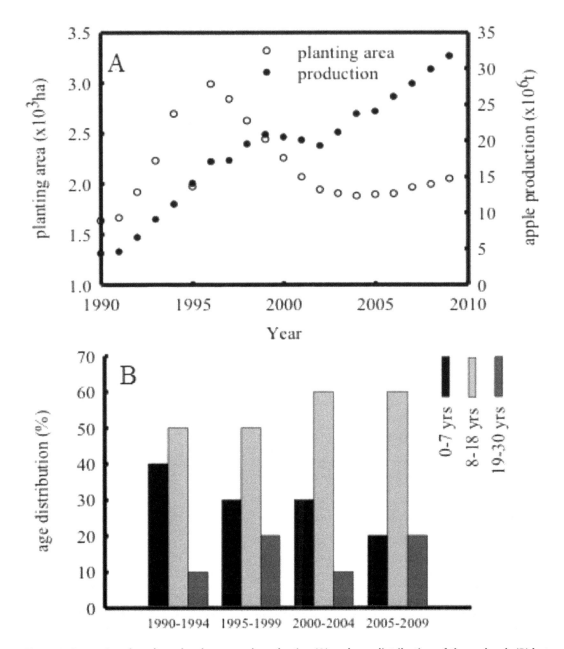

Figure 2. Dynamics of apple orchards area and production (A) and age distribution of the orchards (B) between 1990 and 2010 in China [*data source: Editorial Committee*, **1991;1996;2001;2006;2010**].

exchange system (LiCor 6400 Portable Photosynthesis System with 6400-09 soil CO_2 flux chamber; LiCor, Lincoln, NE, USA). At all three orchards, 27 replicate LiCor soil collars in total were installed. The collars remained in place throughout the experiment period, allowing repeated measurements. All measurements were carried out between 10:00 to 11:30 am. Measurements were not made on days of rain.

Litterfall Decomposition

At the beginning of September 2010, 20 g fresh leaves collected from the apple trees was put into a nylon bag with a mesh size of 2 mm. Six bags were randomly placed on the surface of the orchard soil (after removing the litter layer). At the end of each month for the following three months, two litterbags were collected from each orchard to calculate weight loss and C

Table 2. Allometric equations for dry matter (kg) for different parts of subject trees using the square of the stem diameter (cm) at 20 cm from the ground as an independent variable (sample number is 9 for each component).

	Formula	a	B	c	Adjusted R^2
Fruit	ax^2+bx+c	−1.170	0.260	−0.001	0.809
Branch	ax^b	0.124	1.234		0.984
Stem		0.178	1.101		0.997
Leaf		0.160	0.656		0.951
Root		0.159	0.999		0.946

content of litterfall. The data was used to determine decomposition rate by fitting the exponential function [28]:

$$X_t = X_0 \times e^{-kt} \qquad (2)$$

where X_0 is initial C content of the litterfall (g C), X_t is C content at time t and k is the decomposition rate (d^{-1}). Data processing and allometric equations.

Other Data Sources

There are three geographical regions for apple production in China: the western, the central and the eastern regions. This study covered major apple regions in China (Figure 1), biomass data of apple trees at various ages from geographic locations was used to validate the allometric equations described in 2.2. The dynamics of the net C sink and C storage of apple orchards in China was calculated based on cultivation areas and tree age groups (Figure 2). The groups were set 0–7, 8–18 and 19–30 years old at national level.

Results

Carbon Storage, Emission and Turnover of Apple Trees

C capture of long residence woody, leaf, fruit. Carbon storage from a part of an apple tree is estimated based on dry matter and C content of the part. The average stem diameters at 20 cm from the ground were 3.2 ± 0.2, 12.9 ± 0.8 and 14.2 ± 0.6 cm for the 5-, 18- and 22-year-old trees. The parameter values for the allometric equations to estimate dry matter from different parts of an apple tree are presented in Table 2.

In order to validate the allometric equations which can be applied to the apple production regions in China, biomass data of an apple tree were collected dated from 1990 in major apple production regions (Table 3). The observed biomass and estimated values from the equations were compared (Table 4 and Figure 3). For the trees with stem diameter between 4 and 6 cm, the equations underestimated biomass of trees due to a large standard deviation in the collected data. For the other two groups at stem diameters of 0–4 cm and 10–15 cm, the estimated values confirm to the actual ones. These results indicated that the allometric

Table 4. Comparison of the modelled and observed total dry matter content of an apple tree (kg).

stem diameter range	0–4 cm	4–6 cm	10–15 cm
Modelled	7.5±1.0	16.9±0.2	154.1±31.9
Observed	6.6±1.2	29.2±12.1	138.0±57.6

equations should be suitable for biomass estimation of apple trees which were subjected to managed agricultural production systems.

The annual biomass increments of living organs of apple trees at different ages were calculated using the equations. The increments of an individual tree were converted into the increments per unit land area through multiplying tree density. The results showed that the 5-year-old tree has a much higher growth rate than the other two groups although its standing biomass is low (Figure 4). The growth rates of all parts (except stem) for the trees older than 18 years began to slow down, which may be caused by reduced growth of a mature tree. As fruit trees get older, the proportion of the long residence woody biomass in total standing biomass production also increases.

C capture and turnover of fine roots. The analysis of minirhizotron images indicated that the net growth rate of fine roots had apparent seasonal changes for the trees of all the ages. Two growth peaks appeared, in early summer and late fall for the 5- and 18-year-old trees throughout the observed soil profile. However, no significant seasonal change in growth rates was observed in the 22-year-old trees. The active growing zone of fine roots fell between 20 and 60 cm of the soil profile (Figure 5). Based on the observation and the regression analysis between fine root weight and surface area ($W_{fr} = 0.634 + 0.7689 S_{area}$, $R^2 = 0.8119$ and n = 5), the annual growth rate of fine roots was calculated as 33.4 ± 16.8, 41.7 ± 19.0, and 17.7 ± 6.8 g DM m^{-2} for the 5-, 18- and 22-year-old trees, respectively.

The indices of annual fine root turnover were 7.7, 6.8 and 1.5 for the 5-, 18- and 22-year-old trees, respectively. The proportion of appeared and disappeared fine roots in the minirhizotron window for the 5-year-old trees remained at a similar value for the

Table 3. Biomass (kg tree^{-1}) of various parts of an apple tree with various ages in different geographical areas in China.

Site	Age (yr)	Stem diameter (cm)	Fruit	Branch	Stem	Coarse root	Leaf	Source*
North China	5	3.19	0.98	1.44	2.23	3.21	0.70	1
		3.11	1.58	2.25	2.38	1.10	0.76	1
		3.42	1.64	3.18	2.32	1.19	0.76	1
	18	13.22	26.06	83.29	52.47	23.84	5.73	1
		12.02	17.31	68.02	40.64	30.77	3.61	1
		13.42	14.60	92.00	52.21	29.48	7.11	1
	22	13.69	12.76	67.44	61.39	22.93	3.574	1
East China	7	5.55	2.42	7.82	7.10	4.91	1.62	2, 3
	8	5.51	9.74	5.22	24.24	12.19	3.47	2, 4
		5.51	6.52	5.81	20.20	7.71	4.35	5
	17	11.58	9.00	8.12	26.86	19.47	3.12	2, 6
		11.58	12.23	10.35	12.64	13.68	4.09	2, 5
Northwest China	5	3.58	0.54	1.73	1.69	0.94	0.55	7

Source*: 1 survey by the authors; 2–6 from per. comm. (2 Y. Jiang; 3 L. Zhao; 4 J. Fang; 5 D. Zhang; 6 N. Ding); 7 Zhang et al. [2009].

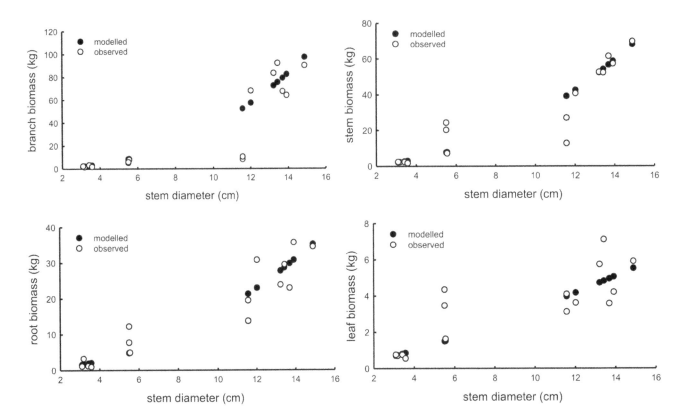

Figure 3. Comparison between the modelled and observed dry matter of branch, stem, fine root and leaf with different stem diameters.

longest period compared to those with other two ages (Figure 6), which indicated that the 5-year-old trees not only produced a large amount of fine roots, but their root systems also had a high metabolic rate.

C emissions of pruning and soil respiration. In China, each individual orchard follows its own guidelines for the disposal of pruned branches based on the age of trees. In general, more branches are pruned with tree ages. Biomass in pruned branches at different ages was shown in Table 5.

Soil respiration from the 18-year-old orchard showed the strongest seasonal changes among the orchards, and two peaks in a year existed (Figure 7). We observed that the highest rate for the 5-year-old orchard occurred in spring and early summer, and that for the 22-year-old orchard occurred in the middle of summer. Based on the samples, annual soil respiration rates were 1.3 ± 0.3, 1.6 ± 0.6 and 1.2 ± 0.5 kg C m^{-2} in the 5-, 18- and 22-year-old orchards, respectively.

C emissions of litterfall. The data from the litterfall bags showed that the decomposition of litterfall followed an exponential equation:

$$X_t = X_0 \times e^{-0.0032t} \qquad \left(R^2 = 0.8463,\, n = 18\right) \qquad (3)$$

The total decomposition rate was equivalent to 312 days of turnover time. i.e., complete decomposition of litterfall would occur within a year either through emission to the atmosphere as CO_2, or transformation into a stable organic matter pool in the soil.

Carbon Sequestration Capability with Various Ages of Apple Trees

Contribution of the components to carbon sequestration. Annual C increment rate for each part of the trees in an orchard was treated as a C sink. The value of the assessment elements related to tree biomass was calculated by the allometric equations, plant density and the conversion fraction of dry matter to C content. The fraction was 0.46 g C g^{-1} DM based on the measurement of this study. In an annual cycle, leaves stay on trees during the growing season and then fall to the ground as litterfall for a certain period. During the year, some of litterfall decomposes releasing C to the atmosphere. Considering the length of the period and the measured decomposition rate, the emitted fraction of C in leaves is set at 0.284. Pruned branches, soil respiration and orchard managements are considered as sources of C emissions. The matrix of assessment elements for each age group of the orchard was shown in Table 6.

Weighting factors for the components. Because there are no substantial changes in long-term residence woody losses within one year, its weighting factor was set to 1. All leaves from apple trees will emerge at the beginning of the growing season and fall after the growing season of the year, so the weighting factor was also set to 1. An apple tree normally sets more fruits than they can be supported to grow and it usually becomes necessary to thin fruits or buds in order to improve the average size of each fruit remaining on the tree. Therefore, the fallen fruit or buds would be decomposed. Because harvested fruits usually grow for half a year, the weighting factor for harvested fruits in an orchard was given a score of 0.5. The annual fine root turnover index derived from the minirhizotron images is 7.69, 6.84 and 1.48 for the 5-, 18- and 22-year-old trees, respectively (details in 3.1.2). The orchard would

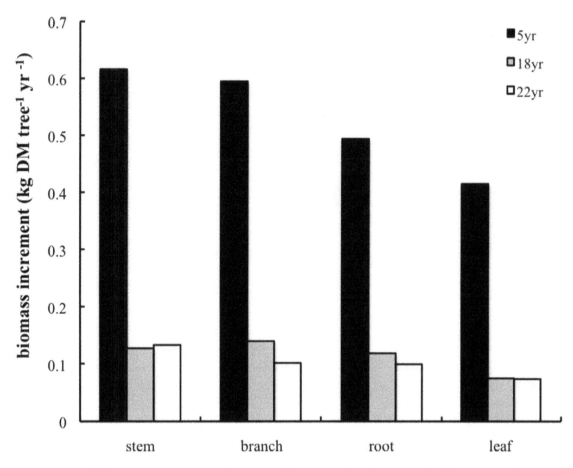

Figure 4. Annual biomass increment rates in various living parts of apple trees with different ages between Oct. 2009 and Oct. 2010.

be pruned once a year, its rate is also to be 1. C emissions from all types of orchard management practices were given a total score of 1. Therefore, the vector of weighting factors for the components from each orchard can be expressed as:

$$B_1 = \begin{bmatrix} B_1 \\ B_2 \\ B_3 \end{bmatrix} = \begin{bmatrix} 1, & 1, & 0.5, & 7.69, & 1, & 1, & 1, & 1 \\ 1, & 1, & 0.5, & 6.84, & 1, & 1, & 1, & 1 \\ 1, & 1, & 0.5, & 1.48, & 1, & 1, & 1, & 1 \end{bmatrix} \quad (4)$$

Assessment of carbon sequestration capability. The final evaluation matrix for each orchard (in a row) was produced through the product of U and B:

$$V = B \cdot U = \begin{bmatrix} 0.76, & -1.47 \\ 4.19, & -2.27 \\ 3.72, & -2.00 \end{bmatrix} \quad (5)$$

The first column in the matrix represents gross C input to an orchard and the second is C emission from the orchard. The net C sink by the end of the experimental period is −0.7, 1.9, and 1.7 kg C m^{-2} for the 5-, 18- and 22-year-old orchards, respectively. There is a transition point from net C emission to net C sink with the tree ages between 5 and 18 years old. The relationship between tree age (X_{age}) and C sequestration (C_{seq}) could be expressed with a parabola function:

$$C_{seq} = -0.0159X_{age}^2 + 0.05853X_{age} - 3.5194 \quad (n=3) \quad (6)$$

The equation showed that an eight-year-old orchard may reach the balance between the source and the sink. The apple trees older than 8 years could be considered as the C sink. When the trees are 18 years old, they reach the peak of C sequestration capability which then begins to decline with their ages.

The proportions of C stocks in short-lived tissues and long residence woody and C emissions from orchard management practices in C sequestration were analyzed (Table 7). In the 5-year-old orchard, the proportion of C emission from the orchard management practices far exceeds the portion of C stocks from both short-lived tissues and long residence woody. As the orchard enters the aging process, the proportion of short-lived tissues is declining, which could explain the lower C sequestration capability in the 22-year-old orchard.

Carbon Sequestration Capability in Apple Orchards in China

There was a great development in apple tree production in the 1980s in China because of government incentives. As a consequence, most of seedling apple trees were planted then. Apple trees younger than 8 years old accounted for the largest percentage in the apple orchards by 1990. Therefore apple orchard inventory

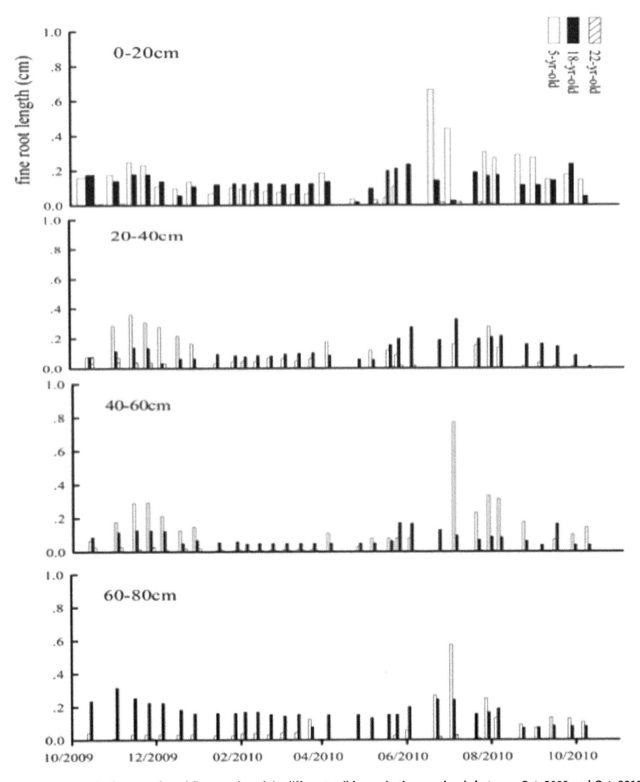

Figure 5. Dynamic changes of total fine root length in different soil layers in three orchards between Oct. 2009 and Oct. 2010.

data from 1990s were analyzed in this paper. The evaluation matrix for the 5-, 18- and 22-year-old apple orchard generated from above session were used to estimate net C sink from 1990–2010 for the age group categories: 0–7 yr., 8–18 yr., and 19–30 yr., respectively. C storage in the living organs of apple trees at national level was assessed using the allometric equations. The

estimated net C sink from apple orchards and the C storage of the apple trees in China were shown in Figure 8. It is noted that the C sequestration capability declined from 1995 to 2000. This is because the market price of apple decreased during the period, and a large number of tress were cut by farmers, especially in the

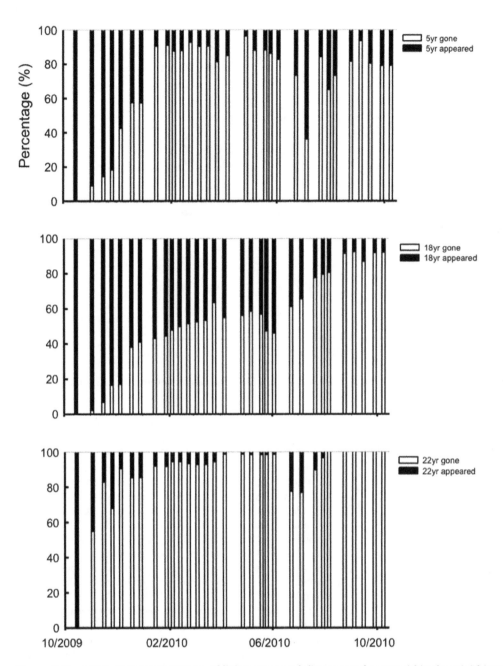

Figure 6. Dynamics of the proportions of living roots and disappeared roots within the minirhizotron window between Oct. 2009 and Oct. 2010.

Table 5. Dry matter and carbon content in pruned branches of an apple tree with different ages assuming 3.75 m² of land is covered by an 18- or 22-year-old tree and 2 m² for a 5-year-old tree.

tree age (yr)	5	18	22
Dry matter (kg tree^{-1})	0.587	3.457	4.630
C content (kg C tree^{-1})	0.270	1.590	2.130
C content (kg C m^{-2})	0.135	0.424	0.568

eastern and northeast China. As a result, the total orchard area was reduced during this period.

The contribution from each region to total C sink from apple orchards is different and variable (Figure 9a). In the eastern and western regions, the change of net C sink from 2000 to 2005 had a similar trend as the national level shown in Figure 8. The sink in the eastern region has decreased, whilst it has kept rising in both the western and the central regions since 2005. The net C sink is 13.3, 5.24 and 9.15 Tg year^{-1} in the western, central and east regions in 2010, respectively. The estimation indicated that the apple orchards in the western region are the major contributor to the C sequestration (Figure 9b). The total net C sink from apple orchards in China is about 27 Tg C year^{-1} in 2010, which is equivalent to sequester 14 t C ha^{-1} year^{-1}.

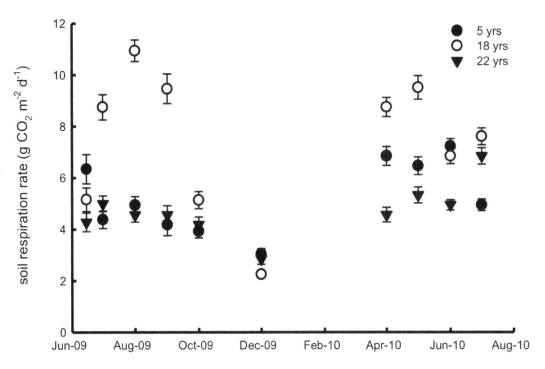

Figure 7. Soil respiration rates in the observed months in the 5-, 18- and 22-year-old apple orchards during the observation period.

Discussion

Monitoring the dynamics of fine roots and soil respiration, litterfall decomposition and destructive sampling on apple trees were made in three apple orchards with 5, 18 and 22 years old, respectively. The results showed that the 5-year-old tree had the highest growth rate among the trees with different ages and the rates of branches and roots for the trees >18 years began to slow down. The analysis of minirhizotron images indicated that the annual fine root turnover index is 7.69, 6.84 and 1.48 for the 5-, 18- and 22-year-old trees, respectively, but the 5-year-old tree produced a large amount of fine roots and its root system also had a high metabolic rate. Soil respiration from the orchard with 18 years old had the strongest seasonal changes among the orchards and two peaks in a year existed. Blanke (1996) [29]suggested Soil and grass respiration occupied a major contribution to the CO_2 flux in a fruit orchard. Consistently our data show the annual respiration rates are 1.25, 1.59 and 1.21 kg C m^{-2} in the 5-, 18- and 22-year-old orchards, respectively. That means there's no significantly relationship between apple trees and respiration. Soil respiration in 18 yr orchard is higher than the respiration in the

other orchards (5-, 22-year-old orchards). We suppose that more animal manure applied to 18–year-old orchard (Table 1) which accelerate the soil respiration. The litterbag experiment made a conclusion that the residence time for litterfall is about 312 days. It is noted that the conclusion was drawn only from the results of a single year experiment. Ideally the experiment should last longer in order to acquire more data for the analysis of C sequestration capability. But it is a time-consuming, labour intensive and expensive experiment. It would be difficult to purely rely on field experiment for the analysis. Modelling may be an option to estimate the contribution of each component in an apple orchard system to C sequestration based on the obtained results from this experiment.

The results of C sequestration capability matrix suggested that the net C sink is −0.7 (source), 1.9, and 1.7 kg C m^{-2} for the 5-, 18- and 22-year-old orchards, respectively. The apple trees older than 8 years could be considered as a C sink. When the trees are 18 years old, they reach the peak of C sequestration capability which then begins to decline with their ages. Only when apple trees grew till a certain age in an orchard, the orchard could start

Table 6. Carbon capture and emission (kg C m^{-2} yr^{-1}) from the evaluation factors in the orchards with different planting ages.

		Long residence woody	Leaf	Fruit	Fine root	Pruning	Soil respiration	Irrigation	Fertilizers
5 yrs old	Capture (U_{11})	0.47	0.091	0.17	0.015	0	0	0	0
	Emission (U_{12})	0	0.0026	0	0	0.135	1.25	0.017	0.034
18 yrs old	Capture (U_{21})	2.2	0.676	2.37	0.019	0	0	0	0
	Emission (U_{22})	0	0.192	0	0	0.424	1.59	0.017	0.038
21 yrs old	Capture (U_{31})	2.3	0.56	1.7	0.008	0	0	0	0
	Emission (U_{32})	0	0.159	0	0	0.568	1.21	0.017	0.038

Table 7. Percentages of various components in net carbon sequestration in the apple orchards with different ages.

Orchard age (yr)	5	18	22
Long residence woody (%)	22	36	45
Short-lived tissues (%)	12	29	18
Orchard managements (%)	66	35	37

to sequester C. Carbon emission derived from management practices would not be compensated through C storage in apple trees before reaching the mature stage. After an orchard became a net C sink, short-lived tissues turnover rates would be the major factor affecting C sequestration in the orchard. Based on apple orchard inventory data from 1990s and the evaluation matrix results from modelled plot in Beijing, it was estimated that the net C sink in apple orchards could range from 13.8 to 32.6 Tg C and the C storage of the apple trees from 227 to 481 Tg C during the period of 1990–2010. Among the three major growing regions of apple trees in China, the apple orchards in the western region are the major contributor to the C sequestration. The total net C sink from apple orchards in China is about 27 Tg C in 2010, which is equivalent to 14 t C ha^{-1}. If taking the life cycling of apple trees as 25 years, the figure is similar to the reported amount for organic kiwifruit and apple production systems in New Zealand [30]. It is predicted that the capability continues growing substantially in China in the future because of the increase of growing areas and more apple orchards entering the mature stage. Therefore apple production systems can be potentially considered as a C sink apart from the function of fruit production.

Net ecosystem productivity is often used as a parameter for C sink at a regional level [11,31] without considering the impact of orchard management practices. In this study, indirect C emissions from fertilizer applications and irrigations, as well as the direct contribution from soil respiration and tree pruning were included in the estimation of the capability, in addition to the biomass turnover rates. The results suggested that a young orchard would

be a source of C emissions initially, and then it become a C sink from the eighth year when the trees are at the full fruit stage and the capacity of C sequestration gradually increases until it reaches 18 years old. During the full fruit stage, apple trees are more active in photosynthesis and metabolism, which would have more C deposited into their organs. Meanwhile, C emissions from orchard management practices would remain at a constant level. As a result, net C sequestration capability could increase when an orchard is between the ages of 8 and 18 years old. In practice, fruit trees are often forced to enter the full fruit stage as early as possible in China. To achieve this goal, heavier inputs (chemical fertilizer, manure, pesticide) was added to the orchards during the early period of orchard establishment to stimulate tree growth, which would induce more C emitted to the atmosphere through the practices.

The partitioning of photosynthate among metabolic activities, short-lived tissues and long residence woody organs in fruit trees, is an important process for C sequestration, especially quantifying the allocation to fine roots and investigating its residence time in soils [32]. It was reported that the proportion of photosynthate allocated for fine root construction could account for 30–50% of total global terrestrial primary productivity [33,34,35]. *Guo et al.* [36] suggested that the most important channel for fixed C into soils is fine roots because of their short life spans and the quick decomposition of dead roots. Whether the dynamics of fine roots in orchards has such an important role in C sequestration is still controversial, as a considerable portion of photosynthate should be used to support fruit production which is one of the main goals for fruit plantation. In this study, the minirhizotron technique was used to monitor the dynamics of fine roots in order to determine fine root turnover time much accurately. The image analysis showed that the annual fine root turnover index of apple trees is much high, reaching almost 7 for the 5-year-old tree although fine root biomass only accounts for a small percentage of total biomass (<0.1%). The fine roots may not make such a large contribution to C sequestration potential in apple orchards.

The preliminary results from this study showed that the proportion of C emission from the orchard management practices far exceeds the portion of C stocks from both the short-lived tissues and long residence woody tissues in the 5-year-

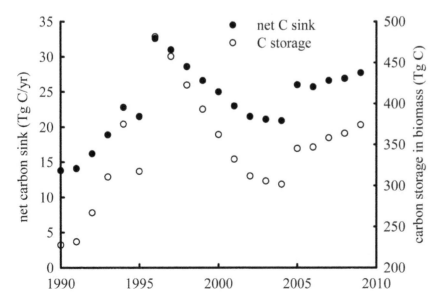

Figure 8. Net carbon sink in apple orchards in China between 1990 and 2010.

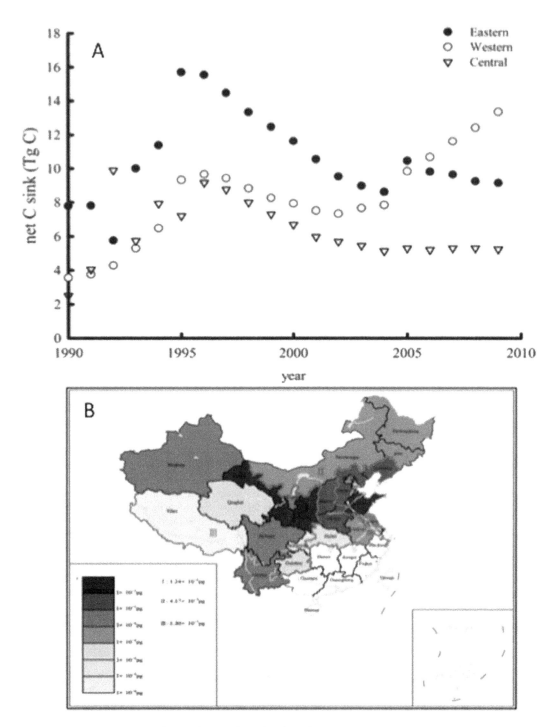

Figure 9. Dynamics of net carbon sink in apple orchards in the western, the central and the eastern regions between 1990 and 2010 (A), and the net carbon sink in each apple growing province in China in 2010 (B).

old orchard. When the orchard enters the aging process, the proportion of short-lived tissues is declining. These conclusions were made from only one experiment in three orchards for one growing season. We caution direct interpretation to other geographic sites where the environmental conditions are significantly different. In order to estimate C sequestration potential from apple orchards in China accurately, more long-term field monitoring experiments at different ages are necessary.

Due to the lack of data on soil organic C pool in apple orchards at a provincial/national level, soil organic C sequestration in the

apple orchards was not included in the net C sink in this study. Continuous applications of organic fertilizer to agroecosystems would result in the accumulation of organic C in cultivated soils, which could have big potential for C sequestration [37]. As a managed system, the production of apple orchards in China will increase the application of organic fertilizers to replace chemical fertilizers in order to produce high quality fruits and ensure the sustainability of the system. C sequestration potential covering trees, management practices and soils in apple orchards could give more accurate estimation in the future.

It was reported that C sequestration potential in Chinese forests was 5.9×10^3 Tg C in 2000, equivalent to 41 t C.ha^{-1} [38]. C sequestration from the apple orchards in 2000 was 332.2 Tg C from this study, which is about 5.6% of forest C sequestration. It was estimated that a mean net C sequestration rate in the terrestrial ecosystems which orchard systems were not explicitly included in China was in the range of 190–260 Tg C yr^{-1} during the 1980s and 1990 [39]. Our results indicated that the net sink from the apple orchards in 1990 was 14.1 Tg C, which equals to 4.5% of the reported total C sequestration from the terrestrial systems in China. The Net C sink has gradually increased since 1990 and reached 27 Tg C in 2010. The estimation from this study suggested that 1.6–3.0% of China's CO_2 emissions from burning fossil fuels in 2000 [40,41] could be compensated by C

accumulation in apple orchards. Therefore, C sequestration in apple orchards can help to offset industrial CO_2 emissions.

Acknowledgments

Authors are grateful to Drs Shilei Lu and David Chadwick for critical comments on the manuscript, and to Huiping Liu for his assistance in selecting model study sites.

Author Contributions

Conceived and designed the experiments: ZH LW TW. Performed the experiments: TW CY RC. Analyzed the data: TW YW. Contributed reagents/materials/analysis tools: TW XZ. Wrote the paper: TW LW ZH.

References

1. Procter JTA, Watson RL, Landsberg JJ (1976) The carbon budget of a young apple tree. Journal of the American Society for Horticultural Science 101: 579–582.
2. Wibbe ML, Blanke MM, Lenz F (1993) Effect of fruiting on carbon budgets of apple tree canopies. Trees - Structure and Function 8: 56–60.
3. Blanke MM, Kappel F (1997) Contribution of soil respiration to the carbon balance of an apple orchard. Acta Horticalture 451: 337–341.
4. Ebert G, Lenz F (1991) Annual course of root respiration of apple-trees and its contribution to the CO 2 -balance. Gartenbauwissenschaft 56: 130–133.
5. Sekikawa S, Kibe T, Koizumi H (2003) Soil carbon budget in peach orchard ecosystem in Japan. Environmental Science 16: 97–104.
6. Sekikawa S, Kibe T, Koizumi H (2003) Soil carbon sequestration in a gape orchard ecosystem in Japan. Journal of the Japanese Agricultural Systems Society 19: 141–150.
7. Sekikawa S, Koizumi H, Kibe T (2002) Diurnal and seasonal changes in soil respiration in a Japanese grape vine orchard and their dependence on temperature and rainfall. Journal of the Japanese Agricultural Systems Society 18: 44–54.
8. Sofo A, Nuzzo V, Palese AM, Xiloyannis C, Celano G, et al. (2005) Net CO 2 storage in Mediterranean olive and peach orchards. Scientia Horticulturae 107: 17–24.
9. Lakso AN, White MD, Tustin DS (2001) Simulation modeling of the effects of short and long-term climatic variations on carbon balance of apple trees. Acta Horticalture 557: 473–480.
10. Lakso AN, Wunsche JN, Palmer JW, Grappadelli LC (1999) Measurement and modeling of carbon balance of the apple tree. Hortscience 34: 1040–1047.
11. Kerckhoffs LHJ, Reid JB (2007) Carbon sequestration in the standing biomass of orchard crops in New Zealand. Hastings, New Zealand: New Zealand Institute for Crop & Food Research Ltd. 1–4 p.
12. Kimmins JP (1997) Forest ecology: a foundation for sustainable management. second edition. New Jersey: Pearson Prentice Hall. 1–596 p.
13. Lauri P-E, Térouanne E, Lespinasse J-M, Regnard J-L, Kelner J-J (1995) Genotypic differences in the axillary bud growth and fruiting pattern of apple fruiting branches over several years - an approach to regulation of fruit bearing. Scientia Horticulturae 64: 265–281.
14. Palmer JW (1988) Annual dry matter production and partitioning over the first 5 years of a bed system of Crispin/M.27 apple trees at four spacings. Journal of Applied Ecology 25: 569–578.
15. Marini RP, Sowers DS (2000) Peach tree growth, yield, and profitability as influenced by tree form and tree density. Hortscience 35: 837–842.
16. Forshey CG, Elfving DC (1989) The relationship between vegetative growth and fruiting in apple trees. Horticultural Reviews 11: 229–287.
17. Grossman YL, Dejong TM (1998) Training and pruning system effects on vegetative growth potential, light interception, and cropping efficiency in peach trees. Journal of the American Society for Horticultural Science 123: 1058–1064.
18. Marsal J, Mata M, Arbones A, Rufat J, Girona J (2002) Regulated deficit irrigation and rectification of irrigation scheduling in young pear trees: an evaluation based on vegetative and productive response. European Journal of Agronomy 17: 111–122.
19. Brady NC, Weil RR (2004) Elements of the nature and properties of soils (Second Edition). New Jersey: Pearson Prentice Hall. 1–624 p.
20. Editorial C (2010) China's Agricultural Yearbook. Beijing: Chinese Agricultural Press. 1–536 (in Chinese) p.

21. Gupta RK, Rao DLN (1994) Potential of wastelands for sequestering carbon by reforestation. Current Science 66: 378–380.
22. Keith H, Mackey BG, Lindenmayer DB (2009) Re-evaluation of forest biomass carbon stocks and lessons from the world's most carbon-dense forests. Proceedings of the National Academy of Sciences of the United States of America 106: 11635–11640.
23. Falloon PD, Smith P, Smith JU, Szabo J, Coleman K, et al. (1998) Regional estimates of carbon sequestration potential: linking the T1 - Rothamsted Carbon Model to GIS databases. Biology and Fertility of Soils 27: 236–241.
24. Laclau P (2003) Biomass and carbon sequestration of ponderosa pine plantations and native cypress forests in northwest Patagonia. Forest Ecology and Management 180: 317–333.
25. Zhang Q, Justice CO (2001) Carbon Emissions and Sequestration Potential of Central African Ecosystems. Ambio 30: 351–355.
26. Engel AB, M.; Lenz, F. (2009) Nutrient translocation from the grass alleyway to the tree strip with mulching in a fruit orchard. Berlin: Springer. 151–161p.
27. West TO, Marland G (2002) A synthesis of carbon sequestration, carbon emissions, and net carbon flux in agriculture: comparing tillage practices in the United States. Agriculture, Ecosystems & Environment 91: 217–232.
28. Neto C, Carranca C, Clemente J (2009) Senescent leaf decomposition in a Mediterranean pear orchard. European Journal of Agronomy 30: 34–40.
29. Blanke MM (1996) Soil respiration in an apple orchard. Environmental and Experimental Botany 36: 339–341.
30. Page G (2011) Modeling carbon footprints of organic orchard production systems to address carbon trading: an approach based on life cycle assessment. Hortscience 46: 324–327.
31. Fang JY, Guo ZD, Piao SL, Chen AP (2007) Terrestrial vegetation carbon sinks in China, 1981–2000. Science in China Series D: Earth Sciences 50: 1341–1350.
32. Joslin JD, Gaudinski JB, Torn MS, Riley WJ, Hanson PJ (2006) Fine-root turnover patterns and their relationship to root diameter and soil depth in a 14 C-labeled hardwood forest. New Phytologist 172: 523–535.
33. Caldwell MM, Camp LB (1974) Belowground productivity of two cool desert communities. Oecologia 17: 123–130.
34. Jackson RB, Mooney HA, Schulze ED (1997) A global budget for the fine root biomass, surface area, and nutrient contents. Proceedings of the National Academy of Sciences of the United States of America 94: 7362–7366.
35. Ruess RW, Hendrick RL, Burton AJ, Pregitzer KS, Sveinbjornsson B, et al. (2003) Coupling fine root dynamics with ecosystem carbon cycling in black spruce forests of interior Alaska. Ecological Monographs 73: 643–662.
36. Guo D, Mitchell R, Hendricks J (2004) Fine root branch orders respond differentially to carbon source-sink manipulations in a longleaf pine forest. Oecologia 140: 450–457.
37. Lal R (2002) Carbon sequestration in dryland ecosystems of West Asia and North Africa. Land Degradation & Development 13: 45–59.
38. Fang JY, Chen AP, Peng CH, Zhao SQ, Ci L (2001) Changes in forest biomass carbon storage in China between 1949 and 1998. Science 292: 2320–2322.
39. Piao S, Fang J, Ciais P, Peylin P, Huang Y, et al. (2009) The carbon balance of terrestrial ecosystems in China. Nature 458: 1009–1013.
40. Boden TA, Marland G, Andres RJ (2001) Global, regional, and national fossil-fuel CO2 emissions. Oak Ridge, Tenn., U.S.A.: Carbon Dioxide Information Analysis Center, Oak Ridge National Laboratory, U.S. Department of Energy. doi 10.3334/CDIAC/00001_V02011 p.
41. Lu AF, Tian HQ, Liu ML, Liu JY, Melillo JM (2006) Spatial and temporal patterns of carbon emissions from forest fires in China from 1950 to 2000. Journal of Geophysical Research 111: 1–12.

11

Estimating Gene Flow between Refuges and Crops: A Case Study of the Biological Control of *Eriosoma lanigerum* by *Aphelinus mali* in Apple Orchards

Blas Lavandero[1]*, **Christian C. Figueroa**[3], **Pierre Franck**[2], **Angela Mendez**[1]

1 Laboratorio de Interacciones Insecto-Planta, Universidad de Talca, Talca, Chile, **2** Plantes et Systèmes de culture Horticoles, INRA, Avignon, France, **3** Facultad de Ciencias, Instituto de Ecología y Evolución, Universidad Austral de Chile, Valdivia, Chile

Abstract

Parasitoid disturbance populations in agroecosystems can be maintained through the provision of habitat refuges with host resources. However, specialized herbivores that feed on different host plants have been shown to form host-specialized races. Parasitoids may subsequently specialize on these herbivore host races and therefore prefer parasitizing insects from the refuge, avoiding foraging on the crop. Evidence is therefore required that parasitoids are able to move between the refuge and the crop and that the refuge is a source of parasitoids, without being an important source of herbivore pests. A North-South transect trough the Chilean Central Valley was sampled, including apple orchards and surrounding *Pyracantha coccinea* (M. Roem) (Rosales: Rosacea) hedges that were host of *Eriosoma lanigerum* (Hemiptera: Aphididae), a globally important aphid pest of cultivated apples. At each orchard, aphid colonies were collected and taken back to the laboratory to sample the emerging hymenopteran parasitoid *Aphelinus mali* (Hymenoptera: Aphelinidae). Aphid and parasitoid individuals were genotyped using species-specific microsatellite loci and genetic variability was assessed. By studying genetic variation, natural geographic barriers of the aphid pest became evident and some evidence for incipient host-plant specialization was found. However, this had no effect on the population-genetic features of its most important parasitoid. In conclusion, the lack of genetic differentiation among the parasitoids suggests the existence of a single large and panmictic population, which could parasite aphids on apple orchards and on *P. coccinea* hedges. The latter could thus comprise a suitable and putative refuge for parasitoids, which could be used to increase the effectiveness of biological control. Moreover, the strong geographical differentiation of the aphid suggests local reinfestations occur mainly from other apple orchards with only low reinfestation from *P. cocinea* hedges. Finally, we propose that the putative refuge could act as a source of parasitoids without being a major source of aphids.

Editor: Jeffrey A. Harvey, Netherlands Institute of Ecology, Netherlands

Funding: This study was funded by the International Foundation for Science Grant No C/4023-2 and FONDECYT (Project Number 11080013). The funders had no role in study design, data collection and analysis, decision to publish, or preparation of the manuscript.

Competing Interests: The authors have declared that no competing interests exist.

* E-mail: blavandero@utalca.cl

Introduction

Natural enemies of insect pests are constantly disturbed in agroecological systems, and classical management practices can severely reduce parasitoid populations. The use of habitat refuges, offering shelter and alternative hosts for these organisms, has been proposed for maintaining high density of parasitoids close to cultivated plants, acting as a constant source to control agricultural pests [1]. At larger scales, landscape heterogeneity has been proposed to have a positive effect on natural enemy populations and parasitism rates in general [2]. Nevertheless, one must have enough evidence that parasitoids do disperse between the refuges and the crop, and that they exert an effect on the herbivore populations.

Ecological specialization of herbivore insects could affect their relationship with the third trophic level. Specialist herbivores that feed on different host plants have been shown to form host-specialized races, evidenced through reduced migration and gene flow [3]. The effect on the next trophic level (the natural enemies) can follow the specialization of their herbivore host, resulting in

the formation of specialized parasitoid races, in a process termed sequential radiation [4]. In fact, as herbivorous insects and their parasitoids interact with their environment on a fine spatial and temporal scale, sequential radiation may be quite common [5]. Thus, parasitoids coming from a refuge may not readily forage on the crop or they may be totally isolated if gene flow between the refuge and the crop is absent, in which case the refuge would not constitute a real source of parasitoids for improving biocontrol.

Genetic markers, particularly highly polymorphic ones such as microsatellites, have been widely used to study several aspects of insect ecology. These DNA markers provide the raw data to estimate genetic diversity and gene flow between insect populations or to reconstruct migration routes and colonization history. Using appropriate bioinformatic tools to analyze DNA marker data, gene flow and genetic diversity within insect species can be quantified, which is critical for explaining population structure and dynamics in time and space (for a review see [6]). For instance, microsatellites in combination with powerful analytical tools [7] have proven to be useful for describing movement of insect pests between continents (for the western corn rootworm see [8]; for the tobacco aphid see [9]), between different production areas (for the codling moth see Fuentes–Contreras et al. [10]; for the woolly apple aphid see [11]), and between native and introduced ranges of parasitoids [12]. To our knowledge, however, there are no studies using neutral genetic variation to estimate natural enemy migration (movement and reproduction) between a putative refuge and the crop.

Here, using neutral genetic variation, we show the existence of geographical natural barriers to aphids in a main apple production area. The level of host specialization of this aphid pest is shown to have no influence on the population differentiation of its most important parasitoid wasp, due to the high gene flow observed among plant species and locations. We argue that the proposed refuge could act as a source of parasitoids without being a major source of the aphid pest.

Results

The aphids

Aphids were found in apple orchards and at four *P. coccinea* hedge sites, irrespective of pest management practices (organic vs. conventional orchards) (Table 1). A total of 581 aphid colonies were sampled and 471 different multilocus genotypes characterized (for a list of multilocus genotypes see Table S1). Twenty six genotypes were found more than once. Frequency of these multicopy genotypes was low in most sites (less than 10%), with the exception of site *Cato* where 44.8% of the colonies belonged to the same genotype. The genotypic diversity was high and similar among all sites as evidenced by the indices of Shannon, Simpson and their evenness (Table 1). Mean standardized allelic richness per site varied from 2.7 to 4.1. Heterozygosity ranged between 0.68 and 0.95 and gene diversity between 0.53 and 0.71 (see Table 1). Significant and frequent departures from Hardy-Weinberg Equilibrium were found in most of the sampled sites due to heterozygote excess. No evidence for null alleles was found (data not shown).

The genetic differentiation of populations (*Phi-pt*) between sites ranged from 2 to 23%. Analyses of Molecular Variance (AMOVA) of the aphid populations revealed different genetic structures that can be explained both by differences among the sites (22%) and differences between the host plants (2%) (p = 0.01). Pairwise comparisons between pairs of neighbouring *Pyracantha* hedges and their corresponding apple orchards showed a significantly high differentiation, ranging between 12.3% for *Colin* (site 9 and C, Figure 1) and 39% for *Cañadilla* (site 3 and A, Figure 1). Further analyses using TESS suggested that the aphid colonies were grouped into seven geographically related clusters, where sites close to each other shared more ancestry than those further apart (represented in Figure 2 and 3 (top) by different colours). The Bayesian clustering method showed different genetic clusters between neighbouring collection sites including samples from different host plants. This was confirmed after analyzing a smaller comparable scale (*P. coccinea* sites A, B, C and D; Apple sites 3, 8 and 9, in Figure 1), revealing a high differentiation between host-plants (5%; *p* = 0.01), although the greater differences among populations were independent of the host (21%; *p* = 0.01). Further analyses using TESS confirmed the AMOVA results by showing almost no admixis between host plants or sites (Figure 4). Analyses using shared allelic distance between individuals at the site *Cañadilla* suggest that aphids from the same host plant are more closely related (Figure 5).

The parasitoids

A total of 1018 parasitoid specimens were obtained (one to three parasitoids emerged from each aphid colony sampled) and 902

individuals were successfully genotyped and considered for analyses. Mean standardized allelic richness per site varied from 3.1 to 4.0. Allelic richness of the parasitoid was independent of the geographical distance between sites (Partial Mantel test; r = −0.1, *p* = 0.46). The proportion of heterozygotes ranged between 0.26 and 0.50, while gene diversity ranged between 0.39 and 0.54 (see Table 2). Slight heterozygote deficiencies were detected in most sites, probably due to null alleles (frequency of null alleles was under 19% for all loci). AMOVA evidenced significant but very low variation between sites (1%) and within host plants (1%), suggesting great gene flow between sites and host plants at the landscape level (see further details of pairwise *Fst* in Table 3). Further analyses using the Bayesian structuring algorithm implemented in TESS and considering all individuals independent of their collection sites, suggested no host or geographically-associated differentiation for the parasitoids (see Figure 3).

Kinship analysis also detected numerous full-sib pairs between parasitoids collected from different aphid colonies sampled from either the same or different trees. Furthermore, parasitoid females emerging from the same aphid colony were usually not full-sibs (Table 4). Parasitism levels ranged from 67.3% to 100%, with no significant differences between organic or conventional orchards (*p* = 0.897). In contrast, parasitism levels were significantly higher on aphids collected from *P. coccinea* than those collected from apples (see Table 2).

Aphid-parasitoid complex

Mean standardized allelic richness for the parasitoids per site were inversely correlated with the parasitism rates per orchard (Spearman r = −0.5, p = 0.038). Parasitism rates were independent of geographical distance when controlling for allelic richness (r = −0.11, p = 0.14). When estimating parasitism rates for the *Malus* sites per genetic cluster according to TESS (Mean ± SE: *Blue* 81.5±4.2; *Dark Yellow* 100±0; *Green* 91.8±3.03; *Pink* 87.7±4.02; *Red* 97.4±1.67 and *Yellow* 81.6±8.59), clusters *Blue* and *Yellow* (Figure 2) had significantly lower parasitism rates (Z values and correspondent p-values for paired comparisons with the *Blue* cluster for the *Dark Yellow* z = 6.266 *Green* z = 5.239 *Pink* z = 2.909; *Red* z = 6.303 and *Yellow* z = 0.001; *p* = 3.70e-10; *p* = 1.61e-07; *p* = 0.00363; *p* = 2.92e-10 and *p* = 0.99951).

Analyses using shared allelic distance between individuals at the site level for the populations from *Cañadilla* suggested that aphids from the same host plant were more closely related; however, the comparable tree for the parasitoids (constructed with individuals emerged from those same aphids), showed no significant grouping of parasitoids per tree or host plant (Figure 5).

Discussion

The very low genetic differentiation among *A. mali* populations suggests that individuals do disperse between sites and host plants, although there is still no clear evidence that this can exert a difference in the herbivore abundances on the crop. The partitioning of molecular variance of the parasitoids revealed very low levels of variation between sites (i.e. orchards), especially considering that parasitoids reproduce sexually. Since no host or geographically-associated structuring was evident for the parasitoid, the natural barriers affecting aphids [11] seem not to be affecting the parasitoids. Moreover, the kinship analysis of parasitoids suggests that oviposition does not occur in a patchy or aggregated fashion. Thus, female parasitoids would lay eggs far away from each other, reducing the endogamy between points by increasing gene flow, at least at the orchard level, thus supporting the idea of a higher dispersal and gene flow between sites. Bayesian

Table 1. Site Number, Location, Host plant, Management conditions (O = Organic, C = Conventional), sample size, Number of Genotypes, Unique vs. Multicopy genotypes (U/M), Shannon diversity (H) and its evenness (VH), Simpson diversity (D) and its evenness (ED), Gene Diversity (1-Q), Inbreeding coefficient (*Fis*) and significance (p-value), Loci under disequilibrium and allelic richness (A) of *Eriosoma lanigerum* females per site.

Site N°	Location	Host plant	Manag.	n	Genotypes	U/M	H	VH	D	ED	(1-Q)	*Fis*	p-value	LD	A
1	Villa Alemana	*Malus*	O	30	28	26/2	3,309	0,993	0,995	0,519	0,867	−0,349	>0,01	2/22	3,6
2	Graneros	*Malus*	C	19	13	11/2	2,347	0,915	0,924	0,481	0,895	−0,642	>0,01	3/22	3
3	Cañadilla	*Malus*	O	13	13	13/0	2,565	1,000	1,000	−1,000	0,890	−0,492	>0,01	1/22	3,1
4	San Fernando	*Malus*	O	29	29	29/0	3,367	0,999	1,000	−1,000	0,823	−0,267	>0,01	4/22	3,4
5	Los Niches	*Malus*	O	51	50	49/1	3,905	0,998	0,999	0,000	0,790	−0,189	>0,01	7/22	3,7
6	Panguilemo	*Malus*	C	28	24	23/1	3,045	0,958	0,974	0,000	0,888	−0,460	>0,01	2/22	3,3
7	Maiten Huapi	*Malus*	O	58	55	53/2	3,980	0,993	0,998	0,453	0,867	−0,316	>0,01	5/22	3,7
8	Las Rastras	*Malus*	C	30	27	25/2	3,245	0,985	0,991	0,462	0,895	−0,351	>0,01	3/22	3,7
9	Colin	*Malus*	C	30	27	25/2	3,245	0,985	0,991	0,462	0,805	−0,256	>0,01	4/22	3,4
10	Las Lomas	*Malus*	C	27	27	27/0	3,296	1,000	1,000	−1,000	0,783	−0,185	>0,01	1/22	3,8
11	Pataguas	*Malus*	C	30	18	13/5	2,691	0,931	0,949	0,824	0,867	−0,554	>0,01	2/22	2,9
12	Miraflores	*Malus*	C	30	28	27/1	3,291	0,988	0,993	0,000	0,810	−0,219	>0,01	8/22	3,8
13	Ancoa	*Malus*	C	37	36	35/1	3,573	0,997	0,998	0,000	0,865	−0,329	>0,01	2/22	3,7
14	Huaquivilo	*Malus*	O	36	35	34/1	3,545	0,997	0,998	0,000	0,679	−0,070	NS	5/22	3,8
15	Mirarios	*Malus*	O	26	26	26/0	3,258	1,000	1,000	−1,000	0,769	−0,113	NS	10/22	4,1
16	Cato	*Malus*	C	29	12	9/3	1,913	0,770	0,786	0,498	0,828	−0,592	>0,01	13/22	2,8
17	Mulchén	*Malus*	O	28	26	25/1	3,214	0,987	0,992	0,000	0,745	−0,245	>0,01	3/22	3,4
			SUBTOTAL	**531**	**474**	**451/26**	**6,072**	**0,984**	**0,999**	**0,930**	**0,830**		**>0,001**	**19/22**	
A	Cañadilla	*Pyracantha*	n/a	12	7	5/2	1,748	0,898	0,864	0,560	0,917	−0,682	>0,001	0/22	2,7
B	Las Rastras	*Pyracantha*	n/a	5	5	5/0	1,609	1,000	1,000	−1,000	0,971	−0,432	>0,001	0/22	3,9
C	Colin	*Pyracantha*	n/a	19	19	19/0	2,944	1,000	1,000	−1,000	0,895	−0,397	>0,001	1/22	3,5
D	Manzano	*Pyracantha*	n/a	8	6	5/1	1,667	0,931	0,893	0,000	0,929	−0,526	>0,001	3/22	3,3
			SUBTOTAL	**44**	**37**	**34/3**	**3,508**	**0,972**	**0,987**	**0,671**	**0,928**		**>0,001**	**1/22**	
			Whole sample	**575**	**511**	**485/29**	**6,146**	**0,985**	**0,999**	**0,941**	**0,879**				

grouping algorithms revealed no geographic or host-driven structuring for the parasitoid, although the aphid host showed seven geographically related groups, where sites close to each other shared more ancestry than those further apart.

As reported before, aphids show low levels of gene flow at the landscape scale, with significant barriers between geographical areas [11]. The high levels of Heterozygosity, and few linked loci, suggest the occurrence of sexual reproduction in *E. lanigerum* in Chile, although this aphid species has not been found on its primary host where sexual reproduction is reported to occur (*Ulmus americana*) [13]. As suggested by Sandanayaka and Bus [14], sexual reproduction could indeed occur on apple, but further studies are necessary to determine the environmental conditions needed to trigger sexual reproduction, and to screen for the presence of sexual morphs in Chile. Interestingly, environmental conditions such as short days and below-zero temperatures (the factors that trigger sexuality in many aphid species [13,15]), could affect parasitism rates through an increased genetic diversity in the aphid host. In any case, this seems not to be enough to affect the parasitoids genetic structure.

The genetic diversity of the woolly apple aphid is clearly geographically structured; however, some of the genetic variation can be also be explained by the different host plants used by the aphids. Analyses comprising only those sites where neighbouring *Pyracantha* hedges are found, suggest a higher differentiation between host plants. Interestingly, the genetic clusters at each *Malus* site were different compared to their corresponding *Pyracantha* hedge. Evidence obtained from TESS, AMOVA and the neighbour-joining tree analyses, clearly separate individuals coming from different host plants. When the survival and preference of females were compared in reciprocal-transference experiments, *E. lanigerum* from *M. domestica* showed a stronger preference for its own natal branch as compared with other *M. domestica* or *P. coccinea* trees (Lavandero, unpublished data). In contrast, aphids born on *P. coccinea* had no significant preference for its natal host, showing a lower rejection for the *M. domestica* host. This could be the case for *E. lanigerum* aphids coming from *Malus*, which are not able to disperse into neighbouring *P. coccinea* hedges, although some individuals from *P. coccinea* may successfully colonize apple trees. This suggests that although *P. coccinea* could potentially become a source of some recolonizing aphids, it should not act as a significant source, as there seems to be a restricted and biased migration between both host plants. Hence, our results are indicative of no sequential radiation in this aphid-parasitoid system; however, aphids still exhibit geographical and some host-driven genetic structure.

Parasitism rates varied greatly among the studied sites; however, the management of the orchards (organic or conventional) did not explain these differences as expected. The literature suggests that the main explanation for parasitism decrease and aphid popula-

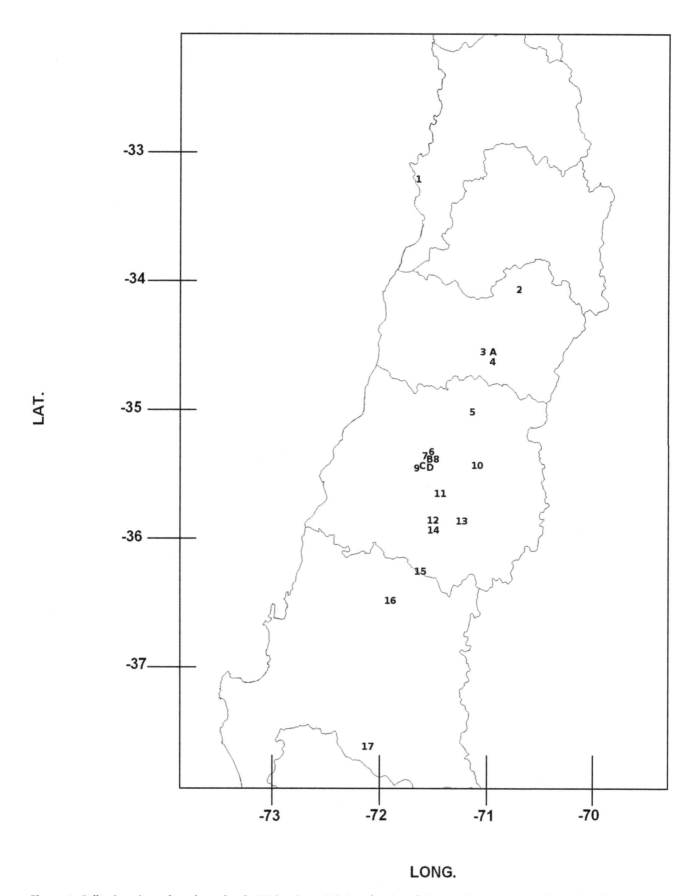

Figure 1. Collection sites of apple orchards (*Malus domestica*) (numbers) and *Pyracantha coccinea* sampling sites (letters). 1 Villa Alemana, 2 Graneros, 3 Cañadilla, 4 San Fernando, 5 Los Niches, 6 Panguilemo, 7 Maiten Huapi, 8 Las Rastras, 9 Colin, 10 Las Lomas, 11 Pataguas 12 Miraflores, 13 Ancoa, 14 Huaquivilo, 15 Miraríos, 16 Cato, 17 Mulchén, A Cañadilla, B Las Rastras, C Colin, D Manzano.

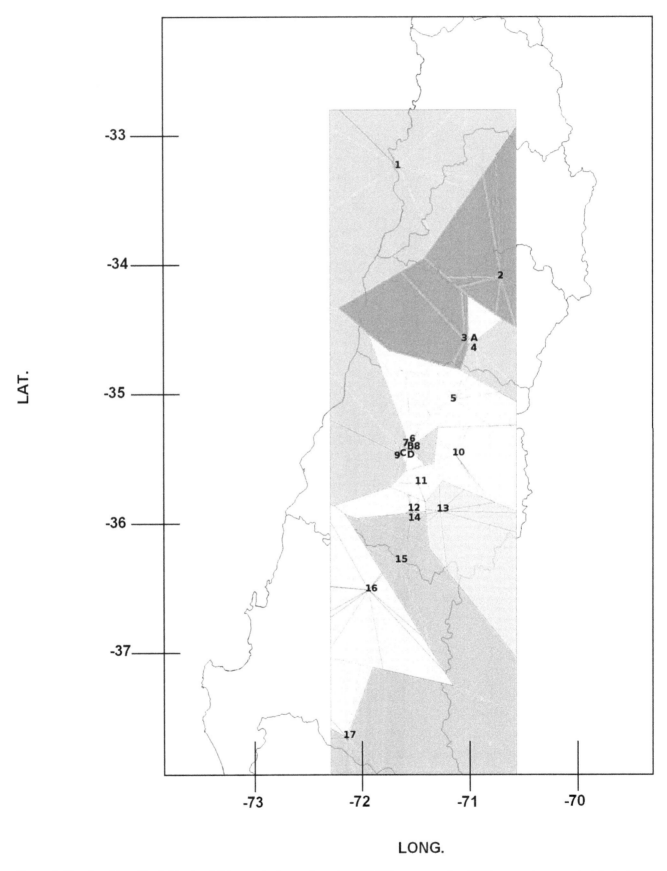

Figure 2. Membership of individuals of *Eriosoma lanigerum* **based on 50,000 sweeps using TESS, assuming no admixis, between sites.** Tessellation is ordered from North to South.

Figure 3. Average assignment probability of individuals of *Eriosoma lanigerum* **(aphid host), independent of sampling origin.** Assignment is based on 100 repetitions of 50,000 sweeps using TESS showing K = 7 genetic clusters and the correspondent structure for its parasitoid *Aphelinus mali*. Individuals (bars) are from North to South.

tion outbreaks are due to the susceptibility of the parasitoids to pesticides (organophosphates and pyrethroids), sulphur and kaoline [16–18]. In both management systems, however, management practices alone cannot account for the differences found (67.3% to 100% rates of parasitism). Indeed, parasitism rates were not related with geographical distance between sites, even considering allelic richness, which could be used as an estimator of effective population sizes [19]. In our study, the allelic richness of the parasitoids was negatively correlated with the parasitism rates per site, which suggest inverse density dependence, meaning that parasitoids are effectively controlling the aphid populations up to a threshold where the rate of increase of aphid populations is greater than the parasitoid ability to exert control. The thermal biology of these organisms could explain this pattern, as the parasitoid has a greater thermal developmental threshold than its aphid host, translating into a lower growth rate (GR) compared to its host (GR = 0.1 parasitoid, 0.14–0.27 for the aphid at 20°C) [20]. On the other hand, aphid populations showed different genetic structures, some genetic clusters showing more susceptibility to *A. mali* parasitism than others, with no significant effect of management practices (i.e. genetic cluster grouped aphids coming from both conventional and organic orchards). Other factors such as land use and nectar availability for parasitoids, among others, need to be further analyzed, as well as the possible interaction between aphid and defense endosymbiont bacteria as found for other aphid species [21].

In conclusion, the lack of genetic differentiation of the parasitoids suggest the existence of a single large and panmictic population, which could parasitise aphids on apple orchards and on *P. coccinea* hedges, the latter being a suitable and putative refuge for parasitoids to increase their effectiveness in biological control. Moreover, the strong geographical differentiation of the aphid suggests that local reinfestations occur mainly from other apple orchards, with little reinfestation occurring from *P. coccinea* hedges. Further mark-recapture studies should be conducted to quantify dispersal, frequency and intensity of aphid infestations in apple orchards coming from both host plants. Quantification of the actual effect of this putative refuge on the population dynamics of the pest across several seasons will be critical if any effort for improving biocontrol is attempted using *P. coccinea*. Overall, we have shown that neutral genetic variation is a useful tool for

addressing population dynamics between host plant species of pests and their parasitoids, determining potential refuges for natural enemies.

Materials and Methods

Study system

Aphids are important pests and disease vectors for a variety of crops, and parasitoids are often introduced for aphid biological control. The woolly apple aphid (*Eriosoma lanigerum* (Haussman)) (Hemiptera: Aphididae) native to North America, is a globally-important pest of apple orchards (*Malus domestica* Borkh). This aphid forms colonies on roots, trunks, branches and shoots, with greatest damage occurring at the shoot level [22]. Other associated damage is cosmetic, as fruits become covered with honeydew leading to subsequent fungus colonization, which reduces their commercial value. Although *M. domestica* is its most common host, this aphid also attacks other Rosacea species, notably *Pyracantha coccinea* (M. Roem) (Rosales: Rosacea), which is a very common plant distributed along farm hedges.

The wooly apple aphid (*E. lanigerum*) was first introduced into Chile during the 19th century, most probably as root colonies from plant material. As the damage to apple orchards in Chile reached dramatic levels, in 1920 the chalcidoid parasitoid *Aphelinus mali* (Hymenoptera: Aphelinidae) was introduced. Although this parasitoid is the main species controlling *E. lanigerum* in Chile, it has been determined that under the current management conditions (conventional agriculture), aphid population outbreaks still occur [23]. There are several reasons for aphid population outbreaks, the most important probably being organophosphates, pyrethroids, sulfur and even kaolin treatments that affect its main parasitoid, *A. mali* [16–18]. In order to improve the effectiveness of the parasitoid, the use of host-plant refuges such as *Pyracantha coccinea* is proposed to attract and maintain parasitoid populations. Indeed, *E. lanigerum* is frequently observed at high densities on *P. coccinea*, with high parasitism rates by *A. mali*. This proposed refuge could be a source of parasitoids when the pest is not present in the orchard or as protection after pesticide use. However, evidence is required that the parasitoids are able to move between the refuge (*P. coccinea*) and the crop (apple), thereby determining its suitability as a source or sink for both aphids and parasitoids.

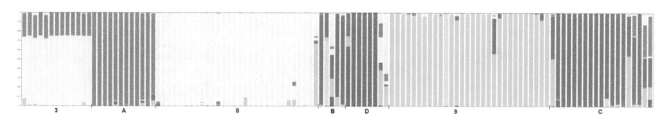

Figure 4. Average assignment probability of a subsample of individuals of *Eriosoma lanigerum* **(aphid host), independent of sampling origin, based on 100 repetitions of 50,000 sweeps using TESS showing K = 7 genetic clusters, ordered showing neighbouring sites between both host plants (***Malus domestica* **and** *Pyracantha coccinea***).** 3 = Cañadilla-*Malus*, A = Cañadilla-*Pyracantha*, 8 = Las Rastras-*Malus*, B = Las Rastras-*Pyracantha*, D = Los Manzanos–*Pyracantha*, 9 = Colin-*Malus*, C = Colin-*Pyracantha*.

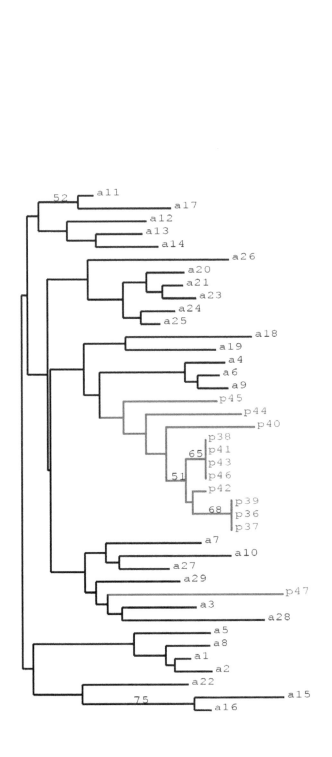

Eriosoma

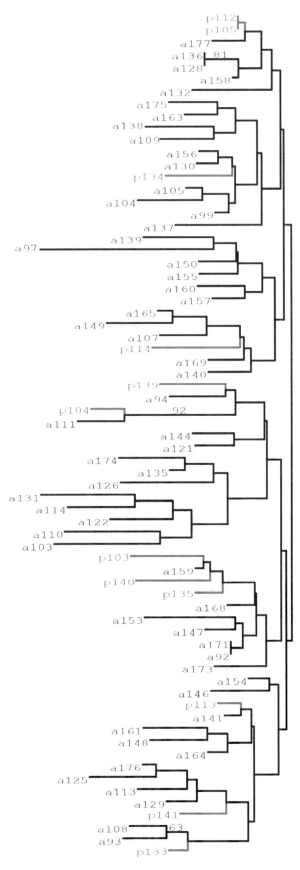

Aphelinus

Figure 5. A neighbor-joining tree constructed using the shared allele distance between individuals of a single site (*Cañadilla*, site 3 Figure 1) among individuals collected on different host plants *Pyracantha* (in red) and *Malus* (black). Bootstrap values were computed over 2000 replications resampling the microsatellite loci. On the left side the tree for the aphid *Eriosoma lanigerum* and on the right the tree for the emerged parasitoids (*Aphelinus mali*).

A 700 Km. North-South transect was sampled, including 17 apple orchards and surrounding *Pyracantha coccinea* hedges at four of the 17 chosen sites (33.19 S 71.733 W to 37.721 S 72.244 W). Orchards were all over 30 ha in size, planted with the Granny Smith apple cultivar. Permissions for entering and taking samples at conventional orchards were issued as part of an ongoing center within Universidad de Talca, Centro de Pomaceas (Stone fruit center), which gives the university the faculty of sampling in their farms (more details at http://pomaceas.utalca.cl/html/index. html). All orchards sampled are members of this center. Permission for entering and using materials of organic orchard were issued as part of an ongoing agreement between Comercial Greenvic Ltda and Universidad de Talca, through their branch Huertos Organicos de Chile S:A: (more details at http://www. huertosorganicosdechile.cl/). All organic orchards sampled are members of this industry-university research agreement. At each orchard, up to 40 colonies of *E. lanigerum* were collected on different apple trees, while all available colonies on the *P. coccinea* hedges were sampled. Each aphid colony was georeferenced and taken back to the laboratory to determine parasitism rates under controlled conditions (20±1°C, 65±10% RH y 16:8 hrs. day/

night cycle). Parasitism rates per orchard were assessed for 10 trees (one colony per tree) per orchard. Colonies taken from the field were individually caged and reared under controlled conditions for two weeks. The number of aphids per colony and emerged parasitoids were registered from each cage. A single wingless adult aphid female per colony was preserved in 95% alcohol for subsequent DNA extraction. Parasitism rates were assessed by rearing aphids on 9 cm long shoots placed on a damp tissue paper inside plastic boxes with top ventilation. At emergence, parasitoids were identified to the species level, and up to three *A. mali* females per colony were preserved in 95% alcohol for DNA extraction. Genomic DNA was obtained following the 'salting out' protocol from [24]. Aphid and parasitoid individuals were genotyped using seven (aphids) and six (parasitoids) microsatellite (SSR) markers described in [25] and [26], respectively. The reverse primer for each pair of primers was fluorescently labeled, and PCR products analyzed on a MegaBASE 1000 automatic DNA Sequencer.

The microsatellite data were checked for null alleles and technical artifacts like stuttering bands and large allele dropout using the MICRO CHEKER v.2.2.3 software [27]. Deviations from the Hardy-Weinberg equilibrium (HWE) and linkage

Table 2. Management conditions, Host plant, Parasitism rates, allelic richness, Observed Heterozygocity (Ho) and Gene Diversity of *Aphelinus mali* females per site.

Site N°	Location	Management	Host plant	Parasitism rates mean	SD	Samples N	Allelic Richness Mean	SD	Ho Mean	SD	Gene diversity Mean	SD
1	Villa Alemana	O	*Malus*	70%	48%	16	4,1	0,9	0,28	0,07	0,52	0,11
2	Graneros	C	*Malus*	95%	31%	15	3,4	1,2	0,31	0,13	0,42	0,21
3	Cañadilla	O	*Malus*	79%	47%	27	3,4	0,8	0,33	0,14	0,52	0,21
4	San Fernando	O	*Malus*	73%	26%	61	3,8	1,0	0,35	0,15	0,49	0,16
5	Los Niches	O	*Malus*	82%	42%	60	3,7	1,1	0,38	0,15	0,53	0,23
6	Panguilemo	C	*Malus*	96%	9%	70	3,7	1,1	0,39	0,14	0,49	0,19
7	Maiten Huapi	O	*Malus*	95%	9%	66	3,7	1,2	0,40	0,16	0,53	0,19
8	Las Rastras	C	*Malus*	88%	18%	21	3,1	1,1	0,38	0,19	0,39	0,15
9	Colin	C	*Malus*	94%	14%	48	4,0	1,0	0,38	0,18	0,49	0,19
10	Las Lomas	C	*Malus*	91%	31%	65	4,0	1,0	0,37	0,15	0,51	0,20
11	Fundo Pataguas	C	*Malus*	96%	13%	57	3,7	1,1	0,36	0,13	0,50	0,16
12	Miraflores	C	*Malus*	67%	50%	29	3,9	1,4	0,34	0,11	0,52	0,17
13	Ancoa	C	*Malus*	100%	/	72	3,6	1,2	0,31	0,11	0,46	0,15
14	Huaquivilo	O	*Malus*	86%	40%	49	3,7	1,2	0,37	0,12	0,48	0,18
15	Miraríos	O	*Malus*	84%	32%	24	4,1	0,8	0,50	0,18	0,51	0,17
16	Cato	C	*Malus*	94%	15%	56	4,0	1,3	0,42	0,20	0,51	0,22
17	Mulchén	O	*Malus*	93%	23%	73	3,9	1,3	0,43	0,17	0,54	0,16
A	Cañadilla	-	*Pyracantha*	100%	/	12	4,0	1,0	0,26	0,09	0,48	0,12
B	Las Rastras	-	*Pyracantha*	100%	/	18	3,5	1,3	0,33	0,12	0,54	0,18
C	Colin	-	*Pyracantha*	100%	/	29	3,7	1,2	0,39	0,16	0,45	0,22
D	Manzanos	-	*Pyracantha*	100%	/	22	3,8	0,9	0,33	0,15	0,46	0,20
			Pyracantha									
	Mean			90%			3,7	1,1	0,36	0,14	0,49	0,18

Table 3. Pairwise Population *Fst* Values, *Aphelinus mali*.

	Pop1	Pop2	Pop3	Pop4	Pop5	Pop6	Pop7	Pop8	Pop9	Pop10	Pop11	Pop12	Pop13	Pop14	Pop15	Pop16	Pop17	Pop18	Pop19	Pop20	Pop21
Pop1	0.000	0.460	0.110	0.010	0.010	0.430	0.270	0.450	0.300	0.140	0.050	0.290	0.030	0.430	0.290	0.290	0.170	0.070	0.510	0.260	0.380
Pop2	0.005	0.000	0.270	0.010	0.160	0.210	0.030	0.010	0.030	0.050	0.030	0.030	0.020	0.300	0.040	0.020	0.390	0.010	0.280	0.150	0.390
Pop3	0.034	0.002	0.000	0.010	0.160	0.010	0.020	0.010	0.140	0.040	0.050	0.060	0.010	0.060	0.120	0.110	0.460	0.010	0.060	0.010	0.370
Pop4	0.022	0.034	0.031	0.000	0.020	0.010	0.020	0.010	0.010	0.040	0.040	0.010	0.020	0.020	0.010	0.030	0.010	0.010	0.020	0.010	0.010
Pop5	0.000	0.008	0.006	0.042	0.000	0.010	0.010	0.010	0.030	0.010	0.010	0.030	0.010	0.010	0.010	0.010	0.430	0.010	0.170	0.010	0.160
Pop6	0.004	0.002	0.014	0.043	0.038	0.000	0.310	0.420	0.030	0.040	0.200	0.180	0.520	0.390	0.180	0.020	0.040	0.100	0.200	0.510	0.060
Pop7	0.000	0.014	0.009	0.029	0.047	0.000	0.000	0.170	0.010	0.140	0.470	0.470	0.010	0.260	0.450	0.040	0.010	0.030	0.020	0.020	0.040
Pop8	0.002	0.012	0.017	0.035	0.043	0.000	0.000	0.000	0.010	0.040	0.060	0.080	0.010	0.190	0.410	0.020	0.010	0.010	0.160	0.090	0.040
Pop9	0.006	0.006	0.003	0.054	0.022	0.002	0.000	0.000	0.000	0.010	0.010	0.030	0.010	0.020	0.310	0.040	0.490	0.010	0.020	0.010	0.450
Pop10	0.010	0.007	0.007	0.022	0.024	0.004	0.000	0.005	0.024	0.000	0.120	0.040	0.090	0.310	0.010	0.210	0.010	0.010	0.110	0.050	0.070
Pop11	0.002	0.015	0.015	0.027	0.060	0.004	0.000	0.003	0.019	0.008	0.000	0.170	0.470	0.050	0.500	0.070	0.020	0.010	0.020	0.050	0.060
Pop12	0.010	0.010	0.009	0.044	0.027	0.002	0.000	0.005	0.009	0.011	0.007	0.000	0.430	0.160	0.070	0.010	0.050	0.030	0.080	0.010	0.100
Pop13	0.000	0.016	0.015	0.034	0.036	0.010	0.000	0.018	0.021	0.005	0.000	0.000	0.000	0.040	0.030	0.020	0.010	0.020	0.030	0.020	0.020
Pop14	0.000	0.001	0.007	0.038	0.026	0.010	0.003	0.003	0.010	0.001	0.012	0.004	0.008	0.000	0.200	0.380	0.040	0.440	0.240	0.230	0.160
Pop15	0.001	0.012	0.006	0.025	0.045	0.005	0.000	0.000	0.003	0.012	0.000	0.009	0.012	0.004	0.000	0.320	0.020	0.020	0.030	0.020	0.240
Pop16	0.008	0.011	0.005	0.034	0.026	0.011	0.009	0.012	0.009	0.003	0.012	0.012	0.010	0.001	0.002	0.000	0.060	0.010	0.050	0.030	0.080
Pop17	0.012	0.000	0.000	0.065	0.001	0.019	0.031	0.029	0.000	0.035	0.042	0.021	0.032	0.019	0.025	0.017	0.000	0.010	0.260	0.010	0.400
Pop18	0.000	0.025	0.029	0.075	0.062	0.015	0.010	0.018	0.028	0.025	0.031	0.012	0.018	0.000	0.025	0.020	0.047	0.000	0.080	0.050	0.030
Pop19	0.005	0.002	0.010	0.050	0.012	0.007	0.026	0.008	0.015	0.015	0.036	0.009	0.022	0.004	0.026	0.012	0.006	0.020	0.000	0.150	0.030
Pop20	0.000	0.007	0.033	0.050	0.062	0.000	0.018	0.009	0.032	0.019	0.022	0.023	0.020	0.003	0.026	0.021	0.051	0.017	0.010	0.000	0.060
Pop21	0.000	0.000	0.002	0.058	0.010	0.015	0.034	0.026	0.000	0.027	0.023	0.019	0.025	0.010	0.009	0.014	0.000	0.035	0.002	0.024	0.000

Fst Values below diagonal.
Probability values based on 9999 permutations are shown above diagonal.

Table 4. Percentage of parasitoid full sibs from the same aphid colony, from aphid colonies collected on the same host plant (for *Pyracantha* and *Malus*) and on different host plants at the Colin and Cañadilla sites.

	% full sibs same Colony	% full sibs *Pyracantha*	% full sibs *Malus*	% full sibs other Host
Colin	7.25	18.84	33.33	47.83
Cañadilla	18.32	2.29	68.70	29.01

disequilibrium (LD) were tested using GENEPOP v.3.2a software [28]. To analyze genotypic data and test for clonality in the aphid populations, the number of genotypes, the rate of unique vs./ multicopy genotypes, Shannon diversity and its evenness, Simpson diversity and its evenness, gene diversity, inbreeding coefficient (Fis) and significance (p-value), loci under disequilibrium and allelic richness per site were estimated using the GenClone 2.0 software [29]. Observed heterozygocity, gene diversity and allelic richness of *A. mali* per site were estimated using HP-RARE 1.0 [30]. Population structure of both species (parasitoids and aphids) was examined first using a hierarchical analysis of molecular variance (AMOVA) assuming asexuality for the aphids (*Phi-pt*; significant deviations from HWE) and sexuality for the parasitoids (*Fst*) as implemented in Genalex v 6.41 [31], with two levels (host plants and total effect). In addition, the population-genetic structure was assessed for aphids and parasitoids using the aggregation Bayesian algorithm implemented in TESS 2.3 [32]. The admixture model was compared with a non-admixture model as suggested by [33], because admixture models are robust to an absence of admixture in the sample, but non-admixture models are robust when admixture is present between some individuals. The TESS algorithm was run with 10,000 sweeps, discarding the first 5,000 with 20 independent iterations for each model for maximum clusters (Kmax) varying from 2 to 12 for both aphids and parasitoids. The highest likelihood runs were selected based on the Deviance Information Criterion (DIC) and graphed against Kmax (as suggested by [32]), allowing selection of the number of hypothetical clusters (K). Then the program was run 100 times for the selected Kmax with 50,000 sweeps discarding the first 10,000. The 10 highest likelihood runs were then averaged. Population genetic structure was assessed again on a subsample consisting of sampling sites with neighboring *P. coccinea* (sites 3, 8 and 9 for *Malus* and A, B, C and D for *P. coccinea* in Figure 1) using the aggregation Bayesian algorithm implemented in TESS, as described before. At the *Cañadilla* site (site 3 on Figure 1) a neighbor-joining tree [34] was constructed using the shared allele distance [35] between individuals, in order to visualize the genetic similarity among individuals collected on different host plants (*P. coccinea* and apple), as site 3 was the only site where aphids were found on hedges of *P. coccinea* inside an apple orchard. Bootstrap values were computed over 2000 resamplings of the microsatellite loci. In order to assess the ability of a parasitoid female to lay eggs grouped or dispersed among the aphid colonies, parasitoids that emerged from the same aphid colony were tested for being

daughters from a single or many females. This was done using a kinship analysis on parasitoids that emerged from aphids sampled at sites where neighboring *P. coccinea* hedges are found (sites *Cañadilla* and *Colin*, 3 and 9 in Figure 1, respectively). Analyses were carried out using the full likelihood method [36,37], as implemented in the software COLONY v 2.0, with data from six SSR loci.

In order to test the hypothesis that parasitoids respond to aphid population structure independently from geographical or sampling effects, a series of partial and simple Mantel tests were carried out. The significance of these correlations were assessed using *zt* version 1.0 [38], with 10.000 permutations [39]. The tested variables were parasitoid allelic richness as an estimate of population sizes, geographical distance between sites, sample size between sites, and parasitism rates per site. Spearman correlation was also carried out between parasitism rates per site and allelic richness of the parasitoids, using R version 2.10.1. Once the number of genetic clusters was estimated for the aphids, parasitism rates per cluster were estimated to asses the influence of the aphid's genetic background on the efficiency of the parasitoid. A generalized linear model (GLM) assuming a Poisson distribution was carried out [40] with the *glm* function in the base package of R version 2.10.1 written by Simon Davies. Mean values per cluster were then compared to the lowest mean value in a series of paired comparisons, and significances were estimated.

Acknowledgments

The authors would like to thank Jason Tylianakis for useful comments and help with English and Marcos Dominguez for assistance during laboratory and field work. P.F. would like to thank Jérôme Olivares for his help in the laboratory.

Author Contributions

Conceived and designed the experiments: BL. Performed the experiments: BL PF. Analyzed the data: BL PF CCF AM. Contributed reagents/ materials/analysis tools: BL PF. Wrote the paper: BL PF CCF.

References

1. Thies C, Tscharntke T (1999) Landscape structure and biological control in agroecosystems. Science 285: 893–895.

2. Bianchi F, Booij CJH, Tscharntke T (2006) Sustainable pest regulation in agricultural landscapes: a review on landscape composition, biodiversity and natural pest control. Proceedings of the Royal Society B-Biological Sciences 273: 1715–1727.

3. Forbes AA, Powell THQ, Stelinski LL, Smith JJ, Feder JL (2009) Sequential Sympatric Speciation Across Trophic Levels. Science 323: 776–779.

4. Abrahamson W, Blair C (2008) Sequential radiation through host-race formation: herbivore diversity leads to diversity in natural enemies. In: Tilmon K, ed.

Specialization, Speciation, and Radiation: The Evolutionary Biology of Herbivorous Insects. BerkeleyCalifornia: University of California Press. pp 188–202.

5. Feder JL, Forbes AA (2010) Sequential speciation and the diversity of parasitic insects. Ecological Entomology 35: 67–76.

6. Behura SK (2006) Molecular marker systems in insects: current trends and future avenues. Molecular Ecology 15: 3087–3113.

7. Guillemaud T, Beaumont MA, Ciosi M, Cornuet JM, Estoup A (2010) Inferring introduction routes of invasive species using approximate Bayesian computation on microsatellite data. Heredity 104: 88–99.

8. Ciosi M, Miller NJ, Kim KS, Giordano R, Estoup A, et al. (2008) Invasion of Europe by the western corn rootworm, Diabrotica virgifera virgifera: multiple transatlantic introductions with various reductions of genetic diversity. Molecular Ecology 17: 3614–3627.

9. Zepeda-Paulo FA, Simon JC, Ramirez CC, Fuentes-Contreras E, Margaritopoulos JT, et al. (2010) The invasion route for an insect pest species: the tobacco aphid in the New World. Molecular Ecology 19: 4738–4752.

10. Fuentes-Contreras E, Espinoza JL, Lavandero B, Ramirez CC (2008) Population genetic structure of codling moth (Lepidoptera : Tortricidae) from apple orchards in central Chile. Journal of Economic Entomology 101: 190–198.

11. Lavandero B, Miranda M, Ramirez CC, Fuentes-Contreras E (2009) Landscape composition modulates population genetic structure of Eriosoma lanigerum (Hausmann) on Malus domestica Borkh in central Chile. Bulletin of Entomological Research 99: 97–105.

12. Hufbauer RA, Bogdanowicz SM, Harrison RG (2004) The population genetics of a biological control introduction: mitochondrial DNA and microsatellie variation in native and introduced populations ofAphidus ervi, a parisitoid wasp. Molecular Ecology 13: 337–348.

13. Blackman RL, Eastop VF (2000) Aphids on the world's crops: an identification and information guide. Aphids on the world's crops: an identification and information guide. pp x+466.

14. Sandanayaka WRM, Bus VGM (2005) Evidence of sexual reproduction of woolly apple aphid, Eriosoma lanigerum, in New Zealand. Journal of Insect Science 5.

15. James BD, Luff ML (1982) Cold-Hardiness and Development of Eggs of Rhopalosiphum-Insertum. Ecological Entomology 7: 277–282.

16. Cohen H, Horowitz AR, Nestel D, Rosen D (1996) Susceptibility of the woolly apple aphid parasitoid, Aphelinus mali (Hym: Aphelinidae), to common pesticides used in apple orchards in Israel. Entomophaga 41: 225–233.

17. Penman DR, Chapman RB (1980) Woolly Apple Aphid Homoptera, Aphididae Outbreak Following Use of Fenvalerate in Apples in Canterbury, New-Zealand. Journal of Economic Entomology 73: 49–51.

18. Marko V, Blommers LHM, Bogya S, Helsen H (2008) Kaolin particle films suppress many apple pests, disrupt natural enemies and promote woolly apple aphid. Journal of Applied Entomology 132: 26–35.

19. Wang J (2005) Estimation of effective population sizes from data on genetic markers. Philosophical Transactions of the Royal Society of London, Biological Sciences Series B 360: 1395–1409.

20. Asante SK (1997) Natural enemies of the woolly apple aphid, Eriosoma lanigerum (Hausmann) (Hemiptera: Aphididae): A review of the world literature. Plant Protection Quarterly 12: 166–172.

21. Castaneda L, Sandrock C, Vorburger C (2010) Variation and covariation of life history traits in aphids are related to infection with the facultative bacterial endosymbiont Hamiltonella defensa. Biological Journal of the Linnean Society 100: 237–247.

22. Weber DC, Brown MW (1988) Impact of Woolly Apple Aphid (Homoptera, Aphididae) on the Growth of Potted Apple-Trees. Journal of Economic Entomology 81: 1170–1177.

23. Moreno V (2003) Dinámica poblacional y enemigos naturales del Pulgón Lanígero (Eriosomalanigerum) en manzano Braeburn/Franco sometidos a manejo tradicional y confusiónsexual para la polilla de la manzana. Talca: Universidad de Talca.

24. Sunnucks P, Hales D (1996) Numerous transposed sequences of mitochondrial cytochrome oxidase I-II in aphids of the genus Sitobion (Hemiptera: Aphididae). Molecular Biology and Evolution 13: 510–524.

25. Lavandero B, Dominguez M (2010) Isolation and characterization of nine microsatellite loci from Aphelinus mali (Hymenoptera: Aphelinidae), a parasitoid of Eriosoma lanigerum (Hemiptera: Aphididae). Insect Science 17: 549–552.

26. Lavandero B, Figueroa CC, Ramirez CC, Caligari PDS, Fuentes-Contreras E (2009) Isolation and characterization of microsatellite loci from the woolly apple aphid Eriosoma lanigerum (Hemiptera: Aphididae: Eriosomatinae). Molecular Ecology Resources 9: 302–304.

27. Van Oosterhout C, Hutchinson WF, Wills DPM, Shipley P (2004) MICRO-CHECKER: software for identifying and correcting genotyping errors in microsatellite data. Molecular Ecology Notes 4: 535–538.

28. Raymond M, Rousset F (1995) Genepop (Version-1.2) - Population-Genetics Software for Exact Tests and Ecumenicism. Journal of Heredity 86: 248–249.

29. Arnaud-Haond S, Belkhir K (2007) Genclone: a computer program to analyse genotypic data, test for clonality and describe spatial clonal organization. Molecular Ecology Notes 7: 1471–8286.

30. Kalinowski S (2005) HP-Rare: a computer program for performing rarefaction on measures of allelic diversity. Molecular Ecology Notes 5: 187–189.

31. Peakall R, Smouse PE (2006) GENALEX 6: genetic analysis in Excel. Population genetic software for teaching and research. Molecular Ecology Notes 6: 288–295.

32. Chen C, Durand E, Forbes F, Francois O (2007) Bayesian clustering algorithms ascertaining spatial population structure: a new computer program and a comparison study. Molecular Ecology Notes 7: 747–756.

33. Francois O, Durand E (2010) Spatially explicit Bayesian clustering models in population genetics. Molecular Ecology Resources 10: 773–784.

34. Saitou N, Nei M (1987) The Neighbor-Joining Method - a New Method for Reconstructing Phylogenetic Trees. Molecular Biology and Evolution 4: 406–425.

35. Chakraborty R, Jin L (1993) Determination of Relatedness between Individuals Using DNA-Fingerprinting. Human Biology 65: 875–895.

36. Wang JL (2004) Sibship reconstruction from genetic data with typing errors. Genetics 166: 1963–1979.

37. Wang J, Santure AW (2009) Parentage and Sibship Inference From Multilocus Genotype Data Under Polygamy. Genetics 181: 1579–1594.

38. Bonnet E, Van de Peer Y (2002) zt: a software tool for simple and partial Mantel tests. Journal of Statistical Software 7: 1–12.

39. Frantz AC, Pourtois JT, Heuertz M, Schley L, Flamand MC, et al. (2006) Genetic structure and assignment tests demonstrate illegal translocation of red deer (Cervus elaphus) into a continuous population. Molecular Ecology 15: 3191–3203.

40. Dobson AJ (2002) An Introduction to Generalized Linear Models CRC Press.

12

Illumina Amplicon Sequencing of 16S rRNA Tag Reveals Bacterial Community Development in the Rhizosphere of Apple Nurseries at a Replant Disease Site and a New Planting Site

Jian Sun, Qiang Zhang, Jia Zhou, Qinping Wei*

Institute of Forestry and Pomology, Beijing Academy of Agriculture and Forestry Sciences, Beijing, China

Abstract

We used a next-generation, Illumina-based sequencing approach to characterize the bacterial community development of apple rhizosphere soil in a replant site (*RePlant*) and a new planting site (*NewPlant*) in Beijing. Dwarfing apple nurseries of 'Fuji'/SH6/Pingyitiancha trees were planted in the spring of 2013. Before planting, soil from the apple rhizosphere of the replant site (*ReSoil*) and from the new planting site (*NewSoil*) was sampled for analysis on the Illumina MiSeq platform. In late September, the rhizosphere soil from both sites was resampled (*RePlant* and *NewPlant*). More than 16,000 valid reads were obtained for each replicate, and the community was composed of five dominant groups (Proteobacteria, Acidobacteria, Bacteroidetes, Gemmatimonadetes and Actinobacteria). The bacterial diversity decreased after apple planting. Principal component analyses revealed that the rhizosphere samples were significantly different among treatments. Apple nursery planting showed a large impact on the soil bacterial community, and the community development was significantly different between the replanted and newly planted soils. Verrucomicrobia were less abundant in *RePlant* soil, while *Pseudomonas* and *Lysobacter* were increased in *RePlant* compared with *ReSoil* and *NewPlant*. Both *RePlant* and *ReSoil* showed relatively higher invertase and cellulase activities than *NewPlant* and *NewSoil*, but only *NewPlant* soil showed higher urease activity, and this soil also had the higher plant growth. Our experimental results suggest that planting apple nurseries has a significant impact on soil bacterial community development at both replant and new planting sites, and planting on new site resulted in significantly higher soil urease activity and a different bacterial community composition.

Editor: Gabriele Berg, Graz University of Technology (TU Graz), Austria

Funding: This work was supported by China Agriculture Research System, CARS-28, Minister of Agriculture (PRC), (http://english.agri.gov.cn/) (QPW), and Beijing Technology Foundation for Selected Overseas Chinese Scholar, Human Resource and Social Security Bureau of Beijing (http://www.bjld.gov.cn/) (JS). The funders had no role in study design, data collection and analysis, decision to publish, or preparation of the manuscript.

Competing Interests: The authors have declared that no competing interests exist.

* Email: qinpingwei@gmail.com

Introduction

Apple trees are among the most important fruit trees in the world. China has the world's highest apple tree acreage (2.060 million hectares, 42.54% of the world's supply; FAOSTAT, 2012) and production (37.00 million tons, 48.44% of the world's supply; FAOSTAT, 2012). However, apple trees in most of China's dominant production areas experience a full fruit period and senescence phase; currently, approximately 70% apple orchards in China are over 20 years old. The need for apple orchard renewal is more than 140 thousand hectares per year. Complicating this renewal is apple replant disease (ARD), which, because of a lack of land resources, is becoming a serious problem in fruit tree nurseries and old orchards.

The term ARD refers to the poor growth of young apple trees, which occurs after replanting on a site that was previously planted with apple. The phenomenon is common to all major apple growing regions of the world, including Asia, Europe, North America, Africa etc. Compared with new planting sites, directly replanting nurseries on old sites can result in decreased tree growth

and a significantly lower seedling survival rate [1]. Trees may take several years to recover from the initial growth depression and eventually reach the size and annual yields of unaffected trees, and the cumulative yields and profitability in ARD-affected orchards are usually much lower than in unaffected orchards [2]. In China, the traditional method to avoid ARD is allowing a fallow period of over three years, while using plants from the grass family such as wheat to 'clean' the soil [3]. The efficacy of growing wheat for reducing apple root infection by species of *Rhizoctonia* and *Pythium* was confirmed by greenhouse and field trials [4,5]. However, the lost productivity is unaffordable for famers because land is limited and expensive.

Typically, in ARD-affected orchards, the root systems of apple seedlings are small, with discolored feeder roots and few functional root hairs [6]. In a previous field trial, leaf analysis for macro- and micronutrients showed most elements in ARD-affected and ARD-unaffected orchards exhibited no significant difference; thus, the observed growth responses were not associated with any nutritional effect [7]. Replant disease of fruit trees has been studied for

Table 1. The invertase, urease and cellulase activity of soil of different treatments and plant growth mass of *RePlant* and *NewPlant*.

	Invertase	Urease	Cellulase	New shoots (DW, g)	New roots (DW, g)
RePlant	8.93±0.93[c]	0.73±0.06[a]	0.164±0.032[b]	150.5±22.9	22.3±6.2
NewPlant	6.59±0.14[b]	1.02±0.22[b]	0.053±0.015[a]	261.4±34.6	44.1±9.7
ReSoil	8.87±0.22[c]	0.55±0.18[a]	0.182±0.064[b]	NA	NA
NewSoil	5.16±0.65[a]	0.60±0.13[a]	0.073±0.040[a]	NA	NA

"DW" means dry weight. Averages of replicates ± standard error; means followed by different letters are significantly different at $P<0.05$.
NA: not applicable.

many decades and although reported to be attributed to certain abiotic elements including phytotoxins [8], nutrient imbalance, low or high pH, and lack or excess of moisture [9], the preponderance of evidence indicates that the disease in large part is due to biotic factors [10–12]. Soil sterilization and fumigation with methyl bromide were effective in treating ARD [13,14], thus demonstrating the role of biotic components of soil as prominent determinants of ARD in orchards.

Soil microorganisms are recognized as the key factor inducing ARD, but it is a challenging task to fully characterize soil microbial communities. The information from culture-dependent methods is limited because only a small fraction of soil microorganisms is culturable [15]. Recently, culture-independent methods have been developed for investigating microbial communities, including molecular analyses of nucleic acids extracted from soil, including denaturing gradient gel electrophoresis (DGGE) and terminal restriction fragment length polymorphism (T-RFLP) after PCR [16,17], as well as community profiling based on fatty acid methyl esters (FAME) and phospholipid fatty acid (PLFA) [18]. Next-Generation Sequencing (NGS) is a new DNA sequencing method, which relies on the detection of pyrophosphate release upon nucleotide incorporation, rather than chain termination with dideoxynucleotides. Compared with DGGE and T-RFLP, NGS may provide more detailed information about the community because each DNA molecule is sequenced as an individual read and because the identification of individual species group is more accurate [19]. Of NGS, both Pyrosequencing and Illumina Miseq are used for characterize soil microbial community structure including rhizosphere microbiome [16,20], however in recent studies, Illumina Miseq platform has been more frequently used

since the around 10-fold increase in read depth, similar sequencing quality together with much lower cost [21].

Growth of fruit plants is affected by soil enzyme activity [22,23], and soil enzyme activity is highly related to soil microbes [24], which are involved in nutrient cycling and plant nutrient availability, and are in turn influenced by plant species. Sun et al. [25] used pyrosequencing to characterize the bacterial community structure of apple rhizosphere soil with different manure ratios, finding that certain levels of manure treatment resulted in significantly higher soil enzyme activity and a more diverse bacterial community composition. Past surveys of microbial communities of apple replant sites in specialized growing areas in Europe [13], Australia [26] and the USA [2] have confirmed a complex of biotic pathogens as causal agents of this etiology. It is also suggested that genotype-specific interactions with soil microbial consortia are linked with apple rootstock tolerance or susceptibility to ARD [27], while some CG series dwarfing rootstock showed high tolerance to replant disease compared to M26 [28]. However, apple orchards in China have significant differences compared with other areas in the world, including soil with low organic matter content and the specific rootstock, such as Pingyitiancha, which is commonly used in the Bohai Bay region, one of the two leading apple-producing areas in China. Little is known about the status of soil microbial communities in ARD orchards in China. For this reason, a specific survey of bacterial community development in the rhizosphere of apple trees at a replant disease site and a new planting site was conducted with the following objectives: (i) to compare tree growth rate and soil enzyme activity between a replant disease site and a new planting site, (ii) to reveal bacterial community development under ARD and non-ARD conditions and (iii) to define the relative importance of biotic components in replant disease etiology.

Methods

Ethics statement

The experiment was carried out in our scientific research field for pomology studies which is owned by our institute, therefore, no specific permissions were required for these locations/activities, and the field studies did not involve endangered or protected species.

Soil sampling

The test soil was collected from an orchard operated by the Institute of Pomology and Forestry, Beijing Academy of Agricultural Sciences. The soil type of local area was sandy loam soil, while the pH was 6.0–6.5. The orchard site was originally planted with apple trees approximately 1985 with trees grafted on *Malus Robusta* rootstock. In September 2012, the old trees were removed, and in April 4, 2013, the soil from a depth of 0–20

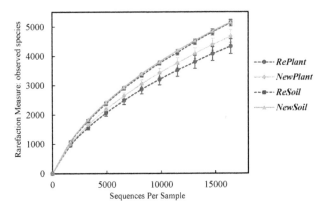

Figure 1. Rarefaction on species-abundance data. Average value of 3 replicates and error bar were showed.

Table 2. Comparison of the estimated operational taxonomic unit (OTU) richness, diversity indexes and Pielous evenness of the 16S rRNA gene libraries for clustering at 97% identity as obtained from the pyrosequencing analysis.

Treatments	Observed OTUs	Shanon	Chao1	Pielous evenness (%)
RePlant	4328.1[a]	10.53[a]	9010.8[a]	87.17[a]
NewPlant	4509.9[a]	10.67[a]	9942.7[ab]	88.09[a]
ReSoil	5125.8[b]	11.02[b]	11371.8[c]	89.45[a]
NewSoil	5176.9[b]	11.12[b]	10614.9[bc]	88.53[a]

Averages of replicates ± standard error; means followed by different letters are significantly different at *P*<0.05.

cm was collected and mixed as replant soil sample (*ReSoil*). Another soil sample from alongside the orchard intensively cultivated with vegetables was collected as a new plant soil sample (*NewSoil*). Soil was put into 64 liter cubic Plexiglas boxes with rainproof shelter, and the nursery was planted right after the soil sampling. In late September, soil from a depth of 0–20 cm in three different locations at 20-cm distances from the center of trunk was collected from both sites and the youngest part of roots and the adjacent soil were resampled as *RePlant* and *NewPlant*.

Rootstock variety and tree growth

The two-year-old apple saplings were planted on April 4, 2013. Trees were of the scion variety 'Fuji' and were first grafted onto SH6 inter stock and then onto *Malus hupehensis* Var. Pingyiensis Jiang rootstock. The rootstock Pingyiensis Jiang has been widely

used in Chinese orchards since the 1970 s. Trees were planted into either the replant soil or the new plant soil, with 10 replicates/ treatment, and trunk heights were handed to 1.0–1.2 m. On September 20, 2013, new shoots and new roots of these trees were collected separately, dried at 105°C for 30 min and then dried at 70°C until a constant weight was reached in a forced-air oven.

Soil enzyme activity characterization

Soil samples were collected from depths of 0–20 cm in three different locations at 20-cm distances from the center of nurseries using a 5-cm diameter soil auger and transferred on ice to the laboratory both before nursery planting, and at the beginning of autumn 2013, just after the growth of autumn-shoot ceased. The soil samples were sieved through a 2-mm screen and homogenized prior to the analysis. One portion of the composite soil was stored

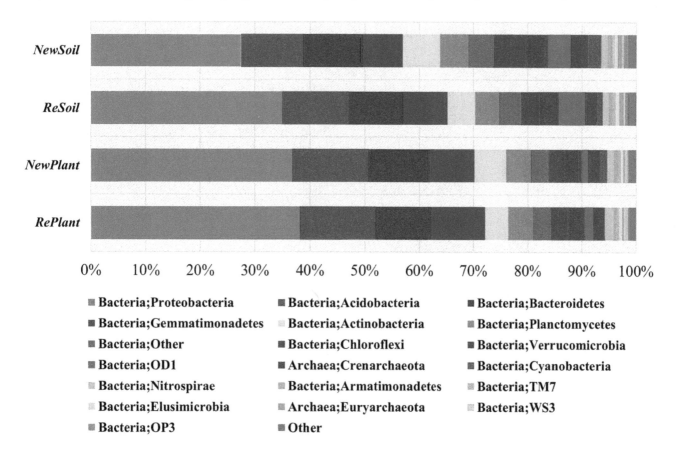

Figure 2. Comparison of the bacterial communities at the phylum level. Relative read abundance of different bacterial phyla within the different communities. Sequences that could not be classified into any known group were labeled "Other".

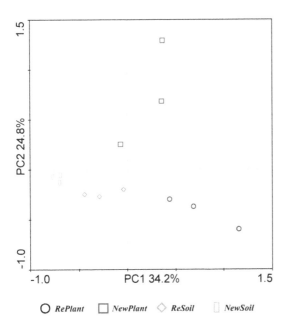

O *RePlant*   □ *NewPlant*   ◇ *ReSoil*   ⬜ *NewSoil*

Figure 3. Principal component analysis (PCA) on the relative abundance of bacterial genera using Canoco 4.5. Principal components (PCs) 1 and 2 explained 34.2% and 24.8% of the variance, respectively.

in DNA-free polythene bags and kept on dry ice for the molecular analysis, while another portion was used for enzyme activity measurements.

Soil urease, invertase and cellulase activities were estimated according to a previous report [25]. Soil urease activity was detected using improved sodium phenate and sodium hypochlorite colorimetry. Invertase and cellulase activities were estimated colorimetrically by determining the reduction of 3,5-dinitrosalicylic acid from reducing sugars after the soil was incubated with a buffered sucrose and sodium carboxymethylcellulose solution and toluene at 37°C for 24 h and 72 h, respectively.

Soil DNA extraction

Three replicate samples were randomly picked for one treatment and used for DNA extraction. Soil DNA was extracted from the 1 g of soil after sieving using a Fast DNA SPIN Kit for soil (MP Biomedicals, Santa Ana, CA), according to the manufacturer's instructions. The extracted soil DNA was dissolved in 100 μl TE buffer, quantified by NanoDrop and stored at −70°C before use.

Bacterial 16S rRNA gene amplification and Illumina Sequencing

To determine the diversity and composition of the bacterial communities in each of these samples, we used the protocol described in Caporaso et al. [29]. PCR amplifications were conducted in with the 515f/806r primer set that amplifies the V4 region of the 16S rDNA gene. The primer set was selected as it exhibits few biases should yield accurate phylogenetic and taxonomic information. The reverse primer contains a 6-bp error-correcting barcode unique to each sample. DNA was amplified following the protocol described previously [30]. Amplicon pyrosequencing was performed on the Illumina MiSeq platforms at Novogene Bioinformatics Technology Co., Ltd,

Beijing, China. Complete data sets are submitted to the NCBI Short Read Archive under accession no. SRX337490.

Pairs of reads from the original DNA fragments were merged by using FLASH [29] -a very fast and accurate software tool which was designed to merge pairs of reads when the original DNA fragments were shorter than twice the length of reads. Sequencing reads was assigned to each sample according to the unique barcode of each sample. Sequences were analyzed with the QIIME [31] software package (Quantitative Insights Into Microbial Ecology) and UPARSE pipeline [32], in addition to custom Perl scripts to analyze alpha (within sample) and beta (between sample) diversity.

First, the reads were filtered by QIIME quality filters. Default settings for Illumina processing in QIIME was used (r = 3 p = 0.75 total read length; q = 3; n = 0).

(p) min_per_read_length: minimum number of consecutive high-qualitybase calls to retain read(as percentage of totalread length).

(r) max_bad_run_length: maximum number of consecutive low-quality base calls allowed before truncating a read.

(n) sequence_max_n: maximum number of ambiguous (N) characters allowed in a sequence.

(q) phred_quality_score: last quality score considered low quality.

Then we use UPARSE pipeline to picking operational taxonomic units (OTUs) through making OTU table. Sequences were assigned to OTUs at 97% similarity. We picked a representative sequences for each OTU and used the RDP classifier [33] to assign taxonomic data to each representative sequence. In order to compute Alpha Divesity, we rarified the OTU table and calculated three metrics: Chao1 metric estimates the richness, the Observed OTUs metric was simply the count of unique OTUs found in the sample, and shannon index. Rarefaction curves were generated based on these three metrics.

Statistical and bioinformatics analysis

Heatmap figures were generated using custom R scripts. Canoco 4.5 was used to run principal component analysis (PCA). Analysis of variance and Spearman's rank correlations were performed using SPSS Statistics 18 (IBM, Armonk, New York, USA). The community richness index, community diversity index, data preprocessing, operational taxonomic unit-based analysis and hypothesis tests were performed using mothur (http://www.mothur.org/). The histogram was created using Microsoft Excel 2010 (Microsoft, Redmond, Washington, USA). Significance was accepted at p<0.05, unless otherwise noted.

Results

Seedling biomass accumulation

In the spring of 2013, dwarfing apple nurseries of 'Fuji'/SH6/Pingyitiancha trees were planted a replant site (*RePlant*) and a new planting site (*NewPlant*). Before planting, soil from the replant site (*ReSoil*) and from the new planting site (*NewSoil*) was sampled. After 1 year of growth, significant differences (P<0.05) were found between *RePlant* and *NewPlant* sites in dry mass accumulation (**Table 1**). Seedling growth was significantly inhibited in replant soil comparing with non-replant soil; the inhibition levels on root dry weight and shoot dry weight were 49.4% and 42.3%, respectively.

Soil enzyme activity

Soil urease activity was highest in *NewPlant* soil, whereas there was no significant difference in urease activity in soil from

Table 3. The genera showing significant differences among the samples.

Taxon	RePlant (%)	NewPlant (%)	ReSoil (%)	NewSoil (%)
Pseudomonas	0.207±0.048[b]	0.120±0.041[a]	0.063±0.031[a]	0.057±0.009[a]
Lysobacter	0.941±0.260[b]	0.471±0.065[a]	0.276±0.061[a]	0.277±0.037[a]
Phenylobacterium	0.337±0.007[c]	0.258±0.045[b]	0.317±0.022[c]	0.135±0.013[a]
Ramlibacter	0.276±0.128[a]	0.728±0.109[b]	0.345±0.061[a]	0.336±0.046[a]
Chthonomonas	0.051±0.013[a]	0.171±0.048[b]	0.12±0.055[ab]	0.202±0.038[b]
Flavisolibacter	0.217±0.060[a]	0.847±0.350[b]	0.364±0.095[a]	0.354±0.048[a]
Opitutus	0.378±0.022[a]	0.400±0.057[a]	0.644±0.120[b]	0.524±0.060[ab]
Bdellovibrio	0.096±0.014[a]	0.161±0.058[ab]	0.256±0.025[b]	0.239±0.031[b]
Novosphingobium	0.185±0.136[ab]	0.266±0.049[b]	0.063±0.018[a]	0.058±0.009[a]
Planctomyces	0.167±0.025[a]	0.152±0.032[a]	0.195±0.032[ab]	0.239±0.002[b]
Kaistobacter	2.463±0.555[b]	2.302±0.461[b]	2.17±0.311[b]	0.978±0.083[a]
Cellvibrio	0.124±0.173[a]	0.100±0.064[a]	0.144±0.051[a]	0.378±0.06[b]
Kribbella	0.012±0.012[a]	0.031±0.011[a]	0.026±0.004[a]	0.081±0.025[b]
Pseudonocardia	0.035±0.014[a]	0.045±0.004[ab]	0.045±0.028[ab]	0.095±0.034[b]
Rubrobacter	0.118±0.066[a]	0.226±0.082[ab]	0.144±0.023[a]	0.311±0.092[b]
Luteimonas	0.388±0.071[b]	0.268±0.158[ab]	0.325±0.052[ab]	0.120±0.029[a]
Steroidobacter	0.902±0.136[ab]	0.701±0.085[a]	0.996±0.104[b]	0.761±0.065[ab]
Candidatus Nitrososphaera	1.282±0.400[a]	2.054±0.372[ab]	2.04±0.357[ab]	3.044±0.683[b]
Cenarchaeaceae;g__	0.476±0.180[b]	0.043±0.043[a]	0.061±0.064[a]	0.017±0.010[a]
Hyphomonadaceae;g__	0.721±0.078[b]	0.360±0.069[a]	0.646±0.055[b]	0.382±0.117[a]
Rhodospirillaceae;g__	0.721±0.104[c]	0.348±0.012[a]	0.577±0.099[bc]	0.455±0.129[ab]
Haliangiaceae;g__	1.520±0.475[b]	0.793±0.028[a]	0.811±0.143[a]	0.879±0.108[a]
Syntrophobacteraceae;g__	0.914±0.046[b]	0.614±0.212[a]	0.563±0.095[a]	0.587±0.051[a]
Rhodothermaceae;g__	0.033±0.018[a]	0.102±0.037[b]	0.033±0.013[a]	0.031±0.016[a]
Saprospiraceae;g__	0.366±0.114[a]	0.671±0.142[b]	0.378±0.054[a]	0.807±0.093[b]
Erythrobacteraceae;g__	0.553±0.150[ab]	0.469±0.080[a]	0.815±0.229[b]	0.309±0.042[a]
Polyangiaceae;g__	0.033±0.015[a]	0.061±0.044[a]	0.073±0.016[a]	0.145±0.012[b]
NB1-j;g__	0.341±0.056[b]	0.157±0.113[a]	0.083±0.044[a]	0.108±0.013[a]
Chromatiales;g__	0.106±0.025[b]	0.020±0.020[a]	0.047±0.029[a]	0.027±0.003[a]
Sphingomonadales;g__	0.159±0.048[ab]	0.240±0.015[c]	0.220±0.043[a]	0.099±0.009[a]
Sphingomonadaceae;g__	0.679±0.076[a]	1.026±0.234[b]	0.689±0.064[a]	0.423±0.051[a]
Micrococcaceae;g__	0.183±0.062[a]	0.563±0.167[b]	0.228±0.013[a]	0.673±0.105[b]

"g__" represents genus not grouped into any known genera within these families/groups. Averages of replicates ± standard error; means followed by different letters are significantly different at $P < 0.05$.

RePlant, ReSoil and NewSoil samples, which were 28.4%~46.1% lower than NewPlant soil (**Table 1**). Cellulase activities in NewPlant and NewSoil were lower than in RePlant and ReSoil, but planting apple nurseries did not significantly change soil cellulase activity. Regardless of whether a new apple nursery had yet been planted, the soil from the old orchard had about three-fold higher cellulase activity than soil from a site that had never been planted. Invertase activity was lower in NewSoil and NewPlant samples than in RePlant and ReSoil samples.

Richness

More than 16,000 valid reads were obtained for each replicate through a sequence optimization process, and the bacterial community richness index was calculated as shown in **Table 2**. After quality filtering, median sequence length of each read was 252 bp. In ReSoil and NewSoil samples, more than 600 additional OTUs were observed compared with RePlant and NewPlant soil

(**Fig. 1**). Higher Shannon and Chao 1 indices before planting indicated that planting of apple nurseries reduced diversity within the bacterial community. However, Pielou's evenness values were indicating approximately equally distributed OTU abundances among community members (**Table 2**).

Taxonomic coverage

All of the sequences were classified into 26 phyla or groups by the mothur program. The overall bacterial composition of the different samples was similar, while the distribution of each phylum or group varied (**Fig. 2**). In all samples, Proteobacteria, Acidobacteria, Bacteroidetes, Gemmatimonadetes and Actinobacteria were the five most dominant phyla, accounting for >60% of the reads. Significantly more unclassified species were detected in ReSoil and NewSoil samples, which was in accordance with their higher diversity indices. Compared with other samples, NewSoil had a significantly higher percentage of GN02 (2.7–5.3-fold), OP3

(1.8–3.1-fold), Chloroflexi (1.8–2.4-fold), and Verrucomicrobia (1.1–1.3-fold), and a lower percentage of Proteobacteria (*NewSoil*: 27.2%, *ReSoil*: 34.9%, *RePlant*: 38.2% and *NewPlant*: 36.8%). The percentages of CyanoBac and Verrucomicrobia were lowest in *RePlant*. More Firmicutes were detected in *NewPlant* than *RePlant* or *ReSoil* samples, and the lowest level was found in *NewSoil*. The OD1 group was approximately two- to three-fold higher in *RePlant* and *NewPlant* compared with *ReSoil* and *NewSoil*.

On a genus level, all 260 detected genera were shared by the four samples, except for *Chryseobacterium*, which was not detected in *ReSoil*, and *Marinobacter*, which was not detected in *RePlant*. Several cold-tolerant species belong to *Chryseobacterium* [34], while *Marinobacter* has been reported to be halophilic and is found in seawater [35].

To further compare the microbiota among the different samples, we performed PCA on the relative abundance of bacterial genera using Canoco 4.5 (**Fig. 3**). Data are presented as a 2D plot to better illustrate the relationship. *ReSoil* and *NewSoil* were relatively similar, but planting of an apple nursery had a significant impact on the soil microbial community. *RePlant* had a significantly higher PC1 value, and *NewPlant* had a higher PC2 value. 32 genera showing significant differences among the samples are listed in **Table 3**.

Pseudomonas, *Lysobacter* and *Phenylobacterium* were significantly higher in *RePlant* compared with *NewPlant*, while *Ramlibacter*, *Chthonomonas* and *Flavisolibacter* were higher in *NewPlant*. *Phenylobacterium* and *Kaistobacter* were higher in *ReSoil* compared with *NewSoil*, while *Cellvibrio*, *Kribbella* and *Rubrobacter* were higher in *NewSoil*. On both the replant site and the new plant site, apple nurseries showed significant impacts on the bacterial community structure. On the new plant site, *Phenylobacterium*, *Ramlibacter*, *Flavisolibacter*, *Novosphingobium* and *Kaistobacter* had increased abundance, and *Planctomyces*, *Cellvibrio* and *Kribbella* decreased in abundance. On the replant site, only *Pseudomonas* and *Lysobacter* increased, while *Opitutus* and *Bdellovibrio* decreased. Compared with the *NewPlant* sample, planting in the replant site also resulted in a greater abundance of certain genera of Cenarchaeaceae, Hyphomonadaceae, Rhodospirillaceae, Haliangiaceae and Syntrophobacteraceae families, however these genera weren't grouped into known genera within these families.

Discussion

Replanted soil significantly inhibited root and shoot development and exhibited different soil enzyme activity and a different bacterial community pattern. The levels of inhibition on root dry weight and shoot dry weight were 49.4% and 42.3%. Rosette disease and decreasing photosynthetic efficiency were also observed in the replant site, as were fewer main branches (data not shown). This observation is in accordance with the negative impact of replanting on apple growth that is widely reported [9,13,36].

Soil enzymes are involved with biological cycling and the development of fertility, so they are crucial indicators of the soil biochemistry. Urease catalyzes the hydrolysis of urea to produce ammonia and carbamate, and it is thus recognized as an important indicator of soil health. In this study, rhizosphere soil of *NewPlant* showed significantly increased urease activity, but rhizosphere soil of RePlant did not, indicating that in new planting sites, the root exudates might support a new and different functional microbial community which was responsible for this apparent increase in mineralization and result in a better supply of

available nutrients. This result was supported by data on plant dry mass (**Table 1**), where *NewPlant* exhibited more than 40% greater plant growth compared with *RePlant*. Xun et al. [22] reported that soil urease activity increased during apple orchard maturation, and in our previous study on manure refinement of apple orchards, urease was the key indicator of soil health and highly correlative to tree growth, no matter whether the soil type was sandy [24] or loam [25]. Therefore, soil urease is an ideal indicator of apple orchard maturation and the lack of a significant increase in urease activity at the replant site may explain the decreased growth and late fruiting. However, soil invertase, which is an important factor affecting hydrolysis of sucrose into glucose and fructose, and cellulase, which is involved in breaking down cellulase, were higher in replant soil rather than new soil, and planting of apple nurseries had no further impact on these two enzyme activities. This could be related to the residual small root tissues of previous trees. Invertase also increased with an overdose of manure refinement in apple orchards, but was not closely associated with tree growth [37].

The soil microbial community composition of the replant site and the new plant site were distinct, and planting of an apple nursery significantly increased the difference. The diversity within the bacterial community was reduced after planting an apple nursery in both replanted and newly planted soil, which was in accordance with the fewer unclassified species observed after nurseries had been growing for one year. It has been reported that many plant species reduce the microbial diversity of rhizosphere soil compared with surrounding sites, including maize [38] and switchgrass [39]. Such a reduced bacterial richness in the plant rhizosphere is known as the 'rhizosphere effect'. This is typically characterized by a selective enrichment of root specialized guilds and reduction of rhizosphere bacterial richness in comparison to unplanted soil. However, although the diversity at both sites decreased after planting, there was no significant difference in bacterial diversity between the replant and new plant sites, either before or after the nursery had been planted, which means the rhizosphere effect of apple trees is the critical factor determining bacterial community diversity.

Although Proteobacteria, Acidobacteria, Bacteroidetes, Gemmatimonadetes and Actinobacteria predominated in all of the samples, *NewSoil* had a unique phyla composition compared with the other samples. Higher percentages of GN02, OP3, Chloroflexi and Verrucomicrobia and lower percentages of Proteobacteria were observed. This could also be explained by the rhizosphere effect' because there must be some root tissue left in the *ReSoil* site, even without a new nursery being planted. It is worth noticing that the OD1 group was approximately two- to three-fold higher in the *RePlant* and *NewPlant* samples as compared with *ReSoil* and *NewSoil*, and the *RePlant* soil had a higher percentage of WS3 compared with the other treatments. Verrucomicrobia were less common in the *RePlant* sample compared with *NewPlant*. Cultivation-independent approaches detect representatives of the Verrucomicrobia phylum in a wide range of environments, including soils, water and human feces [40], suggesting the phylum is widespread, but it is still poorly characterized. A few species that are extremely acidophilic [41] or ectosymbionts of protists [42] belong to this phylum. In our previous study [25], on loam soil of apple orchards with manure refinement, an optimal manure ratio resulted in an increase of Verrucomicrobia compared with soil with no manure applied, with the abundance increasing from 1.10% to 1.51%. However, on sandy soil [24], Verrucomicrobia decreased monotonically from 2.62% to 1.71%, 1.36% and 0.97% following application of 5%, 10% and 15% (which was optimal for tree growth) manure, respectively, and increased back to 1.28% when 20% manure applied. More research is required to determine whether this phylum is related to the rhizosphere effect of apple trees. Because most species of GN02, OP3, OD1 and WS3 groups are known only from metagenomics study and remain uncultivated, little information is available for functional discussion.

Principal component analysis on the genus level showed that apple nurseries had significant impacts on the soil microbial community, and the changes from *ReSoil* to *RePlant* and from *NewSoil* to *NewPlant* differed significantly. The bacterial community difference between *RePlant* and *NewPlant* was much greater than between *ReSoil* and *NewSoil*.

Phenylobacterium and *Kaistobacter* were higher in *ReSoil* compared with *NewSoil*, while *Cellvibrio, Kribbella* and *Rubrobacter* were higher in *NewSoil*. The genus *Phenylobacterium* comprises a single species called *P. immobile*, which has previously been described as growing optimally only on artificial compounds such as chloridazon [43]. More *P. immobile* in replant soil could be related to herbicide applied in the orchard. Unfortunately, little information is available in the literature concerning the genus *Kaistobacter*. *Cellvibrio* is a genus of gamma proteobacteria, which can oxidize cellulose to oxycellulose, but *NewSoil* showed lower cellulase than *ReSoil*, probably because in an aerobic environment, soil cellulase mainly derives from fungi, rather than bacteria. Some species of *Kribbella* have been isolated from the rhizosphere of *Typhonium giganteum* [44] and from tissues such as apricot leaves [45] and roots of *Lupinus angustifolius* [46]. *Kribbella antibiotic* [47] of the genus was reported to have a strong inhibitory activity toward *Botrytis* sp., *Rhizoctonia solani* and *Pyricularia oryzae*. *Rubrobacter* is a genus of Actinobacteria, which are radiotolerant [48], and a novel DNA repair enzyme was isolated from *Rubrobacter radiotolerans* [49], but little information is available regarding its presence in orchard soil.

On the replant site, *Lysobacter* and *Pseudomonas* increased in abundance, while *Opitutus* and *Bdellovibrio* decreased. *Pseudomonas* and *Lysobacter* were significantly higher in *RePlant* compared with *NewPlant*, while *Ramlibacter, Chthonomonas* and *Flavisolibacter* were more abundant in the *NewPlant* sample. *Pseudomonas* has been proposed to play a role in replant disease etiology of peach [50] and apple [51] trees through the production of hydrogen cyanide (HCN). However, *Pseudomonas* also includes several soil bacterial species with plant growth promoting activity,

including *P. fluorescens* [10,52], and *Pseudomonas* and *Bacillus* are two of the most common biocontrol agent sources [53]. *Lysobacter* [54] has been a rich source for novel antibiotics, and some species have potential as biological control agents for plant diseases. Considering the previous study of manure refinement of apple orchard, in which an optimal manure ratio for nursery growth resulted in a decrease of *Pseudomonas* and *Lysobacter* [24,37], we speculate that in the *RePlant* sample, the higher percentage of *Pseudomonas* was related to replant disease and the increase of *Lysobacter* and was probably induced by the high percentages of Pathogenic fungi. However, further culture-based experiments would be needed to confirm this. There was only one species of *Chthonomonas*, *G. calidirosea*, which is an aerobic, pigmented, thermophilic micro-organism [55], while two species of *Flavisolibacter* were isolated from ginseng cultivating soil [56]. However, little was known about their function in soil or their relationship with planting until now.

Compared with *NewPlant*, planting in the replant site also resulted in more of certain genera of Cenarchaeaceae, Hyphomonadaceae, Rhodospirillaceae, Haliangiaceae and Syntrophobacteraceae families; however, these genera weren't grouped into any known genera within these families. This is in accordance with the fact that pyrosequencing can detect many uncultured microbes.

Conclusions

Our study has documented that replanting has a large negative impact on growth of apple nurseries in Beijing, China. Planting apple nurseries raised soil urease activity at the new planting site but not the replant site, while no significant impact on invertase and cellulase was observed. Apple nurseries had a significant impact on the soil bacterial community. *Lysobacter* and *Pseudomonas* were increased at the replant site, and the bacterial communities of the new and replant sites responded differently, resulting in more distinct community patterns.

Author Contributions

Conceived and designed the experiments: JS QPW. Performed the experiments: JS QZ JZ. Analyzed the data: JS QPW. Contributed reagents/materials/analysis tools: QZ JZ QPW. Contributed to the writing of the manuscript: JS QPW.

References

1. Mazzola M, Granatstein DM, Elfving DC, Mullinix K, Gu Y (2002) Cultural management of microbial community structure to enhance growth of apple in replant soils. Phytopathology 92: 1363–1366.

2. Mazzola M (1998) Elucidation of the microbial complex having a causal role in the development of apple replant disease in Washington. Phytopathology 88: 930–938.

3. Fan H, Zhao Z, Liu H, Zhao G, Zhang X, et al. (2008) Changes of Soil Nutrition in Root Zone and Their Effects on Growth of the Replanted Apple. Acta Horticulturae Sinica 12: 3.

4. Mazzola M, Gu Y (2000) Impact of wheat cultivation on microbial communities from replant soils and apple growth in greenhouse trials. Phytopathology 90: 114–119.

5. Mazzola M, Mullinix K (2005) Comparative field efficacy of management strategies containing *Brassica napus* seed meal or green manure for the control of apple replant disease. Plant Dis 89: 1207–1213.

6. Wilson S, Andrews P, Nair TS (2004) Non-fumigant management of apple replant disease. Sci Hortic-Amsterdam 102: 221–231.

7. Hofmann A, Wittenmayer L, Arnold G, Schieber A, Merbach W (2012) Root exudation of phloridzin by apple seedlings (*Malus* x *domestica* Borkh.) with symptoms of apple replant disease. Journal of Applied Botany and Food Quality 82: 193–198.

8. Utkhede RS, Smith EM (1993) Biotic and abiotic causes of replant problems of fruit trees. III International Symposium on Replant Problems 363. 25–32.

9. Rumberger A, Yao S, Merwin IA, Nelson EB, Thies JE (2004) Rootstock genotype and orchard replant position rather than soil fumigation or compost

amendment determine tree growth and rhizosphere bacterial community composition in an apple replant soil. Plant Soil 264: 247–260.

10. Mazzola M, Manici LM (2012) Apple replant disease: role of microbial ecology in cause and control. Annu Rev Phytopathol 50: 45–65.

11. Yim B, Smalla K, Winkelmann T (2013) Evaluation of apple replant problems based on different soil disinfection treatments-links to soil microbial community structure? Plant Soil 1–15.

12. Covey Jr RP, Benson NR, Haglund WA (1979) Effect of soil fumigation on the apple replant disease in Washington [USA]. Phytopathology 69.

13. Manici LM, Kelderer M, Franke-Whittle IH, Rühmer T, Baab G, et al. (2013) Relationship between root-endophytic microbial communities and replant disease in specialized apple growing areas in Europe. Appl Soil Ecol 72: 207–214.

14. Cardenas E, Tiedje JM (2008) New tools for discovering and characterizing microbial diversity. Curr Opin Biotech 19: 544–549.

15. Smalla K, Oros-Sichler M, Milling A, Heuer H, Baumgarte S, et al. (2007) Bacterial diversity of soils assessed by DGGE, T-RFLP and SSCP fingerprints of PCR-amplified 16S rRNA gene fragments: do the different methods provide similar results? J Microbiol Meth 69: 470–479.

16. Hirsch PR, Mauchline TH, Clark IM (2010) Culture-independent molecular techniques for soil microbial ecology. Soil Biology and Biochemistry 42: 878–887.

17. Drenovsky RE, Elliott GN, Graham KJ, Scow KM (2004) Comparison of phospholipid fatty acid (PLFA) and total soil fatty acid methyl esters (TSFAME)

for characterizing soil microbial communities. Soil Biology and Biochemistry 36: 1793–1800.

18. Ronaghi M, Uhlén M, Nyrén P (1998) A sequencing method based on real-time pyrophosphate. Science 281: 363–365.

19. Lee OO, Wang Y, Yang J, Lafi FF, Al-Suwailem A, et al. (2010) Pyrosequencing reveals highly diverse and species-specific microbial communities in sponges from the Red Sea. The ISME journal 5: 650–664.

20. Mazzola M, Strauss SL (2013) Resilience of orchard replant soils to pathogen re-infestation in response to Brassicaceae seed meal amendment. Aspects of Applied Biology 119: 69–77.

21. Gołębiewski M, Deja-Sikora E, Cichosz M, Tretyn A, Wróbel B (2014) 16S rDNA Pyrosequencing Analysis of Bacterial Community in Heavy Metals Polluted Soils. Microbial Ecol 1–13.

22. Qian X, Gu J, Sun W, Li Y, Fu Q, et al. (2014) Changes in the soil nutrient levels, enzyme activities, microbial community function, and structure during apple orchard maturation. Appl Soil Ecol 77: 18–25.

23. Jia T, Chao S, Mingyan Y, Xiaoqi Z (2012) Studies on the activities of three kinds of soil enzyme, organic matters, microbes and the yields and quality of apple in different tree-aged apple orchards in Loess plateau. Journal of Agricultural Science and Technology (Beijing) 14: 115–122.

24. Zhang Q, Sun J, Liu S, Wei Q (2013) Manure Refinement Affects Apple Rhizosphere Bacterial Community Structure: A Study in Sandy Soil. PloS one 8: e76937.

25. Sun J, Zhang Q, Zhou J, Wei Q (2014) Pyrosequencing technology reveals the impact of different manure doses on the bacterial community in apple rhizosphere soil. Appl Soil Ecol 78: 28–36.

26. Dullahide SR, Stirling GR, Nikulin A, Stirling AM (1994) The role of nematodes, fungi, bacteria, and abiotic factors in the etiology of apple replant problems in the Granite Belt of Queensland. Animal Production Science 34: 1177–1182.

27. St. Laurent A, Merwin I, Fazio G, Thies J, Brown M (2010) Rootstock genotype succession influences apple replant disease and root-zone microbial community composition in an orchard soil. Plant Soil 337: 259–272.

28. Atucha A, Emmett B, Bauerle T (2014) Growth rate of fine root systems influences rootstock tolerance to replant disease. Plant Soil 376: 337–346.

29. Caporaso JG, Kuczynski J, Stombaugh J, Bittinger K, Bushman FD, et al. (2010) QIIME allows analysis of high-throughput community sequencing data. Nature methods 7: 335–336.

30. Magoč T, Salzberg SL (2011) FLASH: fast length adjustment of short reads to improve genome assemblies. Bioinformatics 27: 2957–2963.

31. Wang Q, Garrity MG, Tiedje MJ, James RC (2007) Naïve Bayesian Classifier for Rapid Assignment of rRNA Sequences into the New Bacterial Taxonomy. Applied and Environmental Microbiology 73: 5261–5267.

32. Edgar RC (2013) UPARSE: highly accurate OTU sequences from microbial amplicon reads. 10: 996–998.

33. Caporaso JG, Lauber CL, Walters WA, Berg-Lyons D, Lozupone CA, et al. (2011) Global patterns of 16S rRNA diversity at a depth of millions of sequences per sample. Proc Natl Acad Sci U S A 108 Suppl 1: 4516–4522.

34. Kämpfer P, Dreyer U, Neef A, Dott W, Busse H (2003) Chryseobacterium defluvii sp. nov., isolated from wastewater. Int J Syst Evol Micr 53: 93–97.

35. Gauthier MJ, Lafay B, Christen R, Fernandez L, Acquaviva M, et al. (1992) Marinobacter hydrocarbonoclasticus gen. nov., sp. nov., a new, extremely halotolerant, hydrocarbon-degrading marine bacterium. International Journal of Systematic Bacteriology 42: 568–576.

36. Sewell G (1981) Effects of Pythium species on the growth of apple and their possible causal role in apple replant disease. Ann Appl Biol 97: 31–42.

37. Sun J, Zhang Q, Liu S, Wang X, Liu J, et al. (2013) Comparison of soil microbe community structure between organic and conventional apple orchards. Journal of Fruit Science 30: 230–234.

38. García Salamanca A, Molina Henares MA, Dillewijn P, Solano J, Pizarro Tobías P, et al. (2013) Bacterial diversity in the rhizosphere of maize and the surrounding carbonate-rich bulk soil. Microbial biotechnology 6: 36–44.

39. Mao Y, Li X, Smyth EM, Yannarell AC, Mackie RI (2014) Enrichment of specific bacterial and eukaryotic microbes in the rhizosphere of switchgrass (Panicum virgatum L.) through root exudates. Environmental Microbiology Reports.

40. Hou S, Makarova KS, Saw JH, Senin P, Ly BV, et al. (2008) Complete genome sequence of the extremely acidophilic methanotroph isolate V4, Methylacidiphilum infernorum, a representative of the bacterial phylum Verrucomicrobia. Biol Direct 3: 11.

41. Dunfield PF, Yuryev A, Senin P, Smirnova AV, Stott MB, et al. (2007) Methane oxidation by an extremely acidophilic bacterium of the phylum Verrucomicrobia. Nature 450: 879–882.

42. Sato T, Kuwahara H, Fujita K, Noda S, Kihara K, et al. (2013) Intranuclear verrucomicrobial symbionts and evidence of lateral gene transfer to the host protist in the termite gut. The ISME journal.

43. Kanso S, Patel BK (2004) Phenylobacterium lituiforme sp. nov., a moderately thermophilic bacterium from a subsurface aquifer, and emended description of the genus Phenylobacterium. Int J Syst Evol Micr 54: 2141–2146.

44. Xu Z, Xu Q, Zheng Z, Huang Y (2012) Kribbella amoyensis sp. nov., isolated from rhizosphere soil of a pharmaceutical plant, Typhonium giganteum Engl. Int J Syst Evol Micr 62: 1081–1085.

45. Kaewkla O, Franco CM (2013) Kribbella endophytica sp. nov., an endophytic actinobacterium isolated from the surface-sterilized leaf of a native apricot tree. Int J Syst Evol Micr 63: 1249–1253.

46. Trujillo ME, Kroppenstedt RM, Schumann P, Martínez-Molina E (2006) Kribbella lupini sp. nov., isolated from the roots of Lupinus angustifolius. Int J Syst Evol Micr 56: 407–411.

47. Wang D, Zhang Y, Jiang Y, Wu W, Jiang C (2003) A study on polyphasic taxonomy of one antifungal actinomycete strain YIM31530~(T). Journal of Yunnan University (Natural Sciences) 26: 265–269.

48. Kausar J, Ohyama Y, Terato H, Ide H, Yamamoto O (1997) 16S rRNA gene sequence of Rubrobacter radiotolerans and its phylogenetic alignment with members of the genus Arthrobacter, gram-positive bacteria, and members of the family Deinococcaceae. International journal of systematic bacteriology 47: 684–686.

49. Asgarani E, Terato H, Asagoshi K, Shahmohammadi HR, Ohyama Y, et al. (2000) Purification and characterization of a novel DNA repair enzyme from the extremely radioresistant bacterium Rubrobacter radiotolerans. J Radiat Res 41: 19–34.

50. Yang J, Ruegger PM, McKenry MV, Becker JO, Borneman J (2012) Correlations between root-associated microorganisms and peach replant disease symptoms in a California soil. PloS one 7: e46420.

51. Rumberger A, Merwin IA, Thies JE (2007) Microbial community development in the rhizosphere of apple trees at a replant disease site. Soil Biology and Biochemistry 39: 1645–1654.

52. Himani S, Mohinder K, Prakash VP, Sheetal R (2011) Phenotypic and genotypic characterization of plant beneficial fluorescent Pseudomonas species associated with apple to overcome replant problem. Plant Disease Research 26.

53. Santoyo G, Orozco-Mosqueda MDC, Govindappa M (2012) Mechanisms of biocontrol and plant growth-promoting activity in soil bacterial species of Bacillus and Pseudomonas: a review. Biocontrol Sci Techn 22: 855–872.

54. Xie Y, Wright S, Shen Y, Du L (2012) Bioactive natural products from Lysobacter. Nat Prod Rep 29: 1277–1287.

55. Lee KC, Dunfield PF, Morgan XC, Crowe MA, Houghton KM, et al. (2011) Chthonomonas calidirosea gen. nov., sp. nov., an aerobic, pigmented, thermophilic micro-organism of a novel bacterial class, Chthonomonadetes classis nov., of the newly described phylum Armatimonadetes originally designated candidate division OP10. Int J Syst Evol Micr 61: 2482–2490.

56. Yoon M, Im W (2007) Flavisolibacter ginsengiterrae gen. nov., sp. nov. and Flavisolibacter ginsengisoli sp. nov., isolated from ginseng cultivating soil. Int J Syst Evol Micr 57: 1834–1839.

13

Semi-Automatic Normalization of Multitemporal Remote Images Based on Vegetative Pseudo-Invariant Features

Luis Garcia-Torres*, Juan J. Caballero-Novella, David Gómez-Candón, Ana Isabel De-Castro

Institute for Sustainable Agriculture (IAS), Spanish Council for Scientific Research (CSIC), Cordoba, Spain

Abstract

A procedure to achieve the semi-automatic relative image normalization of multitemporal remote images of an agricultural scene called ARIN was developed using the following procedures: 1) defining the same parcel of selected vegetative pseudo-invariant features (VPIFs) in each multitemporal image; 2) extracting data concerning the VPIF spectral bands from each image; 3) calculating the correction factors (CFs) for each image band to fit each image band to the average value of the image series; and 4) obtaining the normalized images by linear transformation of each original image band through the corresponding CF. ARIN software was developed to semi-automatically perform the ARIN procedure. We have validated ARIN using seven GeoEye-1 satellite images taken over the same location in Southern Spain from early April to October 2010 at an interval of approximately 3 to 4 weeks. The following three VPIFs were chosen: citrus orchards (CIT), olive orchards (OLI) and poplar groves (POP). In the ARIN-normalized images, the range, standard deviation (s. d.) and root mean square error (RMSE) of the spectral bands and vegetation indices were considerably reduced compared to the original images, regardless of the VPIF or the combination of VPIFs selected for normalization, which demonstrates the method's efficacy. The correlation coefficients between the CFs among VPIFs for any spectral band (and all bands overall) were calculated to be at least 0.85 and were significant at $P = 0.95$, indicating that the normalization procedure was comparably performed regardless of the VPIF chosen. ARIN method was designed only for agricultural and forestry landscapes where VPIFs can be identified.

Editor: Guy J-P. Schumann, NASA Jet Propulsion Laboratory, United States of America

Funding: All the external funding of this research was provided by the Spanish Commission of Science and Technology (http://www.idi.mineco.gob.es/portal/site/MICINN/), Ministry of Economy and Competition, through the projects AGL2007-60926 and AGL2010-15506. The funders had no role in study design, data collection and analysis, decision to publish, or preparation of the manuscript.

Competing Interests: The authors have declared that the ARIN software was registered, and the ARIN procedure was patented in the Spanish Office for Patents and Trademark (ARIN procedure for the automatic radiometric normalization of multi-temporal series images based in pseudo-invariant features [in Spanish]; Spanish Patent Office, Application date 28 December 2012, Priority number 201232054). Its owner is the author's institution, CSIC (Spanish Council for Scientific Research).

* E-mail: lgarciatorres@ias.csic.es

Introduction

Remote sensing observations are usually instantaneous and are affected by many factors, such as atmospheric conditions, sun angle, viewing angle, dynamic changes in the soil and plant–atmosphere system, and changes in the sensor calibration over time [1,2]. The goal of radiometric corrections is to remove or compensate for all of the above effects. Exceptions to this procedure include corrections for actual changes in the ground target to retrieve surface reflectance (absolute correction) or to normalize the digital counts obtained under the different conditions and to establish them on a common scale (relative correction) [2].

Absolute radiometric corrections (ARC) make it possible to relate the digital counts in satellite image data to radiance at the surface of the Earth. This relation requires sensor calibration coefficients, an atmospheric correction algorithm and related input data among other corrections [2]. A considerable amount of research has been performed to address the problem of correcting images for atmospheric effects. Radiometric normalization of remote imagery requires all of the previously mentioned information at the time of image acquisition. For most historically remote scenes, these data are not available, and for planned acquisitions,

the data may be difficult to obtain [3]. Consequently, absolute surface reflectance retrieval may not always be practical [2].

Relative radiometric normalization (RRN) based on the radiometric information intrinsic to the images themselves is an alternative whenever absolute surface radiances are not required. RRN of imagery is important for many applications, such as land cover change detection, mosaicking and tracking vegetation indices over time, and supervised and unsupervised land cover classification [2,3,4,5]. Several methods have been proposed for the RRN of multitemporal images collected under different conditions at different times [2,4]. All methods operate under the assumption that the relationship between the at-sensor radiances recorded at two different times from regions of constant reflectance is spatially homogeneous and can be approximated by linear functions. The most difficult and time-consuming aspect of these methods is the determination of suitable time-invariant features that will serve as the basis for normalization [3]. Ya'allah and Saradjian [5] developed an automatic normalization method based on regression applied to unchanged pixels in urban areas. The proposed method is based on efficient selection of unchanged pixels through image difference histogram modeling using available spectral bands and calculation of relevant coefficients

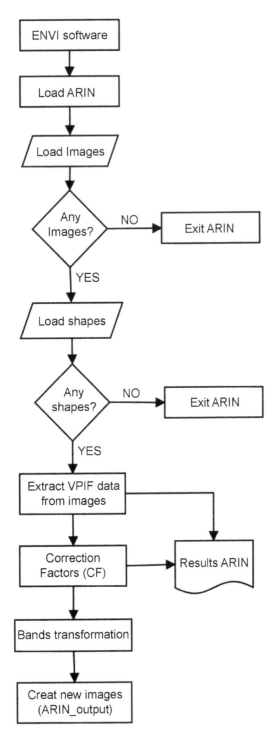

Figure 1. ARIN flowchart.

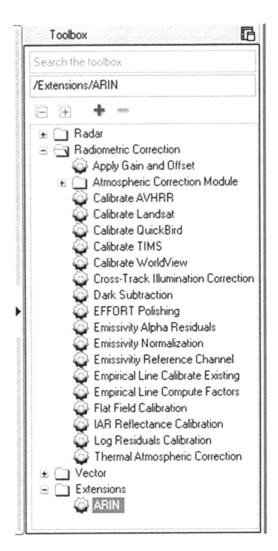

Figure 2. View of radiometric correction available in ENVI5.0 and ARIN software as an extension.

for dark, gray and bright pixels in each band. Yang and Lo [6] studied five methods of RRN applied to Landsat images. This method includes pseudo-invariant features, radiometric control set, image regression, no-change set determined from scatter-grams, and histogram matching, all of which require the use of a reference-subject image pair. Factors that affect the performance of RRN include land-use/land-cover distribution, water-land proportion and topographic relief [6]. Ground reference data are expensive and difficult to acquire for most remotely sensed

(satellite) images, and the selection of PIF is generally subjective [2]. In practice, vegetative targets of absolutely constant reflectance do not exist. Therefore, the concept of PIF is adopted with the assumption that the reflectance is constant over time [2].

Several authors have developed powerful statistical approaches to determine invariant features for the atmospheric normalization of image pairs. Hall et al. (1991) [7] and Coppin and Bauwer (2004) [8] developed radiometric rectification techniques for land cover change detection through the use of landscape elements whose reflectance values are nearly constant over time. Hall et al. (1991) [7] selected PIFs with two sets of data, bright and dark. The two sets were selected in different images by visual inspection. Du et al. (2002) [2] developed a new procedure for radiometric normalization between multitemporal images of the same area. In this method, the selection of PIF is performed statistically, and quality control and principal component analysis (PCA) are used to find linear relationships between temporal images of the same area. Several authors have proposed a change-detection technique called multivariate alteration detection (MAD), which is invariant to linear and affine scaling [3,9]. Thus, if MAD was used for change detection applications, pre-processing by linear radiometric normalization is superfluous. An iteratively re-weighted modification of the MAD transformation (IR-MAD) established a better background of no change upon which significant changes can be examined [10–11]. Some authors have used the IR-MAD transformation for relative radiometric normalization of multi-temporal images and MAD transformation for change detection [12]. Others converted digital number values to reflectance directly by relative radiometric normalization using IR-MAD [13]. Baisantry et al. (2012; [14]) performed RRN using MAD transformation and selected PIFs automatically through the Bin-Division Method. Kim et al. (2012; [15]) developed a method designed to automatically extract pseudo-invariant features for the RRN of hyperion hyperspectral images and used band-to-band linear regression. Philpot and Ansty (2013; [16]) developed an analytical formula that relates pseudo-invariant features (PIFs) to the radiometric properties of the scenes. The formula is then inverted to yield an estimate of the ratio of the transmission spectra of the two images given the path radiance for each scene and a set of invariant features. Sadeghi et al. (2013) [17] proposed an automated RRN to adjust a non-linear based on artificial neural network and unchanged pixels.

QUAC (quick atmospheric correction) and FLAASH (fast line-of-sight atmospheric analysis of spectral hypercubes) are atmospheric correction modules used in ENVI image processing software (Exelis-Visual Information Solutions, Inc. 4990 Pearl East Circle Boulder, CO 80301 USA, http://www.exelisvis.com). QUAC is an on-the-fly method for use in real-time data processing that determines parameters directly from the information contained in the scene using the observed pixel spectra [18]. FLAASH is a physics-based correction method built on MOD-TRAN4 atmospheric correction software [19]. FLAASH allows the user to define all parameters that influence atmospheric absorption and scattering, such as relative solar position, atmospheric, aerosol, and scattering models, and visibility parameters, among others. The advantage of QUAC is that an in-scene approach is easily implemented, while FLAASH uses a very diverse atmospheric ancillary parameter, and the data are therefore highly tunable by the image expert. Hu et al. (2011, [20]) compared the FLAASH and MAD normalization methods on Landsat time-series images. Other authors applied FLAASH to change detection applications [21].

To our knowledge, no information is available on relative image normalization (ARIN) of multitemporal agricultural scenes based on VPIFs or on the development of software to achieve this semi-automatically. Our specific objectives were as follows: 1) to describe the ARIN procedure and implement it in the GeoEye-1 multitemporal image series; 2) to comparatively study the selected VPIFs in relation to the ARIN method efficacy; 3) to compare ARIN-transformed multitemporal images to the original (ORI) and to the FLAASH and QUAC calibrated images; and 4) to develop semi-automatic software to normalize any set of multi-temporal remote imagery by identifying VPIFs.

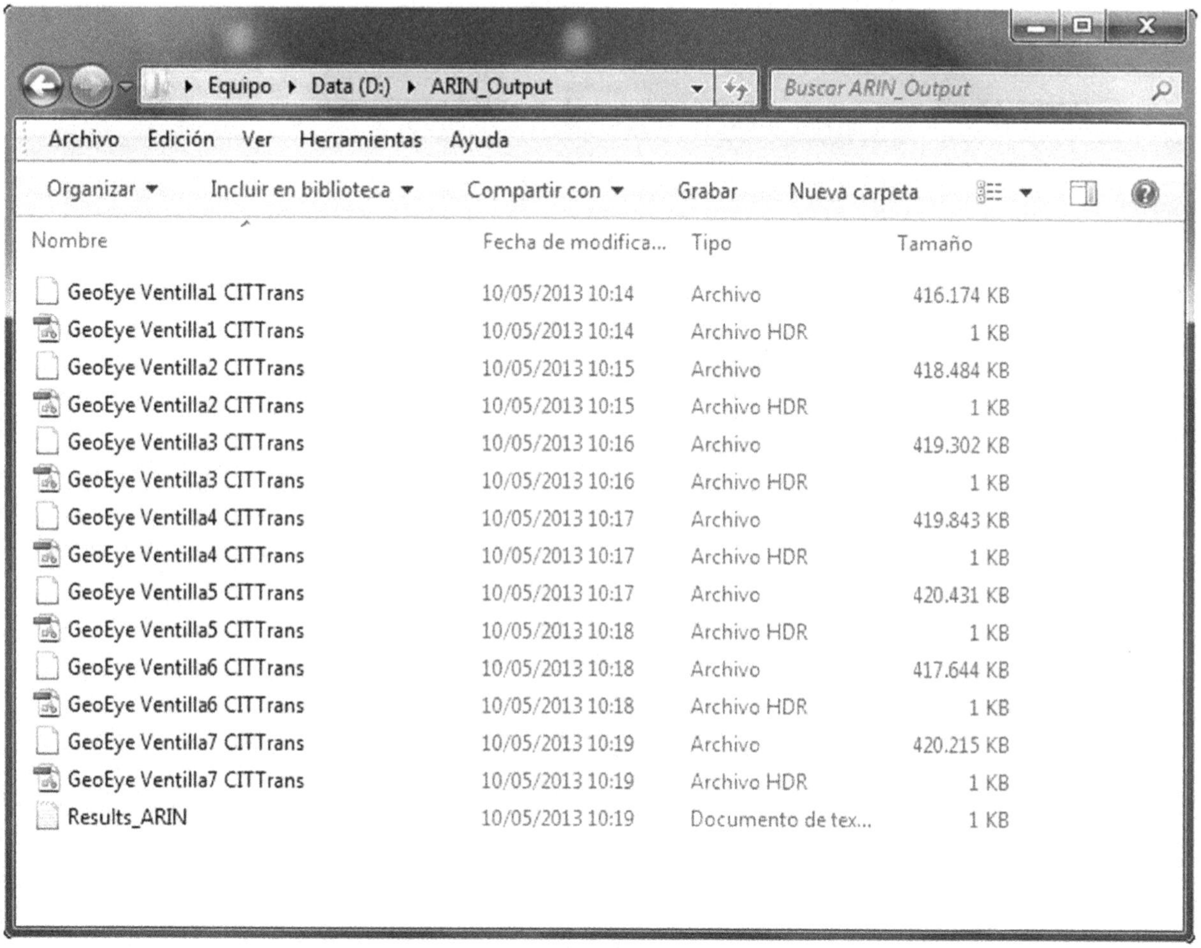

Figure 3. ARIN_output: normalized images ("transf") using CIT VPIF.

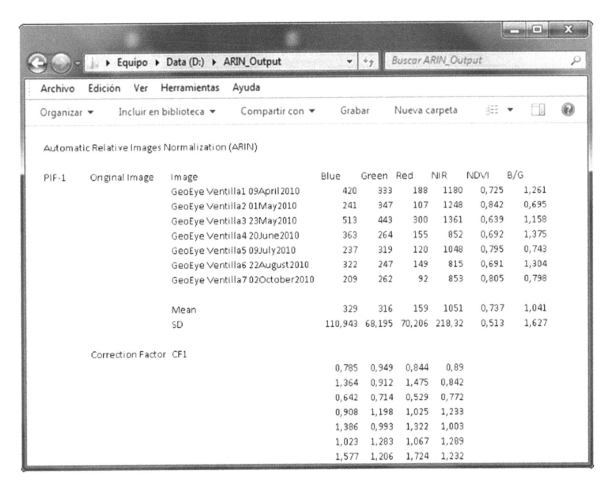

Figure 4. ARIN_results: transposed.txt and Excel files. The VPIF spectral bands of the original images and the corresponding band correction factors are shown.

Materials and Methods

1. ARIN Procedure and Software

The procedure developed for the relative normalization of multitemporal images consists of the following steps: 1) selecting one or several VPIFs; 2) defining the same parcel or parcels for each selected VPIF in each multitemporal image; 3) extracting the VPIF spectral band data for each image; 4) calculating the correction factors (CFs) for each image band to fit each band value to the average value of the image series; and 5) obtaining the normalized images by transforming each band through CF linear functions. Further information of VPIF and of vegetative variant features (VVF) will be given later in this article. Basically they coincide with permanent orchards/mature tree plantations and annual/herbaceous crops, respectively. We select the VPIF parcels at random, among many parcels of very similar characteristics available in our agricultural scene. Main steps of ARIN procedure can be achieved by conventional image processing menus including the parcel definition, the spectral band data extraction, and the image liner transformation through the estimated correction factors (CF), as will be later defined. The CF can be calculate manually or in a excel sheet once the VPIF spectral band values has been extracted.

Environment for Visualizing Images (ENVI 4.8 and ENVI 5.0, Exelis-Visual Information Solutions) software was used to visualize and process the images. Generally, mature non-deciduous tree orchards and permanent lawn green cover are eligible to be

VPIFs, and at least one must be present in the scene for normalization. A parcel of the selected VPIF needs to be drawn through the ROI/SHAPE menu in one image and then moved to the rest of the image series through the VECTOR/SHAPE menu (convert the ROI to a DXF vector). The B, G, R and NIR spectral bands of the selected VPIF can be extracted for each image through the ROI/SHAPE Tool Statistical menu.

The CFs of any VPIF spectral band, for example, the G band of the image i (CF_{Gi}), are defined as the ratio G_m/G_i, where G_m is the average original value of the G band in the original image (ORI), and G_i is the band value of image i. Then, each band of each ORI will be transformed by applying the corresponding linear CF through the Basic Tool-Band Math menu. The normalized image will be composed of the transformed bands through the Layer Stacking menu. To semi-automatically perform the previously described steps (steps 2 through 5), the so-called ARIN software and procedure were developed [22,23]. The ARIN flowchart is shown in Figure 1. The main ARIN software screens are shown in Figures 2, 3, and 4. A partial view of the vegetative pseudo-invariant features (VPIFs) used in this study is shown in Figure 5.

2. ARIN Procedure Validation

2.1 Study location and GeoEye-1 images series. Seven multi-spectral and panchromatic GeoEye-1 satellite images (GeoEye-1, 2012), each covering approximately 100 km^2, were

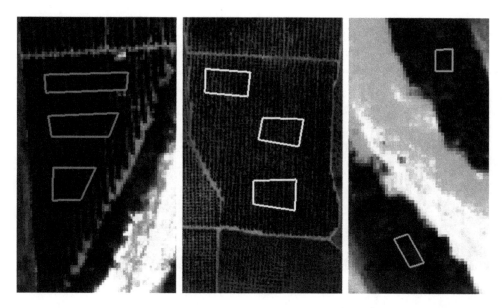

Figure 5. Partial view of the vegetative pseudo-invariant features (VPIFs) used in this study. a) CIT-citrus, b) OLI-olives, and c) POP-poplar groves. The VPIF SHP files are transposed to other multi-temporal images during the ARIN process.

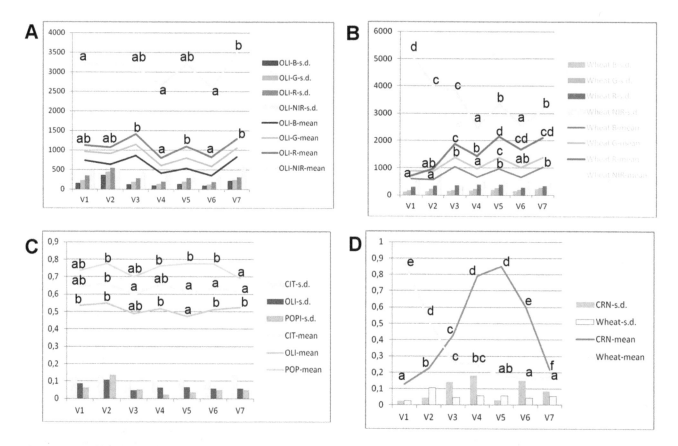

Figure 6. a) and b) Spectral band evolution in pseudo-invariant (OLI, olive) and variant (wheat) features; c) and d) NDVI evolution in pseudo-invariant features (CIT, citrus; OLI, olives, and POP, poplar groves); and d) winter wheat and corn as representative cropping systems that have variant vegetative features. The abscissa is the remote image timing (V1- early April to V7 - early October, interval of approximately one month; data are presented as the means, and vertical bars represent the standard deviation (s. d.) of six parcels of approximately 3 ha each).

Figure 7. Wheat crop evolution. A: early April, grain filling growth stage, showing intense green color and therefore high NDVI values; B: early May, mid-senescence, green-yellowish color, and mid-NDVI values; C: late May/early June, late senescence, predominant yellow color, low NDVI values; D: stubble, typical of mid-June throughout the summer. These growth stages roughly coincide with the V1, V2, V3 and V5 satellite images taken, respectively, and with the wheat NDVI data evolution shown in Figure 6d.

taken over the same area of LaVentilla village (a province of Cordoba, southern Spain) from April to October 2010. The geographic coordinates (Universe Transverse Mercator System, Zone 30 North) in the upper-left corner of the images were $X = 315206$ m/$Y = 4186133$ m. The images were taken on April 9, May 1, May 23, June 20, July 9, August 22 and October 2 and are named V1 to V7, respectively. The panchromatic image was 0.50 m pixels^{-1}, and the multi-spectral-image spatial resolution was 2.00 m pixels^{-1}, providing information on blue (B, 450–510 nm), green (G, 510–580 nm), red (R, 655–690 nm) and near-infrared (NIR, 780–920 nm) spectral bands. The swath width was 15.2 km. The ground was predominantly flat, with an average slope grade of 2.12%. The georeferencing accuracy of the GeoEye-1 images was improved by using ground control points (GCPs) and image-to-image co-registration [24].

2.2 Land uses and selected variant and invariant vegetative features. The LaVentilla area was surveyed approximately every 3 weeks from April to October 2010 to identify the crop of each parcel, its stage of development, and any key agricultural features. A total of 23 land uses were identified in the

Geo-Eyes-1 scenes. Vegetative systems, such as alfalfa, avena, broad beans, citrus orchards, chickpeas, corn, cotton, Mediterranean forest, olive orchards, potatoes, sunflower, rapeseed, poplar groves and winter wheat, among others, were identified. Additionally, non-vegetative land uses, such as rivers, water reservoirs, paved roads, bare soil roads, and civil buildings, were also found. The phenotypes of some herbaceous vegetation parcels, such as wheat, sunflower, corn and cotton, varied considerably throughout the growing season and can thus be designated variant vegetative features (VVFs). The phenotypes of high density adult tree plantation parcels, such as citrus orchards (CIT), olive orchards (OLI), and poplar groves (POP), demonstrate much less vegetation change throughout the cropping season. Thus, they can be designated pseudo-invariant vegetative features (VPIFs). To show the differences between VVF and VPIF parcels, the spectral bands and NDVI vegetative index evolution were determined in four parcels of approximately 0.3–0.5 ha for each selected VPIF (CIT, OLI and POP) and for wheat (WHT) and corn (CRN), as representatives of the winter (autumn-sown) and summer (spring-sown) VVFs.

Figure 8. Corn crop evolution. A: late May and early June, corn at the vegetative growing phase, characterize with increasing NDVI values, and coinciding with the V3–V4 satellite images of this stage; B: flowering stage, July, V5 image; C: senescence period, which take place in the second part of August (NDVI values decrease; satellite image V6); and D: corn stubble, beyond mid-September (satellite image V7). Corn NDVI data evolution shown in Figure 6d.

2.3 Implementation of the ARIN procedure. CIT, OLI and POP were used as VPIFs. Four parcels of approximately 0.3–0.5 ha were drawn for each VPIF using the regions of interest (ROI-VECTOR)/SHAPE menu of ENVI. The ARIN procedure provided the B, G, R and NIR spectral bands, the NDVI and the G/B vegetative indices for the original and transformed/normalized images. Normalization was achieved using a single VPIF and by validating the other two VPIFs or by using two VPIFs consecutively and validating with the other.

2.4. Absolute corrections using QUAC and FLAASH. To allow comparison with ARIN, the original V1 to V7 images were transformed using QUAC and FLAASH software. QUAC software requires the presence of dark and bright pixels in the images to serve as a basis for the implemented corrections. FLAASH was implemented by fitting to GeoEye-1 sensor specifications and to the atmospheric mid-latitude summer geographic area where the satellite images were taken. This area was aerosol rural, the highest image ground elevation was 150 m, the water column retrieval parameter was 2.92, and the scene

visibility parameter was 100 km for all images except for the V7 image (140 km).

2.5. Statistical parameters. VPIF and VVF parcels spectral band and vegetation index data were subjected to analysis of variance and means were separated at the 5% level of significance by the least significance difference (LSD) test with the use of SPSS Statistical-21 software (IBM North America, New York, NY, United States). For any original or transformed image, the VPIF mean, range, standard deviation (s. d.) and root mean standard error (RMSE) of the band spectral values, NDVI and B/G vegetation indices were determined. The root mean square error (RMSE) of the series of images was calculated by the following equation [25]:

$$RMSE = \left[\sum_{i=1}^{n} \left[(X_i - X_m)^2 \right] / n \right]^{1/2},$$

Where n is the number of images, X_i is the values of each image

Figure 9. Canopy structure of citrus (A and B) and olive (C and D) orchards vary very little throughout the year and particularly in the main growing season (April to October). Photographs A and C were taken early May and B and D late September, roughly coinciding with the V2 and V7 satellite images of this study. In photograph D can be appreciated the orange fruit but not the olives in photograph D at the distance where the photograph was taken. Due to the little changes in the trees canopy and soil surface covered the NDVI values of citrus and olive are very similar throughout the growing season (Figure 6d), and can be used as pseudo-invariant features for radiometric normalization as shown in this study.

and X_m is the average of all images. The smaller range, s. d. and RMSE of data in a given series of images indicated more uniformity among images than in the others.

The correlation coefficients and the level of significance between CFs were determined across VPIFs for each image band and for all images.

Results and Comments

1. Evolution of Vegetative Variant and Pseudo-invariant Features

Generally, the evolution of the spectral bands and the NDVI throughout the growing season varied to a greater extent in VVFs than in VPIFs (Figures 6, 7, 8, and 9). For example, the NIR band average, range and s. d. for OLI were 466, 240, and 81 (Figure 6a), respectively, whereas for WHT, these values were 3494, 2683 and 936, respectively (Figure 6b), for the images series. Similarly, the NDVI evolution varied significantly in VVFs, while it was

relatively stable in VPIFs. For example, the NDVI mean, range and s. d. were 0.63, 0.064 and 0.024 for CIT and 0.34, 0.70 and 0.26 for WHT, respectively (Figure 6c & d). These data confirmed that the phenotype, morphology, development, and observable physical characteristics of perennial plantation are very stable throughout the agricultural season for VPIFs (Figure 9), while the opposite is true for the herbaceous cropping systems (VVFs, Figures 7 and 8). This observation has been obvious to field workers and agronomists for many years. This simple, evident finding is very important for land use classification of agricultural scenes through remote sensing.

2. Relative Radiometric Normalization

2.1 Using a single VPIF. The spectral band values of the original GeoEye-1 image series for the given VPIFs varied considerably among the images (Table 1). For example, the CIT B band digital values for the images V2 and V6 were 241 and 322, respectively, and the POP NIR values for the same images were

Table 1. Vegetative pseudo-invariant feature (VPIF) spectral band values of the original (ORIG) and VPIF ARIN-transformed (-transf.) images.

VPIF	Spectral Band	ImagesType	Series of images[1]							Statistical			
			V1	V2	V3	V4	V5	V6	V7	Mean	Range	S. d.[2]	RMSE
CIT[1]	B	ORIG[2]	420e[2]	241b	513f	363b	237a	322d	209a	329 g[3]	304 g	111 g	103 g
		POP-transf.	334b	335b	347b	310ab	341b	324b	307a	328 g	40 h	15 h	14 h
		OLI-transf.	343ab	322a	346b	320a	330a	338ab	283a	326 g	63 h	21 h	20 h
	G	ORIG	333c	347c	443d	264ab	319c	247a	262b	316 g	196 g	68 g	63 g
		POP-transf.	322bc	346cd	350cd	274a	324bc	302ab	281a	314 g	76 h	30 h	28 h
		OLI-transf.	342d	323bc	352cd	289ab	309bc	323bc	266a	315 g	86 h	30 h	28 h
	R	ORIG	188c	107b	300e	155cd	120a	149c	92a	159 g	208 g	70 g	65 g
		POP-transf.	158ab	158ab	197b	133a	175ab	142a	129a	156 g	68 h	24 h	23i
		OLI-transf	163ab	136a	190b	140a	153a	163ab	135a	154 g	55 h	20 h	19 h
	NIR	ORIG	1180c	1248d	1361e	852a	1048b	815a	853a	1051 g	546 g	218 g	202 g
		POP-transf.	1173d	1095cd	1192d	904a	1005b	906a	1031bc	1048 g	298 h	117 h	108 h
		OLI-transf.	1247d	1120d	1088cd	951ab	1000bc	1029bc	895a	1047 g	342 h	117 h	108 h
OLI	B	ORIG	462e	283b	559f	428d	271a	360c	278b	377	288 g	110 g	102
		POP-transf.	367a	393b	378ab	365a	390ab	362a	409b	381	47 h	17i	17 h
		CIT-transf.	362a	386c	359a	388bc	376ab	368ab	438c	382	79i	27 h	26 h
	G	ORIG	349c	384d	450e	327a	368c	273a	352c	358	177 g	54 g	50 g
		POP-transf.	337a	383bc	355ab	340a	375bc	335a	377bc	357	48 h	21 h	19 h
		CIT-transf.	331ab	350bc	321a	392c	366b	351b	425d	362	104i	36i	34i
	R	ORIG	312d	214b	427e	300	212ab	248c	185a	271	242 g	83 g	77 g
		POP-transf.	261ab	314c	281b	259ab	309c	237c	259ab	274b	77 h	28 h	27 h
		CIT-transf.	263a	316c	226a	308c	280ab	264a	319c	282	93 h	34 h	34 h
	NIR	ORIG	731c	861d	966e	692b	809c	612a	736bc	772	354 g	117 g	108 g
		POP-transf.	727b	756b	846c	735ab	776b	680a	890c	773	210 h	72 h	67 h
		CIT-transf.	651a	725a	746a	854b	812b	789a	907b	783	256 h	85 h	80 h
POP	B	ORIG	428c	245b	504d	399c	237a	339b	231a	340	273 g	108 g	100 g
		CIT-transf.	336ab	335ab	324a	362ab	329a	346ab	365b	342	41 h	16 h	15 h
	.	OLI-transf.	350ab	327ab	340ab	352ab	330ab	355b	314a	338	41 h	15 h	14 h
	G	ORIG	307c	298cd	376d	286bc	292bc	243ac	277b	297	133 g	40 g	37 g
		CIT-transf.	292a	272a	268a	342a	290a	311a	335a	301	74 h	29 h	27 h
		OLI-transf.	315ab	277a	299ab	313ab	283ab	317b	282ab	298	40i	17i	16i
	R	ORIG	186bc	106a	237d	181b	107a	163b	111a	156 g	131 g	50 g	46 g
		CIT-transf.	157ab	157ab	125a	186ab	141ab	174ab	192b	162 g	67 h	24 h	23 h
		OLI-transf.	162a	135a	151a	164a	137a	179a	163a	156 g	44 h	16i	15i
	NIR	ORIG	879b	995b	997b	823ab	911b	785ab	722a	873 g	275 g	104 g	96 g

Table 1. Cont.

VPIF	Spectral Band	ImagesType	Series of images[1]							Statistical			
			V1	V2	V3	V4	V5	V6	V7	Mean	Range	S. d.[2]	RMSE
		CIT-transf.	782a	838a	770a	1015b	913ab	1013b	890a	889 g	245 g	100 g	94 g
		OLI-transf.	928b	893ab	797a	918b	869ab	992b	758a	879 g	234 g	80 g	74i

[1]Series of images: from V1, early April, to V7, October.
[2]Abbreviations: ORI, original images; CIT, citrus orchards; OLI, Olive orchards; POP, poplars grove; -transf., transformed images; B, blue; G, Green; R, red; NIR, near infra-red; S. d., standard deviation; RMSE, Root Mean Square Error.
[3]For each VPIF, spectral band and image type the data of the multitemporal images followed by the same letter are not significantly different at P≥0.05.
[4]For each VPIF and spectral band statistical data of image types followed by a different letter are significantly different at P≥0.05.

995 and 785, respectively (Table 1). Consequently, the vegetation indices varied considerably among the original images at any given VPIF (Table 2). For example, the CIT NDVI values varied from 0.64 at V3 to 0.81 at V7, and the POP NDVI values ranged from 0.62 at V3 to 0.81 at V2. This wide variation in the VPIF band digital counts and vegetation indices among images indicates that the radiometric normalization process is highly recommended for a comparative follow-up of cropping systems and other environmental features in any multitemporal image series.

Generally, the range and s. d. of the spectral bands and vegetation indices in the ARIN-transformed images were considerably reduced compared to those of the original images (Table 1 and 2), regardless the VPIF considered. First, it should be noted that for a single transformation (using just one VPIF), the spectral bands and vegetation indices of the VPIF taken as reference produce exactly the same value for any transformed image. This value is the average of the image series, for example, 329 and 1051 for the CIT B and NIR spectral bands, respectively. Additionally, the corresponding range and s. d. are negligible.

The ARIN process was an efficient normalization process regardless of the single VPIF or the combination of VPIFs chosen. The selection of the single VPIF used for the ARIN normalization process only slightly affected the normalization results of other VPIFs. For example, the range, s. d. and RMSE of the CIT VPIF B and NIR bands at the POP transformed images series were 40, 15 and 14 and 288, 117 and 108, respectively, whereas these values were 304, 111 and 103 and 546, 218 and 202 in the original images (Table 1). Similarly, the range, s. d. and RMSE values for the R band of the CIT VPIF in the OLI transformed images were 55, 20 and 19, compared to 208, 70 and 65 for the original images, respectively (Table 1).

2.2 Using two VPIFs consecutively. For each selected VPIF, the normalization effect caused by the ARIN procedure was also determined using two VPIFs consecutively; thus, the potential slight stationary phenotypic variation could be balanced for each VPIF. Applying the ARIN process using two VPIFs consecutively is also an effective method to normalize the multitemporal series of images. For example, after normalization with the pseudo-invariants OLI+POP, the CIT NDVI range, s. d. and RMSE of the image series were 0.08, 0.03 and 0.02, respectively, in comparison to 0.20, 0.07 and 0.07 for the original images (Table 2). Similarly, the consecutive implementation of ARIN CIT+POP resulted in OLI NDVI range, s. d. and RMSE values of 0.14, 0.05 and 0.04 in comparison to 0.21, 0.10 and 0.09 for the original images, respectively.

2.3 VPIF correlation factors. The VPIF spectral band CFs used to implement the ARIN linear normalization procedure varied greatly among the spectral bands for any given image (Figure 4) and among images for any given spectral band. Furthermore, the correlation coefficients between the CFs among the VPIFs for any spectral band and for all bands were found to be at least 0.85 and were significant at P = 0.95 or higher (Table 3). This finding also demonstrates that the ARIN normalization process is efficient regardless of the VPIF selected.

3. ARIN vs. QUAC and FLAASH

Generally, for the series of GeoEye-1 images studied, ARIN was more efficient than QUAC and as efficient as FLAASH, varying slightly with the VPIF selected. For example, if we consider the VPIF CIT, the NDVI s. d. of the original, OLI+POP-transf., QUAC and FLAASH images were 0.07, 0.03, 0.05 and 0.08, respectively, and the same statistics for the B/G index were 0.29, 0.06, 0.11 and 0.07 (Table 2). Considering the VPIF POP and the

s. d. and RMSE statistical, the results are better for ARIN CIT+ OLI-transf. (0.03 and 0.03) and FLAASH (0.03 and 0.03), followed by QUAC (0.05 and 0.04) compared to the original images (0.08 and 0.07) (Table 2). Considering the B/G index, the ARIN-transf. procedure was also more effective than QUAC and FLAASH in any VPIF studied.

Discussion

In our study, the VPIF spectral band and vegetation index values of the original GeoEye-1 images varied considerably among the images, indicating that the calibration or radiometric normalization process is highly recommended for comparative follow-up of cropping systems. In fact, the calibration and normalization of multitemporal images has been a challenge in remote sensing for decades [1–16].

ARC relates image digital counts to radiance at the surface of the Earth and requires sensor calibration coefficients, an atmospheric correction algorithm and related input data, among other corrections [2]. For most historical images, such data are not available, and for planned acquisitions, they may be difficult to obtain [3]. Consequently, ARC retrieval is not often a practical method [2]. RRN is based on the radiometric information intrinsic to the images themselves and is an alternative whenever absolute surface radiances are not required, as in change detection applications or for supervised land cover classification [3,4].

To our knowledge, no RRN methods for multitemporal remote images using VPIF as a reference have previously been developed. The concept of PIF is adopted with the assumption that the reflectance is constant over time [2]. Moreover, any individual plant, cropping system, or vegetative feature varies with time, and therefore, there is a unanimous agreement that the reflectance is not an absolute invariant. As we have shown, most agricultural areas have vegetative parcels that change drastically throughout the growing season, such as annual herbaceous crops (VVFs), while others features, such as dense forest, permanent lawn or dense non-deciduous orchard plantations, remain comparatively invariant throughout the growing season (VPIFs). We have shown drastic differences between the selected VVFs and VPIFs in the spectral bands and vegetation index evolution. The phenotypic or morphological aspects of VPIFs are well known, and therefore, the light reflectance changes very little throughout the annual growing season, which is a key factor for the land use classification of agricultural scenes through remote sensing. Additionally, the pseudo-invariability of VPIF is the characteristic used in the ARIN procedure to normalize a set of images of a common scene.

ARIN method was developed for the radiometric normalization of multitemporal images of agricultural and forestry scenes where vegetative pseudo-invariant features (VPIFs) can be identified. In our work, we have shown that selected mature CIT (citrus orchards), OLI (olive orchards) and POP (poplar groves) in a Mediterranean landscape can be chosen efficiently as VPIFs throughout spring and summer. Generally, the ARIN normalization process efficiently produced relatively uniform data for all images, regardless of the single VPIF or the combination of VPIFs chosen. This result is likely because the range and s. d. of the spectral bands and vegetation indices in the transformed images are considerably reduced compared to those of the original images.

Regardless of the single VPIF used to estimate the band CFs for the image series transformation, the results were relatively normalized when compared to those of the original images. The VPIF spectral band correction factors (CFs) used to implement the ARIN linear normalization procedure varied greatly among spectral bands for any given image and among images for any given spectral band. Furthermore, the high and statistically significant correlation coefficients between the CFs among the VPIFs for any spectral band and for all bands suggest that the ARIN normalization process was efficient regardless of the VPIF selected.

Implementing the ARIN procedure consecutively using two VPIF was also an efficient method of normalizing the multitemporal images series. Generally, for the multitemporal series of GeoEye-1 images studied, ARIN was more efficient than QUAC and as efficient as the FLAASH absolute calibration method. The advantage of the ARIN method is that weather calibration parameters are not necessary, whereas they are required for the highly tunable FLAASH methods.

VPIF size is clearly related to the image spatial resolution and should have a sufficient number of pixels to provide a solid average of the selected vegetative feature. With medium to high spatial resolution images from the satellite GeoEye-1 (i.e., <5 m pixel), VPIF size normally coincides with uniform parcels of approximately 0.3 to 0.5 ha or larger. Moreover, ARIN can also be used with very high spatial resolution images (i. e. 3 to 5 cm pixel) such as those provided by unmanned aerial vehicles (UAV) [26]. In UAV images a VPIF size of 2 to 4 m^2 will cover a high number of pixels to provide a solid pseudo-invariant vegetative feature sample, and in practice, it may coincide with a non-deciduous tree of 2 to 4 m^2. UAV images can be normalized through the ARIN procedure avoiding the use of the barium sulfate standard spectralon panel, which is placed in the middle of the field to calibrate data [26].

Step-by-step implementation of the VPIF-based ARIN normalization procedure through the available ENVI image-processing tools is time consuming, and therefore, it is not economically feasible. ARIN software [22] quickly and easily executes ARIN and generates the transformed images; consequently, its development is essential for the practical application of the ARIN method. However ARIN method can be applied only in any agricultural and forestry landscapes where a VPIF can be identified.

Conclusions

A novel method for the radiometric normalization of multitemporal images, named ARIN, was developed to be used in agricultural and forestry scenes where a vegetative pseudo-invariant features (VPIFs) can be identified. This new procedure identifies one common VPIF parcel in all scenes, extracts the spectral band values of each image and transforms them to common band values through linear transformation. We validated ARIN using a series of GeoEye-1 satellite images of one scene. ARIN worked correctly, regardless of the three VPIFs considered (citrus orchards, olive orchards and poplar groves). The ARIN method was slightly more efficient than the absolute calibration QUAC method and as efficient as the highly tunable FLAASH method, which uses solar position and weather calibration parameters. Implementing the ARIN procedure through conventional image processing is time consuming. The software ARIN executes ARIN semi-automatically in an economically feasible manner.

Table 2. Selected vegetative pseudo-invariant feature (VPIF) vegetation indices of the original (ORIG) and VPIF ARIN-, QUAC- and FLAASH-transformed (-transf.) images.

VPIF	Vegetation Index	Images	Series of images[1]							Statistical			
			V1	V2	V3	V4	V5	V6	V7	Mean	Range	S. d.[2]	RMSE.[2]
CIT	NDVI	ORIG[2]	0.73b[3]	0.84d	0.64a	0.69a	0.79c	0.69a	0.81cd	0.74 g[4]	0.20 h	0.07 h	0.07i
		OLI+POP-transf.	0.77ab	0.75a	0.72a	0.75a	0.71a	0.73a	0.78b	0.74 g	0.08 g	0.03 g	0.02 g
		QUAC-transf.	0.79a	0.85ab	0.79a	0.90a	0.79a	0.90b	0.89ab	0.84 h	0.12 g	0.05 g h	0.05 hi
		FLAASH-transf.	0.76b	0.69a	0.80b	0.74a	0.78b	0.78b	0.79bc	0.76 g	0.11 g	0.04 g	0.07i
	B/G	ORIG	1.26dc	0.69a	1.16c	1.38d	0.74a	1.30b	1.08a	1.05 h	0.68i	0.29i	0.45i
		OLI+POP-transf	1.02a	0.96a	0.98a	1.12a	1.04a	1.06a	1.08a	1.04 h	0.16 g	0.06 g	0.06 g
		QUAC-transf	0.44b	0.57c	0.24a	0.43b	0.54c	0.43b	0.47bc	0.45 g	0.33 h	0.11 h	0.17 g
		FLAASH-transf.	0.59b	0.64c	0.55b	0.59bc	0.47a	0.64c	0.66c	0.59 g	0.19 g	0.07 g	0.27 h
OLI	NDVI	ORIG	0.40a	0.60c	0.39a	0.40a	0.58c	0.42a	0.60c	0.48 g	0.21 h	0.10 h	0.09 h
		CIT+POP-transf.	0.46ab	0.40a	0.49b	0.47ab	0.42a	0.48bc	0.54c	0.47 g	0.14 g	0.05 g	0.04 g
		QUAC-transf.	0.37a	0.51bc	0.54c	0.60cd	0.45b	0.70d	0.66d	0.55 g	0.33	0.11 h	0.10 h
		FLAASH-transf.	0.50b	0.42a	0.45a	0.42a	0.49b	0.45a	0.45a	0.45 g	0.08 g	0.03 g	0.04 g
	B/G	ORIG	1.32c	0.74a	1.24c	1.31c	0.74a	1.32c	0.79b	1.07 g	0.59i	0.29i	0.271
		CIT+POP-transf.	1.08a	1.02a	1.05a	1.06a	1.03a	1.07a	1.07a	1.05 g	0.06 g	0.02 g	0.02 g
		QUAC-transf.	0.60c	0.76d	0.33a	0.60b	0.63c	0.56b	0.67d	0.59 g	0.42i	0.13 h	0.12 h
		FLAASH-transf.	0.54a	0.76c	0.68b	0.68b	0.61ab	0.82c	0.79c	0.70 g	0.28 h	0.10 h	0.09 h
POP	NDVI	ORIG	0.65ab	0.81d	0.62a	0.64a	0.79d	0.66b	0.73c	0.70 g	0.19 h	0.08 h	0.07 h
		CIT+OLI-transf.	0.70ab	0.73b	0.67a	0.69ab	0.72b	0.69ab	0.64a	0.69 g	0.09 g	0.03 g	0.03 g
		QUAC-transf.	0.73a	0.81b	0.81b	0.85c	0.80ab	0.88c	0.81b	0.81 g	0.15 h	0.05 g h	0.04 g
		FLAASH-transf.	0.70ab	0.70ab	0.75b	0.76b	0.75b	0.67a	0.76b	0.73 g	0.09 g	0.03 g	0.03 h
	B/G	ORIG	1.39b	0.82a	1.34b	1.40b	0.81a	1.40b	0.83a	1.14 h	0.58 i	0.30 i	0.27i
		CIT+OLI-transf.	1.11a	1.18a	1.14a	1.13a	1.16a	1.12a	1.12a	1.14 h	0.07 g	0.03 g	0.02 g
		QUAC-transf.	0.60b	0.82c	0.33a	0.63b	0.68b	0.58b	0.64b	0.61 g	0.49i	0.15 h	0.14 h
		FLAASH-transf.	0.72c	0.79c	0.67b	0.67b	0.59a	0.78c	0.77c	0.71 g	0.20 h	0.07 g	0.08 h

[1]Series of multitemporal images: from V1, early April, to V7, October.
[2]Abbreviations: ORi; original images; CIT, citrus orchards; OLI, Olive orchards; POP, poplars grove; -transf., transformed images; S. d., standard deviation; RMSE, Root Mean Square Error. Vegetation indexes: NDVI: $(NIR-R)/(NIR+R)$; B/: B and G are spectral bands.
[3]For each VPIF, vegetation index and image type the data followed by the same letter are not significantly different at $P \geq 0.05$.
[4]For each VPIF and vegetation index statistical data of image types followed by a different letter are significantly different at $P \geq 0.05$.

Table 3. Correlation coefficients between the spectral band CFs of VPIF CIT, OLI and POP.

	Spectral band				
VPIF[1]	B	G	R	NIR	Overall
CIT vs. OLI	0.97**	1.0**	0.97**	0.85*	0.96**
CIT vs. POP	0.98**	0.91*	0.92*	0.89*	0.93**
OLI vs. POP	0.99**	0.92*	0.93**	0.85*	0.92**
Overall	0.97**	0.93**	0.92**	0.87**	0.93**

[1]Abbreviations: VPIF; vegetative pseudo-invariant features; CIT, citrus orchards; OLI, olive orchards; POP, poplar groves; B, blue; G, green, R, read, NIR, near-infrared; * and ** Statistically significant at ≥95% and ≥99% probabilities.

Acknowledgments

We appreciate the interesting discussions regarding this paper with Drs. Francisca López-Granados and Jose-Manuel Peña-Barragán.

Author Contributions

Conceived and designed the experiments: LGT DGC. Performed the experiments: DGC JJCN LGT. Analyzed the data: DGC JJCN LGT AIDC. Contributed reagents/materials/analysis tools: DGC JJCN LGT. Wrote the paper: LGT DGC JJCN. Software development: JJCN AIDC.

References

1. Inoue Y (2003) Synergy of Remote Sensing and Modelling for Estimating Eco-physiological Processes in Plant Production. Plant Production Sci 6(1): 3–16.
2. Du Y, Teillet PM, Cihlar J (2002) Radiometric normalization of multitemporal high resolution satellite images with quality control for land cover change detection. Remote Sens Environ 82(1): 123–134.
3. Canty MJ, Nielsen AA, Schmidt M (2004) Automatic radiometric normalization of multi-temporal satellite imagery. Remote Sens Environ 91(3–4): 441–451.
4. Furby SL, Campbell NA (2001) Calibrating images from different dates to "like-value" digital counts. Remote Sens Environ, 77(2): 186–196.
5. Ya'allah SM, Saradjian MR (2005) Automatic normalization of satellite images using unchanged pixels within urban areas. Inf Fusion 6: 235–241.
6. Yang X, Lo CP (2000) Relative radiometric normalization performance for change detection from multi-date satellite images. Photogramm Eng Remote Sens 66: 967–980.
7. Hall FG, Strebel DE, Nickeson JE, Goetz SJ (1991) Radiometric Rectification: Toward a Common Radiometric Response Among Multidate, Multisensor Images. Remote Sens Environ 35(1): 11–27.
8. Coppin PR, Bauwer ME (1994) Processing of multitemporal Landsat TM imagery to optimize extraction of forest cover change features. IEEE Geoscience and Remote Sensing 60(3): 287–298.
9. Nielsen AA, Conradsen K, Andersen OB (2002) A change oriented extension of EOF analysis applied to the 1996–1997 AVHRR sea surface temperature data. Physics and Chem of the Earth 27(32–34): 1379–1386.
10. Nielsen AA (2007) The regularized iteratively reweighed MAD method for change detection in multi- and hyperspectral data. IEEE Transact Image Processing 16(2): 463–478.
11. Canty JM, Nielsen AA (2008) Automatic radiometric normalization of multitemporal satellite imagery with the iterative re-weighted MAD transformation. Remote Sens Environ 112(3): 1025–1036.
12. Broncano-Mateos CJ, Pinilla-Ruiz C, González-Crespo R, Castillo-Sanz A (2010) Relative Radiometric Normalization of Multitemporal images. Intern Journal Artificial Intelligence and Interactive Multimedia 1(3): 53–58.
13. Hu C, Tang P (2011) Converting DN value to reflectance directly by relative radiometric normalization. Proceedings CISP-4th Internat Congress Image and Signal Processing 1(3): 1614–1618.
14. Baisantry M, Negi DS, Manocha OP (2012) Automatic Relative Radiometric Normalization for Change Detection of Satellite Imagery, ACEEE Internat J Information Technology 2(2): 28–31.
15. Kim D, Pyeon M, Eo Y, Byun Y, Kim Y (2012) Automatic pseudo-invariant feature extraction for the relative radiometric normalization of hyperion hyperspectral images. GIScience and Remote Sensing 49(5): 755–773.
16. Philpot W, Ansty T (2013) Analytical Description of Pseudoinvariant Features, IEEE T. Geoscience and Remote Sensing 51(4-1): 2016–2021.
17. Sadeghi V, Ebadi H, Ahmadi F (2013) A new model for automatic normalization of multitemporal satellite images using Artificial Neural Network and mathematical methods Applied Mathematical Modelling 37(9): 6437–6445.
18. Bernstein LS, Xuemin J, Brian G, Adler-Golden SM (2012) Quick atmospheric correction code: algorithm description and recent upgrades, SPIE Optical Engineering 51(11), 111719.
19. Adler-Golden M, Berk A, Bernstein LS, Richtsmeier SC, Acharya PK, et al. (1998) FLAASH, a MODTRAN4 Atmospheric Correction Package for Hyperspectral Data Retrievals and Simulations. 1998 AVIRIS Geoscience Workshop, Jet Propulsion Laboratory, Pasadena, CA.
20. Hu Y, Liu L, Jiao Q (2011) Comparison of absolute and relative radiometric normalization use Landsat time series images. Proceedings of SPIE - The Internat Society Optical Engineer, Vol. 8006, Article 800616.
21. Hien LP, Kim DS, Eo YD, Yeon SH, Kim SW (2012) Comparison of radiometric pre-processing methods to detect change using aerial hyperspectral images. Intern J Advanc Comput Technol, 4(9): 1–9.
22. Garcia-Torres L, Gomez-Candón D, Caballero-Novella JJ, Jurado-Expósito M, Peña- Barragan JM, et al. (2012) ARIN software for the automatic radiometric normalization of multi-temporal series images based in pseudo-invariant features (in Spanish). Spanish Property Registration Office, Public Notary Protocol 1391/2012, 16 November 2012.
23. Garcia-Torres L, Gomez-Candón D, Caballero-Novella JJ, Peña-Barragan JM, Lopez-Granados F (2012) ARIN procedure for the automatic radiometric normalization of multi-temporal series images based in pseudo-invariant features (in Spanish). Spanish Patent Office, Application date 28 December 2012, Priority number 201232054, PCT/ES2013/070873.
24. Gómez-Candón D, López-Granados F, Caballero-Novella JJ, Gómez-Casero M, Jurado-Expósito M, et al. (2011) Geo-referencing remote images for precision agriculture using artificial terrestrial targets. Precision Agricul 12(6): 876–891.
25. ERDAS Inc. (1999). RMS error, page 362–365, in ERDAS Field Guide. 5th Edition. Atlanta, Georgia, 672 pp.
26. Torres-Sanchez J, López-Granados F, De-Castro AI, Peña-Barragán JM (2013) Configurations and specifications of an unmanned aerial vehicle (UAV) for early site weed management. PLOS ONE, 8, 3, e58210.

14

Alicyclobacillus Contamination in the Production Line of Kiwi Products in China

Jiangbo Zhang, Tianli Yue*, Yahong Yuan

College of Food Science and Engineering, Northwest A&F University, Yangling, PR China

Abstract

Alicyclobacillus are spoilage microbes of many juice products, but contamination of kiwi products by *Alicyclobacillus* is seldom reported. This study aims to investigate the whole production line of kiwi products in China to assess the potential risk of their contamination. A total of 401 samples from 18 commercial products, 1 processing plant and 16 raw material orchards were tested, and 76 samples were positive, from which 76 strains of microbes were isolated and identified as 4 species of *Alicyclobacillus*, including *Alicyclobacillus acidoterrestris*, *Alicyclobacillus contaminans*, *Alicyclobacillus herbarius* and *Alicyclobacillus cycloheptanicus*, and another 9 strains as 3 species of *Bacillus* by sequencing of their 16S rDNA. Through phylogenetic tree construction and RAPD-PCR amplification, it was found that there exist genotypic diversities to some extent among these isolates. Four test strains (each from one species of the 4 *Alicyclobacillus* species isolated in this study) could spoil pH adjusted kiwi fruit juice and some commercial kiwi fruit products with producing guaiacol (11–34 ppb).

Editor: Adam Driks, Loyola University Medical Center, United States of America

Funding: This study was financially supported by the China State "12th-Five-Year Plan" Scientific and Technological Schemes Support (2012BAK17B06) (http://www.most.gov.cn/), and the National Natural Science Foundation of China (31071550, 31171721)(http://www.nsfc.gov.cn/).The funders had no role in study design, data collection and analysis, decision to publish, or preparation of the manuscript.

Competing Interests: The authors have declared that no competing interests exist.

* E-mail: yuetl@nwsuaf.edu.cn

Introduction

Alicyclobacillus species are a group of thermo-acidophilic, non-pathogenic rod-shaped, endospore-forming bacteria which were first isolated from Japanese hot springs in 1967 and can cause the spoilage of many fruit juice products [1–4]. Fruit juice are susceptible to the growth of yeasts, mycelial fungi and lactic acid bacteria due to their ability to grow in high-acid environments [5], but the pH of fruit juice had been thought to be too low for the spore-forming bacteria to grow [6]. In 1982, a large-scale spoilage incident caused by a new type of bacterium happened in Germany [7]. The most common characteristic of the spoilage is a medicinal, antiseptic offensive off-odor in commercial pasteurised apple juice [7]. The microbe causing this spoilage incident was a thermo-acidophilic, endospore-forming bacterium that was later identified as *Alicyclobacillus acidoterrestris* [2]. After this spoilage incident the potential dangers of *Alicyclobacillus* species were realized by the juice industries. During the next decade, after many this kind of thermo-acidophilic bacteria, which were distinct from the bacteria of the genus *Bacillus*, have been reported, they were allocated to a new genus, *Alicyclobacillus*, based on comparative sequence analysis of their 16S rRNA genes and the presence of ω-alicyclic fatty acids in their cell membrane [2]. To date, 19 species, 2 subspecies and 2 genomic species that belong to the genus *Alicyclobacillus* have been identified [8], [9].

The thermal processes to inactivate pathogenic foodborne microorganisms and vegetative non-heat resistant spoilage microorganisms in juice are insufficient to inactivate *Alicyclobacillus* [10]. Chang and Kang [11] also proved that the pasteurisation treatments applied to fruit products are not sufficient to inactivate *Alicyclobacillus* in 2004. A 1998 survey by the European National Food Processors Association showed that 35% of juice manufacturing participants experienced *Alicyclobacillus* spp. related problems. In North America, as well as in Europe, a similar survey in the same year conducted by the Grocery Manufacturers Association of the USA also showed that almost 1/3 of the responded companies had experienced spoilage incidents which may be caused by *Alicyclobacillus* [12]. In 1999, Eiroa et al. [5] found that up to 14.7% of 75 orange juice samples from 11 Brazilian companies to contain *Alicyclobacillus*. According to a survey by the European Fruit Juice Association (AIJN) in 2005, 45% of the 68 companies from the fruit processing industry experienced *Alicyclobacillus* related product contaminations during the three years prior to the survey, including 33% experiencing contaminations more than once [13]. These reports provide evidence to support that the problem caused by *Alicyclobacillus* spp. is a major and widespread microbial spoilage concern for the juice and beverage industries, which has not been thoroughly studied [12], [14].

Alicyclobacillus spp. are soil-borne bacteria [15], and there have been reports of *Alicyclobacillus* isolated from orchards [16], [17]. *Alicyclobacillus* have been isolated from many kinds of fruit juices products and concentrated fruit juices products, including apple [18-20], pear [16], [17], [14], orange [21], banana [22], watermelon [21], mango [23], lemon [24], grapefruit and blueberry [14] and so on with apple and pear are the most frequent isolation sources, but there are almost no reports about *Alicyclobacillus* contamination of kiwi fruit products. According to a 14-year survey in Argentina, 8556 samples from 7 Argentinean provinces of 19 different kinds of fruit and vegetable juices were analyzed for the presence of *Alicyclobacillus*, and the result showed that there was no *Alicyclobacillus* found in the only one sample of kiwi fruit [25].

Even so, we believe that there might be a potential risk of *Alicyclobacillus* contamination in kiwi fruit products because we have isolated 1 strain of *Alicyclobacillus* from a kiwi fruit product. As the origin source area of kiwi fruit and the largest production country in the world, China's industry of kiwi fruit products (juice, vinegar, wine, fresh-cut slices) is developing very fast in recent years, and out of 60% of kiwi fruit production are from Shaanxi province [26], especially from Mei county and Zhouzhi county. Gocmen and Pettipher [27] [28] have proved that cell numbers between 10^5 and 10^6 CFU/mL of *A. acidoterrestris* produced sufficient guaiacol (2 ppb) to spoil orange and apple juices, but to our knowledge there is still no reports about the growth and taint (mainly guaiacol) production of *Alicyclobacillus* in kiwi fruit juice or commercial beverages related to kiwi fruit, although vanillin, which is a precursor of guaiacol is contained in kiwi fruit. This research aims to investigate the existence condition of *Alicyclobacillus* in the production line of kiwi products (orchards, processing factories and commercial products) in China's Shaanxi province and study the growth to and taint production of *Alicyclobacillus* in kiwi fruit juice and commercial kiwi fruit products to assess the potential risk of *Alicyclobacillus* contamination in kiwi fruit production line under the situation of a rapid development of this industry.

Materials and Methods

Sampling and Isolation

Samples were collected from three sources: eighteen samples were collected from commercial kiwi products bought from supermarkets, stores in Yangling or online, 102 samples were collected from a Hazard Analysis Critical Control Point accredited fresh-cut and frozen fruit and vegetable producer in Shaanxi province of China (the name of the authority who issued the permission: Baoji DuLe Food Co., Ltd), and 281 samples (fruits, soil and air) were collected from 16 orchards covering main regions of kiwi fruit production in Shaanxi province (Figure 1) (Table 1). The names of the authorities who issued the permissions for each sampling location are also shown in Table 1.

As the most efficient isolation medium, the *Bacillus acidoterrestris* medium (BAM) [29] which can support nearly all species and all spoilage related species of *Alicyclobacillus*, was used as the medium for the enrichment of *Alicyclobacillus* in this study [9]. A hundred milliliter samples of kiwi products were mixed with 200 mL BAM broth in 500 mL flasks followed by a heat shock treatment at 80°C for 10 min [18] to inhibit the growth of yeast and fungi and promote the germination of *Alicyclobacillus* spores, after which the flasks were shaken at 50°C, 120 rpm in a shaker for 5 days to obtain an enrichment. Then 1 mL of these cultures were diluted appropriately in test tubes and obtain a same heat shock treatment and streaked on BAM plate and yeast starch glucose (YSG) [30] plates duplicately followed by a incubation at 45°C (BAM) or 60°C (YSG) for 4 days. After the enrichment, according to their morphology features, colonies which may be *Alicyclobacillus* species were selected randomly from plates and re-streaked to obtain pure colonies [9]. Then all pure colonies were streaked on pH 7.0 Luria-Bertani (LB) plates. The colonies which did not grow on LB plates were selected and observed with a microscope. Colonies of spore-forming rods were selected for further examination and stored at -40°C in corresponding broth supplemented with 30% sterile glycerol.

The number and the enrichment methods of the samples are shown in Table 2. Samples from the orchards were collected

randomly into sterile sample bags from 2 sources: soil and fruits (both on trees and dropped), and the air samples were also collected with BAM plates. The samples were treated immediately after collection with 100 mL sterile water poured into the sample bags directly and then put in a shaker for 10 min at 120 rpm. After settlement the supernatant solution was mixed with 100 mL BAM broth and incubated at 50°C, 120 rpm for 5 days after a heat shock treatment at 80°C for 10 min to obtain an enrichment. The consequent isolation steps of samples from the plant and orchards were the same as the isolation steps of kiwi products samples. The plates containing air samples were incubated at 50°C for 5days directly followed by isolation steps similar to other samples.

16S rRNA Gene Amplification and Sequencing

Selected colonies were grown in corresponding broth, and then their genomic DNA was extracted with the TIANamp Bacteria DNA Kit (TIANGEN, Beijing, China) according to the manufacturer's instructions. Then a portion of their 16S rRNA gene was amplified using the primers 27F (5′-AGA GTT TGA TCC TGG CTC AG-3′) and 1492R (5′-GGY TAC CTT GTT ACG ACT T-3′) (BGI, supplied by BGI, Beijing, China) [31].

PCR amplifications were performed under conditions as below: fifty microliter of a total reaction volume, 0.3 μL of 5 U/μL *Taq* DNA polymerase (Takara, supplied by Takara Biotechnology Co. Ltd., Dalian, China), 5 μL of 10× PCR reaction buffer, 3 μL of 25 mM MgCl₂, 4 μL of deoxyribonucleoside triphosphate (dNTPs) mixture (Takara) with 2.5 mM each, 4 μL of each primers, 3 μL of template DNA. PCR reactions were done in an Alpha Unit Block Assembly for Peltier Thermal Cycler DNA Engine Systems (MJ RESEARCH Inc., Watertown, Massachusetts, USA) under conditions as below: initial denaturation at 94°C for 2 min, 30 cycles of denaturation at 94°C for 45 s, annealing at 50°C for 45 s, elongation at 72°C for 2 min, and final elongation at 72°C for 10 min [17].

The PCR products were checked by 1% agarose (Invitrogen, supplied by Invitrogen, Carlsbad, CA92008, USA) electrophoresis with an electrophoresis apparatus (Liuyi, Beijing Liuyi Instrument Factory, Beijing, China) to confirm they contained a 1.5 kb fragment each. Then these fragments were sent to BGI for purification and sequencing.

Phylogenetic Tree Construction and RAPD-PCR Amplification

The sequence data of 16S rDNA of these isolates were compared to sequences in GenBank using BLAST 2.2.27 and then submitted to GenBank using Sequin. Then these 16S rDNA sequences of isolates were aligned with CLUSTAL W to construct a phylogenetic tree with MEGA 5.10 using both the neighbour-joining method [32] based on the p-distance model and the maximum-parsimony method [33].

RAPD-PCR amplifications were performed in 50 μL volume reactions containing 0.3 μL of 5 U/μL *Taq* DNA polymerase (Takara), 5 μL of 10× PCR reaction buffer, 3 μL of 25 mM MgCl₂, 4 μL of dNTPs mixture (Takara) with 2.5 mM each, 4 μL of 2 different primers (A-01:5′-CAGGCCCTTC-3′; AZ-14:5′-CACGGCTTCC-3′) (BGI). The volume of template DNA was adjusted from 1 to 4 μL to get the clearest electrophoresis patterns. The PCR conditions were the same as above, except for annealing at 38°C for 45 s, and elongation at 72°C for 2 min 30 s. The results were checked by 1% agarose electrophoresis.

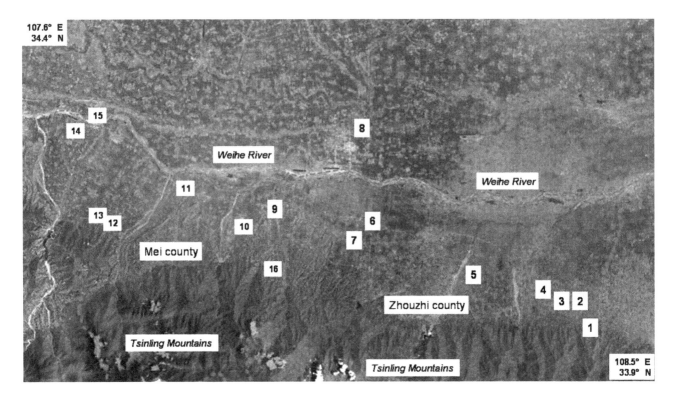

Figure 1. Sampling sites of 16 orchards. The blue dash lines on the map are the administrative boundaries of Mei county and Zhouzhi county which located on the middle of Shaanxi province of China. The plain between the Weihe River and the Tsinling Mountains within these two counties, which is a part of the Guanzhong Plain, is the main region of kiwi fruit production in Shaanxi province.

Enumeration of Growth and GC-MS Analysis for Taint Production

Four strains (C-ZJB-12-32, C-ZJB-12-35, C-ZJB-12-55 and C-ZJB-12-69, each 1 strain for one species of *Alicyclobacillus* isolated in this study) were selected and inoculated into BAM broth and incubated for 12 h to reach the log phase for activation, then these cultures were serially diluted (to make the number of CFU/mL of each sample was below 100 after inoculation) and each inoculated into 150 mL kiwi fruit juice (pH adjusted to 4.2 using 4 M NaOH) and 150 mL kiwi fruit juice (pH 2.5, without adjustment), then all 8 samples were incubated at 45°C for 21 days. Activated C-ZJB-12-35 was also inoculated into 150 mL of some commercial kiwi products mentioned in the sampling and isolation part, including 2 kinds of kiwi fruit juice (pH 3.5, soluble solids (SS) content 10°Brix, raw kiwi fruit juice content 60%), vinegar (pH 2.5, 5°Brix) and wine (pH 3.5, 7°Brix, alcohol 6%) (<100 CFU/mL after inoculation) and incubated at 45°C for 21 days. All 14 samples were plated onto BAM agar every day for the first week of incubation and every 3 days for the remainder of the experiment. Samples of kiwi fruit juice without pH adjustment obtained another test with a heat shock treatment of 80°C for 10 min before plating to check spores in them.

According to some previous reports [27] [28] [34] [35], guaiacol and 2 kinds of halophenols, 2,6-dibromophenol (2,6-DBP) and 2,6-dichlorophenol (2,6-DCP) were chosen as aim taint compounds. Standard solutions were prepared with guaiacol (Fluka Analytical, 2931 Soldier Springs Road, Laramie, WY, USA), 2,6-DBP (SUPELCO, 595 North Harrison Road, Bellefonte, PA, USA) and 2,6-DCP (SIGMA-ALDRICH, Co., 3050 Spruce Street, St. Louis MO, USA) in concentrations of 2.5, 5, 7.5, 10 and 20 ppb in distilled water. A 5 ppb standard solution for guaiacol was also prepared in kiwi fruit juice as a contrast. All

standard solutions were kept at 4°C in the dark until analysis (within 2 days). Ten milliliter of samples which could accumulate >10^6 CFU/mL after incubation and all standard solutions were transferred to 20 mL glass vials. After a 15-min equilibrium at 45°C, a solid phase microextraction (SPME) fiber (50/30mm DVB/Carboxen/PDMS; Supelco Co., Bellefonte, PA, USA) was exposed to the headspace of the vials at 45°C for 1 h to extract their volatile compounds.

After the extraction, the fiber was inserted into a Thermo-Finnigan Trace GC ultra/Trace DSQ (Thermo-Finnigan, San Jose, CA, USA) injection port using a 30 m ×0.25 mm i.d.×0.25 μm DB-Wax column (Agilent, Palo Alto, CA, USA). The GC-MS conditions were as follows: carrier gas He at 1 ml/min, splitless mode, injector temperature 250°C, starting temperature 50°C (2 min), final temperature 230°C (5 min), temperature rate 10°C /min, ion source temperature 230°C, scanning mass range 50–350 m/z. NIST 2002 fragmentation spectra database was used for identifications.

Nucleotide Sequence Accession Numbers

All the 16S rRNA gene (16S rDNA) sequences were deposited in GenBank (accession numbers KC193182 to KC193190 and KC354615 to KC354691, also see details in Figure 2).

Results

Positive Percentages and Properties of Isolates

A total of 401 samples were tested, and 76 samples were positive (*Alicyclobacillus* detected, 19.0%), from which 85 strains of microbes were isolated and identified. The total percentages of positive samples of the 16 orchards and all sampling sections of the producer were 20.3% and 17.6%, respectively (Table 1; Table 2),

Table 1. The number of samples and positive samples of the 16 orchards and their geographic coordinate.

Number of orchards	Locations of orchards	Geographic coordinate of orchards	authority who issued the permission	Number of samples	Positive samples
East					
1	Shaanxi Zhouzhi Bairui Kiwi fruit Experimental Base	108.45°E, 34.05°N	Shaanxi Bairui Kiwi Fruit Research Institute Co. Ltd.	14	5
2	Kiwi fruit Experimental Base of Agricultural Science-technology Demonstration Park (Nanqianhu village in Zhouzhi county)	108.45°E, 34.07°N	the government of Jiufeng township	18	4
3	Gengxi village, Jiufeng township, Zhouzhi county	108.41°E, 34.07°N	p	22	0
4	Liujiabao village, Jixian township, Zhouzhi county	108.38°E, 34.10°N	p	16	4
5	Erhezhuang village, Louguan township, Zhouzhi county	108.27°E, 34.11°N	p	15	6
6	Shangtiantun village, Situn township, Zhouzhi county	108.14°E, 34.17°N	p	19	3
7	Changdong village, Yabai township, Zhouzhi county	108.10°E, 34.15°N	p	14	6
8	Xiajiagou village, Yangcun township, Yangling town	108.11°E, 34.30°N	p	21	7
West					
9	Jinjia village, Hengqu town, Mei county	107.99°E, 34.19°N	p	22	5
10	Tuling village, Tangyu town, Mei county	107.93°E, 34.18°N	p	18	9
11	Taoyuan village, Huaiya town, Mei county (pollution free orchards)	107.85°E, 34.21°N	p	16	3
12	Luoyukou village, Yingtou town, Mei county	107.75°E, 34.16°N	p	8	0
13	Tongyu village, Yingtou town, Mei county	107.74°E, 34.16°N	p	7	2
14	Yuechen village, Diwucun town, Mei county	107.68°E, 34.28°N	p	25	1
15	Chenjiagou village, Diwucun town, Mei county	107.71°E, 34.29°N	p	26	0
16	Kiwi fruit Experimental Station of Northwest A & F University (Qinghua Town)	107.99°E, 34.12°N	Northwest A&F University	20	2
Total				281	57

p: private land, and the owner of the land have given permission to conduct the study on this site.

and only one sample (around 140 CFU/mL) among 18 commercial kiwi fruit products was detected as positive (5%), from which the one isolate was identified as *Alicyclobacillus acidoterrestris* later.

All isolates were Gram-positive, endospore-forming rods, and could form creamy white or yellowish, opaque colonies. Some isolates were motile. Among all 85 isolates, 1 was isolated from commercial products, 16 were isolated from the producer and 68 were isolated from the 16 orchards, including 26 from fruits samples (mainly from drop fruits), 37 from soil samples and 5 from air samples. The data revealed that *Alicyclobacillus* are soil-borne bacteria, as reported by Deinhard et al. in 1987 [15]. All samples from orchards' air were negative except 2 air samples (formed 5, 7 colonies on plates respectively) from the orchard in Xiajiagou village, which might be because of the sprinkle on the day of sampling.

Identification of Isolates

As shown in Table 3, the similarity (max identity of BLAST results) of 16S rDNA among all isolates was beyond 99% except

C-ZJB-12-85, and the number of nucleotides sequenced of all isolates were beyond 1400 (almost complete sequence of 16S rDNA) except C-ZJB-12-53 and C-ZJB-12-55. These mean that the isolates were very likely the same species as their nearest phylogenetic neighbours showed in Table 3. Forty six isolates from 20 sources (all sampling sections of the quick-frozen kiwi fruit slice producer and 7 orchards: the orchard in Xiajiagou village, Bairui Kiwi fruit Experimental Base, Changdong village, Liujiabao village, Kiwi fruit Experimental Station of Northwest A & F University, kiwi fruit Experimental Base of Agricultural Science-technology Demonstration Park, and Yuechen village) of total 85 isolates (54.1%) were identified as *Alicyclobacillus acidoterrestris* through the sequence data compared with nucleotides sequences in GenBank, which means that *Alicyclobacillus acidoterrestris* is the most widespread species of *Alicyclobacillus*.

In this study, 10 strains of *Alicyclobacillus contaminans* (4 orchards: the orchards in Bairui Kiwi fruit Experimental Base, Changdong village, Liujiabao village, and Shangtiantun village), 19 strains of *Alicyclobacillus herbarius* (5 orchards: the orchards in Bairui Kiwi

Table 2. The number and the sampling and enrichment methods of the samples from the producer.

Sample form and sampling sections	Number of samples	Sampling and enrichment methods	Positive samples
Orchards			
Kiwi fruits before peeling	10	Samples were put into sterile sample bags and treated immediately after collection with 300 mL sterile water poured into the sample bags and then put in a shaker for 10 min at 120 rpm. After settlement 100 mL of the supernatant solution was mixed with 100 mL BAM broth and incubated at 50°C, 120 rpm for 5 days after a heat shock treatment at 80°C for 10 min to have an enrichment.	2
Kiwi fruits after peeling	12	s	2
Kiwi fruits after color protection treatment	8	s	2
Washing & fresh-cut shops			
Kiwi fruits before washing	7	s	1
Kiwi fruits after washing	9	s	1
Kiwi fruit slices	10	s	1
Wash water	5	A total 2500 mL of wash water was collected from 5 sites of the washer randomly with five 500 mL sterile flasks. After delivery to the lab all the samples were mixed with BAM broth at the ratio of 1:1 in 500 mL flasks and put in a shaker at 50°C, 120 rpm for 5 days after a heat shock treatment at 80°C for 10 min to have an enrichment.	2
Shop environment (walls)	4	Sampling were carried out with sterile cotton bud smearing on the surfaces of the shop environment and then put into test tubes containing 10 mL BAM broth. The tubes were incubated at 50°C for 5 days after a heat shock treatment at 80°C for 10 min as an enrichment.	1
Shop environment (floor)	8	s	2
Shop environment (raw material bins)	9	s	3
Shop environment (air)	10	s	0
Quick-freezing shop			
Quick-frozen kiwi fruit slices	10	After melting at room temperature, the samples were treated by the same method as used for samples from the orchards.	1
Total	102		18

s: same as above.

fruit Experimental Base, Changdong village, Liujiabao village, kiwi fruit Experimental Base of Agricultural Science-technology Demonstration Park, and Tongyu village) and 1 strains of *Alicyclobacillus cycloheptanicus* (Shangtiantun village) were also isolated.

One thing to be noted was that there were 9 strains of thermo-acidophilic *Bacillus* isolated, including 2 strains of *Bacillus coagulans* from shop environment (raw material bins), 5 strains of *Bacillus fumarioli* from soil of the orchards in 4 locations (Xiajiagou village, Shangtiantun village, Bairui Kiwi fruit Experimental Base, Erhezhuang village), and 2 strains of *Bacillus ginsengihumi* from soil of the orchard in Erhezhuang village and Xiajiagou village.

Phylogenetic Positions and Genotypic Diversities of Isolates

From the neighbour-joining phylogenetic tree of the 85 isolates and 4 reference species (Figure 2) we can see that all strains fell into 7 clusters. All the strains of each of the 7 species (4 *Alicyclobacillus* species and 3 *Bacillus* species) clustered together. The biggest cluster (46 isolate strains and 1 reference strain) including all *Alicyclobacillus acidoterrestris* strains were mainly composed of 2 groups. The bootstrap values of the branches within these 2 groups and the *Alicyclobacillus contaminans* cluster and the *Alicyclobacillus herbarius* cluster were low (some bootstrap values were even below

30), which indicated that the differences among these strains were very small according to their 16S rDNA sequences.

Genotypic diversities of all 85 isolates and 4 reference species were evaluated by RAPD-PCR (Figure 3). Generally, the RAPD-PCR with primer AZ-14 (pattern B in Figure 3) generated more bands after electrophoresis than with primer A-01 (pattern A in Figure 3), which might be caused by the relatively higher affinity of primer AZ-14 to the genomic DNA of *Alicyclobacillus*. Therefore, the genotypic diversity revealed through RAPD-PCR with primer AZ-14 was bigger than with primer A-01. In detail, in pattern A, within the cluster of *Alicyclobacillus acidoterrestris*, there were 4 groups and 2 unique strains which were different from each other, as well as from any strains of the 4 groups. Whereas in pattern B, there were 6 groups and 7 unique strains in the same cluster. But the biggest 2 groups in this cluster were similar. Group I was composed of 22 strains in both pattern A and B, and 16 strains of them were the same, while group III in pattern A and group VI in pattern B were composed of exactly the same 8 strains, which agreed with the result of the phylogenetic tree's corresponding part precisely. Within the cluster *Alicyclobacillus contaminans* there were 2 groups and 3 unique strains in pattern A, but each strain including the reference strain DSM 17975 belong to this cluster in pattern B was different from each other, which suggested there might exist great genotypic diversity within this species of *Alicyclobacillus* in the kiwi fruit production area in Shaanxi province of China. Although

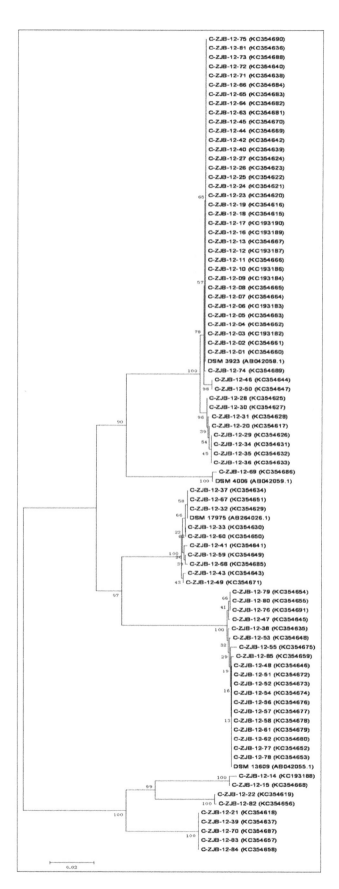

Figure 2. Neighbour-joining phylogenetic tree of isolates and reference species. The 4 reference species are *Alicyclobacillus acidoterrestris* DSM 3923, *Alicyclobacillus cycloheptanicus* DSM 4006, *Alicyclobacillus contaminans* DSM 17975, and *Alicyclobacillus herbarius* DSM 13609. Construction is based on 16S rRNA gene sequence comparisons. Bootstrap percentages based on 1000 replicates are shown. Bar, p-distance (0.02 substitutions per nucleotide position).

there were 3 groups within the cluster *Alicyclobacillus herbarius* in both 2 patterns, but pattern B still showed bigger genotypic diversity, with 6 unique strains, whereas there was only 1 unique strain in pattern A. But the 2 patterns also shared some similarity. There were 2, 5, and 2 same strains between group I, II and III, respectively.

Growth and Taint Production

Changes in microbial populations over incubation are shown in Figure 4. Replicate uninoculated control samples of 2 kinds of kiwi fruit juice and 6 commercial kiwi fruit products had undetectable *Alicyclobacillus* populations (<30 CFU/mL) over incubation (data not shown). Microbial population levels of all pH adjusted kiwi fruit juice samples reached >10^7 CFU/mL within 7 days of incubation, and the growth curves of all 4 test strains were similar. All curves in Figure 4B did not seem extending regularly, which might be caused by the testing errors (with a inoculation volume of 100 μL, there usually were only <5 CFUs on a plate). But it was obvious that there was hardly any growth in all kiwi fruit juice samples without pH adjustment. In all these samples a certain number of spores kept their abilities of reproduction throughout the whole incubation time (Figure 4C). One juice sample of the six commercial kiwi fruit products samples reached population levels >10^7 CFU/mL within 10 days, and the 2 wine samples accumulated *Alicyclobacillus* population levels 10^6 and 10^4, respectively. *Alicyclobacillus* did not grow obviously in the rest samples of this group.

A GC–MS chromatogram, a mass spectrum from the peak thought to be guaiacol in the sample of pH adjusted kiwi fruit juice inoculated with C-ZJB-12-35 and a mass spectrum of standard guaiacol are shown in Figure 5. The compound peaked at 14.47 min should be guaiacol based on the values of SI and RSI given by the NIST 2002 library and comparison with the retention time of standard solutions. Three major mass fragments of guaiacol (mass to charge, m/z, 81,109, 124) can be seen in Figure 5. The GC–MS chromatograms of other chosen samples were similar to the above one, according to which guaiacol were produced by all chosen samples. No 2,6-DBP and 2,6-DCP were detected. A GC–MS chromatogram of all standard solutions and a standard curve generated by Excel are shown in Figure 6. The mass spectrums of the peaks at 14.47 min in the TIC chromatograms of the contrast 5 ppb standard solution for guaiacol in kiwi fruit juice and the sample of commercial kiwi fruit wine A show low SI and RSI values of many compounds through NIST library. The reason might be there were several compounds detected at this retention time or the limit of the instrument precision, and changing a more suitable column or optimizing the conditions of microextraction and GC-MS might obtain better results. The peak areas of guaiacol in samples and the calculated concentrations of guaiacol are shown in Table 4 (see Chromatogram S1-S6 for their raw data files). According to the standard curve all the samples accumulated guaiacol between 11–34 ppb after the 21 days incubation.

Discussion

The decline of percentages of positive samples from orchards to producer shops and commercial products reveals the distribution

Table 3. All isolated strains, their sources, number of nucleotides sequenced, nearest phylogenetic neighbour and similarity.

Isolated strains	Source	No. of nucleotides sequenced	Nearest phylogenetic neighbour (GenBank accession number)	Similarity
C-ZJB-12-01	Kiwi fruit wine (pH 3.5)	1423	*Alicyclobacillus acidoterrestris* (NR_040844.1)	99.9%
C-ZJB-12-02	Kiwi fruits before peeling	1413	*Alicyclobacillus acidoterrestris* (NR_040844.1)	99.8%
C-ZJB-12-03	s	1424	*Alicyclobacillus acidoterrestris* (AB682390.1)	99.6%
C-ZJB-12-04	s	1437	*Alicyclobacillus acidoterrestris* (AJ133631.1)	99.7%
C-ZJB-12-05	s	1428	*Alicyclobacillus acidoterrestris* (NR_040844.1)	99.8%
C-ZJB-12-06	Kiwi fruits after peeling	1428	*Alicyclobacillus acidoterrestris* (NR_040844.1)	99.6%
C-ZJB-12-07	Kiwi fruits after color protection treatment	1430	*Alicyclobacillus acidoterrestris* (NR_040844.1)	99.6%
C-ZJB-12-08	Kiwi fruits before washing	1430	*Alicyclobacillus acidoterrestris* (NR_040844.1)	99.6%
C-ZJB-12-09	s	1433	*Alicyclobacillus acidoterrestris* (NR_040844.1)	99.6%
C-ZJB-12-10	Kiwi fruits after washing	1438	*Alicyclobacillus acidoterrestris* (NR_040844.1)	99.6%
C-ZJB-12-11	Wash water	1432	*Alicyclobacillus acidoterrestris* (NR_040844.1)	99.7%
C-ZJB-12-12	Shop environment (walls)	1440	*Alicyclobacillus acidoterrestris* (AB682390.1)	99.6%
C-ZJB-12-13	Shop environment (floor)	1422	*Alicyclobacillus acidoterrestris* (AB682390.1)	99.9%
C-ZJB-12-14	Shop environment (raw material bins)	1470	*Bacillus coagulans* (AB696800.1)	98.8%
C-ZJB-12-15	s	1441	*Bacillus coagulans* (AB696800.1)	99.9%
C-ZJB-12-16	s	1441	*Alicyclobacillus acidoterrestris* (NR_040844.1)	99.2%
C-ZJB-12-17	s	1435	*Alicyclobacillus acidoterrestris* (NR_040844.1)	99.5%
C-ZJB-12-18	Fruits from the orchard in Xiajiagou village	1409	*Alicyclobacillus acidoterrestris* (NR_040844.1)	99.9%
C-ZJB-12-19	s	1434	*Alicyclobacillus acidoterrestris* (NR_040844.1)	99.9%
C-ZJB-12-20	s	1412	*Alicyclobacillus acidoterrestris* (AB059675.1)	99.8%
C-ZJB-12-21	Soil from the orchard in Xiajiagou village	1450	*Bacillus fumarioli* (AJ581126.1)	99.7%
C-ZJB-12-22	s	1429	*Bacillus ginsengihumi* (NR_041378.1)	99.9%
C-ZJB-12-23	Air from the orchard in Xiajiagou village	1429	*Alicyclobacillus acidoterrestris* (NR_040844.1)	99.7%
C-ZJB-12-24	s	1422	*Alicyclobacillus acidoterrestris* (NR_040844.1)	99.8%
C-ZJB-12-25	s	1432	*Alicyclobacillus acidoterrestris* (NR_040844.1)	99.6%
C-ZJB-12-26	s	1409	*Alicyclobacillus acidoterrestris* (NR_040844.1)	99.8%
C-ZJB-12-27	s	1435	*Alicyclobacillus acidoterrestris* (NR_040844.1)	99.6%
C-ZJB-12-28	Fruits from the orchard in Bairui Kiwi fruit Experimental Base	1423	*Alicyclobacillus acidoterrestris* (AB059675.1)	99.5%
C-ZJB-12-29	s	1417	*Alicyclobacillus acidoterrestris* (AB059675.1)	99.6%
C-ZJB-12-30	s	1419	*Alicyclobacillus acidoterrestris* (AB059675.1)	99.5%
C-ZJB-12-31	s	1423	*Alicyclobacillus acidoterrestris* (AB682383.1)	99.4%
C-ZJB-12-32	s	1430	*Alicyclobacillus contaminans* (NR_041475.1)	99.3%
C-ZJB-12-33	s	1418	*Alicyclobacillus contaminans* (AB264027.1)	99.4%
C-ZJB-12-34	s	1429	*Alicyclobacillus acidoterrestris* (AB059675.1)	99.6%
C-ZJB-12-35	s	1412	*Alicyclobacillus acidoterrestris* (AB059675.1)	99.6%
C-ZJB-12-36	s	1430	*Alicyclobacillus acidoterrestris* (AB059675.1)	99.5%
C-ZJB-12-37	s	1431	*Alicyclobacillus contaminans* (NR_041475.1)	99.5%
C-ZJB-12-38	Soil from the orchard in Bairui Kiwi fruit Experimental Base	1444	*Alicyclobacillus herbarius* (AB681266.1)	99.2%
C-ZJB-12-39	s	1433	*Bacillus fumarioli* (J581126.1)	99.8%
C-ZJB-12-40	s	1428	*Alicyclobacillus acidoterrestris* (NR_040844.1)	99.9%
C-ZJB-12-41	s	1451	*Alicyclobacillus contaminans* (AB264027.1)	99.2%
C-ZJB-12-42	Fruits from the orchard in Changdong village	1412	*Alicyclobacillus acidoterrestris* (NR_040844.1)	99.9%
C-ZJB-12-43	s	1437	*Alicyclobacillus contaminans* (AB264027.1)	99.5%
C-ZJB-12-44	s	1422	*Alicyclobacillus acidoterrestris* (NR_040844.1)	99.6%
C-ZJB-12-45	s	1424	*Alicyclobacillus acidoterrestris* (NR_040844.1)	99.7%
C-ZJB-12-46	s	1451	*Alicyclobacillus acidoterrestris* (AB682390.1)	99.9%
C-ZJB-12-47	s	1455	*Alicyclobacillus herbarius* (AB681266.1)	99.2%

Table 3. Cont.

Isolated strains	Source	No. of nucleotides sequenced	Nearest phylogenetic neighbour (GenBank accession number)	Similarity
C-ZJB-12-48	s	1453	*Alicyclobacillus herbarius* (AB681266.1)	99.5%
C-ZJB-12-49	s	1437	*Alicyclobacillus contaminans* (AB264027.1)	99.6%
C-ZJB-12-50	s	1438	*Alicyclobacillus acidoterrestris* (AB682390.1)	99.8%
C-ZJB-12-51	Soil from the orchard in Changdong village	1449	*Alicyclobacillus herbarius* (AB681266.1)	99.5%
C-ZJB-12-52	s	1426	*Alicyclobacillus herbarius* (AB681266.1)	99.4%
C-ZJB-12-53	s	993	*Alicyclobacillus herbarius* (AB681266.1)	99.7%
C-ZJB-12-54	s	1448	*Alicyclobacillus herbarius* (AB681266.1)	99.1%
C-ZJB-12-55	s	1003	*Alicyclobacillus herbarius* (AB681266.1)	99.0%
C-ZJB-12-56	s	1439	*Alicyclobacillus herbarius* (AB681266.1)	99.2%
C-ZJB-12-57	s	1429	*Alicyclobacillus herbarius* (AB681266.1)	99.4%
C-ZJB-12-58	s	1435	*Alicyclobacillus herbarius* (AB681266.1)	99.2%
C-ZJB-12-59	Soil from the orchard in Liujiabao village	1401	*Alicyclobacillus contaminans* (AB264027.1)	99.5%
C-ZJB-12-60	s	1439	*Alicyclobacillus contaminans* (AB264027.1)	99.0%
C-ZJB-12-61	s	1448	*Alicyclobacillus herbarius* (AB681266.1)	99.5%
C-ZJB-12-62	s	1449	*Alicyclobacillus herbarius* (AB681266.1)	99.2%
C-ZJB-12-63	s	1433	*Alicyclobacillus acidoterrestris* (NR_040844.1)	99.8%
C-ZJB-12-64	s	1426	*Alicyclobacillus acidoterrestris* (NR_040844.1)	99.5%
C-ZJB-12-65	s	1429	*Alicyclobacillus acidoterrestris* (NR_040844.1)	99.7%
C-ZJB-12-66	s	1431	*Alicyclobacillus acidoterrestris* (AB682390.1)	99.6%
C-ZJB-12-67	Fruits from the orchard in Shangtiantun village	1440	*Alicyclobacillus contaminans* (NR_041475.1)	99.2%
C-ZJB-12-68	s	1452	*Alicyclobacillus contaminans* (AB264027.1)	99.5%
C-ZJB-12-69	s	1435	*Alicyclobacillus cycloheptanicus* (AB680830.1)	99.4%
C-ZJB-12-70	Soil from the orchard in Shangtiantun village	1442	*Bacillus fumarioli* (AJ581126.1)	99.9%
C-ZJB-12-71	Soil from the orchard in Kiwi fruit Experimental Station of Northwest A & F University	1422	*Alicyclobacillus acidoterrestris* (NR_040844.1)	99.8%
C-ZJB-12-72	s	1412	*Alicyclobacillus acidoterrestris* (NR_040844.1)	99.9%
C-ZJB-12-73	Fruits from the orchard in kiwi fruit Experimental Base of Agricultural Science-technology Demonstration Park	1434	*Alicyclobacillus acidoterrestris* (AB682390.1)	99.7%
C-ZJB-12-74	Soil from the orchard in kiwi fruit Experimental Base of Agricultural Science-technology Demonstration Park	1407	*Alicyclobacillus acidoterrestris* (NR_040844.1)	99.9%
C-ZJB-12-75	s	1428	*Alicyclobacillus acidoterrestris* (NR_040844.1)	99.6%
C-ZJB-12-76	s	1430	*Alicyclobacillus herbarius* (AB681266.1)	99.2%
C-ZJB-12-77	s	1452	*Alicyclobacillus herbarius* (AB681266.1)	99.3%
C-ZJB-12-78	s	1456	*Alicyclobacillus herbarius* (AB681266.1)	99.2%
C-ZJB-12-79	s	1451	*Alicyclobacillus herbarius* (AB681266.1)	99.0%
C-ZJB-12-80	s	1451	*Alicyclobacillus herbarius* (AB681266.1)	99.3%
C-ZJB-12-81	Soil from the orchard in Yuechen village	1420	*Alicyclobacillus acidoterrestris* (NR_040844.1)	99.7%
C-ZJB-12-82	Soil from the orchard in Erhezhuang village	1460	*Bacillus ginsengihumi* (FJ357590.1)	100%
C-ZJB-12-83	s	1459	*Bacillus fumarioli* (AJ581126.1)	99.9%
C-ZJB-12-84	s	1457	*Bacillus fumarioli* (AJ581126.1)	99.8%
C-ZJB-12-85	Soil from the orchard in Tongyu village	1447	*Alicyclobacillus herbarius* (AB681266.1)	98.8%

s: same as above.

rule of microbes in a whole production line. There seem no clear relations among the percentages of positive samples of the 16 orchards, and the percentages of some orchards (Gengxi village, Luoyukou village, Chenjiagou village) are even 0%, despite their number of samples. This may be because of the sampling locations. Positive samples are likely from drop fruits and the soil around them (data not shown). Therefore, there are probably other species of *Alicyclobacillus* except the 4 species found in this study because of the randomness and limited quantity of sampling.

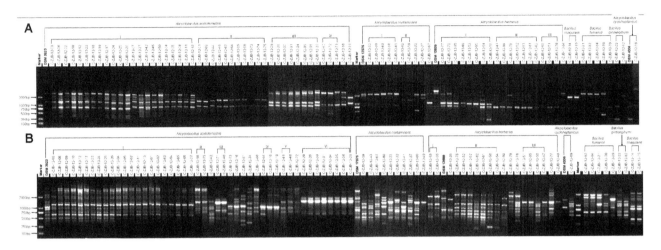

Figure 3. Electrophoresis patterns of RAPD-PCR of the genomic DNA of isolates and reference species. The 4 reference species are *Alicyclobacillus acidoterrestris* DSM 3923, *Alicyclobacillus cycloheptanicus* DSM 4006, *Alicyclobacillus contaminans* DSM 17975, *Alicyclobacillus herbarius* DSM 13609. I–VI indicate groups within a certain species. Marker, DNA marker DL2000 (BGI). A, Primer A-01; B, Primer AZ-14.

The result that *Alicyclobacillus acidoterrestris* is the most widespread species of *Alicyclobacillus* in this study is similar to many previous reports [36], [17], [14]. *Alicyclobacillus acidoterrestris* is the most important spoilage microbe of juice products and off-odour (mainly guaiacol) producer [18], [37], therefore there will be a potential risk of spoilage when more and more kiwi fruits from

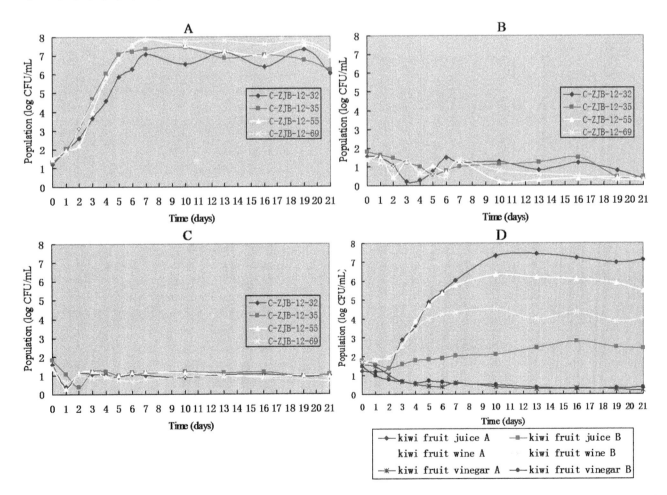

Figure 4. Changes in microbial populations during the 45°C incubation. A, pH adjusted kiwi fruit juice inoculated with 4 strains of *Alicyclobacillus*; B, kiwi fruit juice inoculated with 4 strains of *Alicyclobacillus*; C, kiwi fruit juice inoculated with 4 strains of *Alicyclobacillus* (with a heat shock treatment); D, six kinds commercial kiwi fruit products inoculated with C-ZJB-12-35.G.

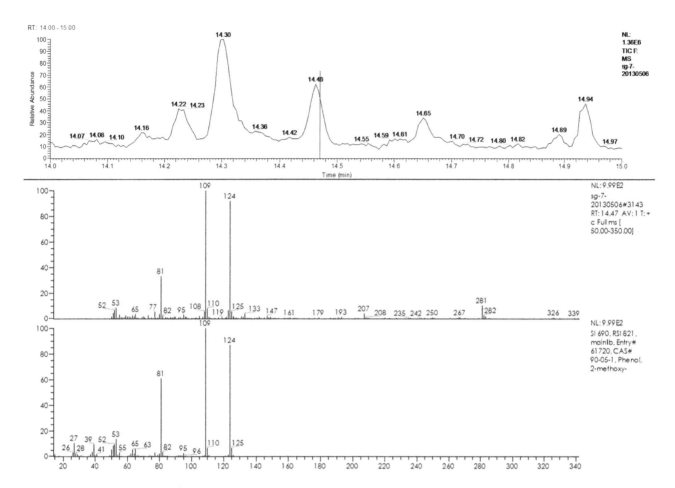

Figure 5. A GC–MS chromatogram of pH adjusted kiwi fruit juice inoculated with C-ZJB-12-35. Mass spectrums from the peak (14.47 min) thought to be guaiacol in this sample and from standard guaiacol are also shown. The compound peaked at 281 m/z may come from the SPME fiber, because it exist in the peak around 14.47min in all samples. Because the scanning mass range was 50–350 m/z, peaks below 50 m/z were not included. Both of these may reduce the SI and RSI values.

these orchards are used as raw materials for juice, wine and other products in the future. Except 2 isolates, all the isolates from all sampling sections of the producer, including the raw material orchard, processing shops and final products, are identified as *Alicyclobacillus acidoterrestris*. This result indicates that the potential risk mentioned above truly exists, which is also proved by the fact that the only one isolate from commercial kiwi fruit products is identified as *Alicyclobacillus acidoterrestris* too. If more producers' shops are tested, other species of *Alicyclobacillus* may be found.

To our knowledge, this is the first report on the isolation of *Alicyclobacillus contaminans* and *Alicyclobacillus herbarius* from China's orchards. *Alicyclobacillus herbarius* and *Alicyclobacillus cycloheptanicus* has also been proved as taint producers [27] [36]. Therefore, kiwi fruit products coming from these orchards may be contaminated by these non-*acidoterrestris Alicyclobacillus*.

Alicyclobacillus acidoterrestris and *Alicyclobacillus contaminans* were widely distributed among various fruit orchards in Japan [36]. According to the results of this study, it is thought that *Alicyclobacillus acidoterrestris*, *Alicyclobacillus contaminans*, and *Alicyclobacillus herbarius* might be the predominant species of *Alicyclobacillus* in kiwi fruit orchards in Shaanxi province of China.

The $D_{80°C}$ of *Bacillus coagulans* is 40 min in a double concentrated tomato paste media according to Sandoval et al. [38], and *Bacillus fumarioli* grows optimally at pH 5.5 and 50°C [39], while *Bacillus ginsengihumi* can also grow in acid media at temperature up to 50°C [40]. These reports proved that these *Bacillus* can be an interference to the detection of *Alicyclobacillus* from samples and cause false positive results, although there are no reports relating them with spoilage of juice products and guaiacol production.

Through the analysis of the 2 patterns it was thought the division of genotypic diversities among a relatively large number of samples (as in this study) though PAPD-PCR with different primers may be different because of the different affinity between the primer and the sample strain's genomic DNA. There seems no large possibility of RAPD-PCR as a traceability method of *Alicyclobacillus* contamination, because the RAPD-PCR results showed that one certain genotypic group within a cluster of a certain species could be isolated from different sources (such as group I of the cluster of *Alicyclobacillus acidoterrestris*, which included strains from 3 sources in pattern A and 4 sources in pattern B, with the plant of the producer of frozen kiwi fruit slice as one source), and strains from one source could be divided into different groups within a cluster of a certain species (such as strains from Xiajiagou village [C-ZJB-12-18 ~ C-ZJB-12-27] which belonged to *Alicyclobacillus acidoterrestris* were in 2 groups in pattern A and 3 groups in pattern B). These results show some differences from Groenewald's [17] report.

From Figure 4A,B and C we can see that the main reason for *Alicyclobacillus* to unable to grow in raw kiwi fruit juice is not its

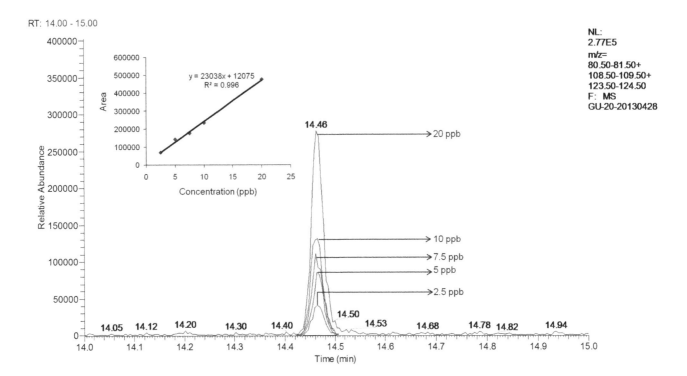

Figure 6. A GC–MS chromatogram of all standard solutions and a standard curve generated by Excel. For comparison purposes only peaks between the mass range of 81, 109, 124 m/z are displayed.

containing some bacteriostatic factors, but its relatively low pH (2.5) compared with other fruit juices (apple and orange, pH 3.5–4.0), and their spores can retain viability in raw kiwi fruit juice and grow and produce taint like in apple and orange juices when the pH rises while dilution or mixing with other relatively high pH juices to make palatable commercial products. Figure 4D proves this conclusion further. The growth curve in the commercial kiwi fruit juice A products are similar to the ones in pH adjusted kiwi fruit juice, only showing longer lag phase and longer time (about 8–10 days) to reach the population levels of 10^7 CFU/mL, which may be because of the lower pH and different media contents compared with pH adjusted kiwi fruit juice. The reason why C-ZJB-12–35 is unable to grow in the 2 vinegar samples may lie in their low pH (2.5). The alcohol in the 2 wine samples might have negative influence on the growth of *Alicyclobacillus*, but according to the GC-MS chromatogram of wine A, a small amount of guaiacol should exist in the sample, although guaiacol might not be separated well with other compounds. As for another commercial

juice which unable to support the growth of C-ZJB-12-35, it is inferred that it does not contain enough raw kiwi fruit juice as described on its packaging (no antiseptics found).

Guaiacol has been found in all 5 samples proved that it is the most important taint compound, which is similar to Pettipher's report [28]. It has been proved that the best estimated threshold (BET) value for odour of guaiacol was around 2 ppb in apple juice [28], and similar threshold values for odour of guaiacol in apple juice were even lower in other reports [41] [42]. Although there are still no reports about such values in kiwi fruit juice, it is thought they might not be much higher than in apple juice, therefore the guaiacol levels in this study are high enough to make a consumer aware of the spoilage. The different amount of guaiacol production among the 5 samples might be attributed to the differences among species, cell numbers. Goto et al. [36] has proved *Alicyclobacillus herbarius* to produce guaiacol, although in the same study, they did not detect guaiacol from *Alicyclobacillus contaminans*, which may be because of the limit of the test method

Table 4. The peak areas of guaiacol of samples and the calculated concentrations.

Samples	Start Retention Time (min)	End Retention Time (min)	Peak area	Concentration (ppb)
C-ZJB-12-32 in A	14.44	14.54	280389.50	11.65
C-ZJB-12-35 in A	14.44	14.53	435151.81	18.36
C-ZJB-12-55 in A	14.43	14.52	678012.37	28.91
C-ZJB-12-69 in A	14.44	14.50	812313.79	34.74
C-ZJB-12-35 in B	14.43	14.50	650602.09	27.72

According to the standard curve, concentrations were calculated based on the peak areas (mass range: 81, 109, 124 m/z).
A: pH adjusted kiwi fruit juice; B: commercial kiwi fruit juice A.

and the culture media. In this study, we first proved that *Alicyclobacillus contaminans* could produce guaiacol (11.65 ppb) in pH adjusted kiwi fruit juice. There were no halophenols (2,6-DBP and 2,6-DCP) detected in all samples in this study, although *Alicyclobacillus acidoterrestris* has been proved to produce guaiacol and 2,6-DBP, and *Alicyclobacillus cycloheptanicus* has been proved to produce guaiacol and both 2,6-DBP and 2,6-DCP in Gocmen's study [27]. This might be attributed to the media used in this study.

Although there was only one *Alicyclobacillus* strain isolated from commercial kiwi fruit products, *Alicyclobacillus* strains were isolated from kiwi fruit product processing plant and raw material orchards, and there existed genotypic diversities to some extent. Furthermore, the test strains from all 4 species isolated in this study can spoil the pH adjusted kiwi fruit juice and some commercial products, and their spores can retain viability in raw kiwi fruit juice. Therefore, there do exist a potential risk of *Alicyclobacillus* contamination, even spoilage incidents, when a large quantity of kiwi fruits were used as raw materials to make juice or wine or other products, although there is no report about such incidents yet in China. Because of the similarity of 16S rDNA between one species and its subspecies could beyond 99% in genus *Alicyclobacillus* (e.g. between some strains of *Alicyclobacillus acidocaldarius* and *Alicyclobacillus acidocaldarius* subsp. *Rittmannii* after searching through NCBI), the genotypic diversities (through RAPD-PCR) of these isolates indicate that there might be some new subspecies among them. Therefore, further research will be focused on the growth conditions and off-odour production of all isolated *Alicyclobacillus* in kiwi fruit juice and commercial products to further assess their capability of contamination, and looking for new subspecies to expand our knowledge about *Alicyclobacillus*.

References

1. Uhino F, Doi S (1967) Acido-thermophilic bacteria from thermal waters. Agricultural Biology and Chemistry 31: 817–822.
2. Wisotzkey JD, Jurtshuk P, Fox GE, Deinhard G, Poralla K (1992) Comparative sequence analysis on the 16S rRNA (rDNA) of *Bacillus acidocaldarius*, *Bacillus acidoterrestris*, and *Bacillus cycloheptanicus* and proposal for creation of a new genus. *Alicyclobacillus* gen. nov. International Journal of Systematic Bacteriology 42: 263–269.
3. Borlinghaus A, Engel R (1997) *Alicyclobacillus* incidence in commercial apple juice concentrate (AJC) supplies –method development and validation. Fruit Processing 7: 262–266.
4. Walls I, Chuyate R (2000) Spoilage of fruit juices by *Alicyclobacillus acidoterrestris*. Food Australia 52: 286–288.
5. Eiroa MNU, Junqueira VCA, Schmidt FL (1999) *Alicyclobacillus* in orange juice: occurrence and heat resistance of spores. Journal of Food Protection 62: 883–886.
6. Blocher JC, Busta FF (1983) Bacterial spore resistance to acid. Food Technology 37: 87–99.
7. Cerny G, Hennlich W, Poralla K (1984) Spoilage of fruit juice by *bacilli*: isolation and characterisation of the spoiling microorganism. Zeitschrift feur Lebensmittel Untersuchung und Forschung 179: 224–227.
8. Steyn CE, Cameron M, Witthuhn RC (2011) Occurrence of *Alicyclobacillus* in the fruit processing environment–A review. International Journal of Food Microbiology 147: 1–11.
9. Smit Y, Cameron M, Venter P, Witthuhn RC (2011) *Alicyclobacillus* spoilage and isolation–A review. Food Microbiology 28: 331–349.
10. Splittstoesser DF, Churey JJ, Lee CY (1994) Growth characteristics of aciduric spore forming bacilli isolated from fruit juices. Journal of Food Protection 57: 1080–1083.
11. Chang SS, Kang DH (2004) *Alicyclobacillus* spp. in the fruit juice industry: history, characteristics, and current isolation/detection procedures. Critical Reviews in Microbiology 30: 55–74.
12. Walker M, Phillips CA (2008) *Alicyclobacillus acidoterrestris*: an increasing threat to the fruit juice industry? International Journal of Food Science and Technology 43: 250–260.
13. Howard I (2006) ACB Workshop October 2005–Review. European Quality Control System (EQCS) Workshop 2006. Available: http://www.eqcs.org/download/Workshop2006/07_ACBworkshop2005.pdf.
14. Durak MZ, Churey JJ, Danyluk MD, Worobo RW (2010) Identification and haplotype distribution of *Alicyclobacillus* spp. from different juices and beverages. International Journal of Food Microbiology 142: 286–291.
15. Deinhard G, Blanz P, Poralla K, Altan E (1987) *Bacillus acidoterrestris* sp. nov., a new thermotolerant acidophile isolated from different soils. Systematic and Applied Microbiology 10: 47–53.
16. Wisse CA, Parish ME (1998) Isolation and enumeration of sporeforming, thermoacidophilic, rod-shaped bacteria from citrus processing environments. Diary, Food and Environmental Sanitation 18: 504–509.
17. Groenewald WH, Gouws PA, Witthuhn RC (2009) Isolation, identification and typification of *Alicyclobacillus acidoterrestris* and *Alicyclobacillus acidocaldarius* strains from orchard soil and the fruit processing environment in South Africa. Food Microbiology 26: 71–76.
18. Walls I, Chuyate R (1998) *Alicyclobacillus* – historical perspective and preliminary characterization study. Dairy, Food and Environmental Sanitation 18: 99–503.
19. Previdi MP, Lusardi E, Vicini E (1999) Resistance of *Alicyclobacillus* spp. spores to a disinfectant. Industria conserve 74: 231–236.
20. Chen S, Tang Q, Zhang X, Zhao G, Hu X, et al. (2006) Isolation and characterization of thermo-acidophilic endosporeforming bacteria from the concentrated apple juice-processing environment. Food Microbiology 23: 439–445.
21. Goto K, Mochida K, Kato Y, Asahara M, Ozawa C, et al. (2006) Diversity of *Alicyclobacillus* isolated from fruit juices and their raw materials, and emended description of *Alicyclobacillus acidocaldarius*. Microbiological Culture Collections 22: 1–14.
22. Baumgart J, Menje S (2000) The impact of *Alicyclobacillus acidoterrestris* on the quality of juices and soft drinks. Fruit Processing 10: 251–254.
23. Gouws PA, Gie L, Pretorius A, Dhansay N (2005) Isolation and identification of *Alicyclobacillus acidocaldarius* by 16S rDNA from mango juice and concentrate. International Journal of Food Science and Technology 40: 789–792.
24. Pinhatti MEMC, Variane S, Eguchi SY, Manfio GP (1997) Detection of acidothermophilic Bacilli in industrialized fruit juices. Fruit Processing 9: 350–353.
25. Oteiza JM, Ares G, Sant'Ana AS, Soto S, Giannuzzi L (2011) Use of a multivariate approach to assess the incidence of *Alicyclobacillus* spp. in concentrate fruit juices marketed in Argentina: Results of a 14-year survey. International Journal of Food Microbiology 151: 229–234.
26. Huang H, Ferguson AR (2001) Review: Kiwifruit in China. New Zealand Journal of Crop and Horticultural Science 29: 1–14.
27. Gocmen D, Elston A, Williams T, Parish M, Rouseff RL (2005) Identification of medicinal off-flavours generated by *Alicyclobacillus* species in orange juice using GC-olfactometry and GC-MS. Letters in Applied Microbiology 40: 172–177.

Supporting Information

Chromatogram S1 The GC-MS raw data file of pH adjusted kiwi fruit juice inoculated with C-ZJB-12-32.

Chromatogram S2 The GC-MS raw data file of pH adjusted kiwi fruit juice inoculated with C-ZJB-12-35.

Chromatogram S3 The GC-MS raw data file of pH adjusted kiwi fruit juice inoculated with C-ZJB-12-55.

Chromatogram S4 The GC-MS raw data file of pH adjusted kiwi fruit juice inoculated with C-ZJB-12-69.

Chromatogram S5 The GC-MS raw data file of commercial kiwi fruit juice A inoculated with C-ZJB-12-35.

Chromatogram S6 The GC-MS raw data file of commercial kiwi fruit wine A inoculated with C-ZJB-12-35.

Chromatogram S7 The GC-MS raw data file of 5-ppb standard solution for guaiacol in kiwi fruit juice.

Author Contributions

Conceived and designed the experiments: TY YY. Performed the experiments: JZ. Analyzed the data: TY JZ. Contributed reagents/materials/analysis tools: TY. Wrote the paper: JZ.

28. Pettipher GL, Osmundson ME, Murphy JM (1997) Methods for the detection and enumeration of *Alicyclobacillus acidoterrestris* and investigation of growth and production of taint in fruit juice and fruit juice-containing drinks. Letters in Applied Microbiology 24: 185–189.

29. International Federation of Fruit Juice Producers (IFU) (2007) Method on the Detection of Taint Producing *Alicyclobacillus* in Fruit Juices. IFU Method No. 12. IFU, Paris, 1–11.

30. Goto K, Tanimoto Y, Tamura T, Mochida K, Arai D, et al. (2002) Identification of thermoacidophilic bacteria and a new *Alicyclobacillus* genomic species isolated from acidic environments in Japan. Extremophiles 6: 333–340.

31. Jiang CY, Liu Y, Liu YY, You XY, Guo X, et al. (2008) *Alicyclobacillus ferrooxydans* sp. nov., a ferrous-oxidizing bacterium from solfataric soil. International Journal of Systematic and Evolutionary Microbiology 58: 2898–2903.

32. Saitou N, Nei M (1987) The neighbor-joining method: a new method for reconstructing phylogenetic trees. Molecular Biology and Evolution 4: 406–425.

33. Kumar S, Tamura K, Nei M (2004) MEGA3: integrated software for molecular evolutionary genetics analysis and sequence alignment. Briefings in Bioinformatics 5: 150–163.

34. Danyluk MD, Friedrich LM, Jouquand C, Goodrich-Schneider R, Parish ME, et al. (2011) Prevalence, concentration, spoilage, and mitigation of *Alicyclobacillus* spp. in tropical and subtropical fruit juice concentrates. Food Microbiology 28: 472–477.

35. Zierler B, Siegmund B, Pfannhauser W (2004) Determination of off-flavour compounds in apple juice caused by microorganisms using headspace solid phase microextraction–gas chromatography–mass spectrometry. Analytica Chimica Acta 520: 3–11.

36. Goto K, Nishibori A, Wasada Y, Furuhata K, Fukuyama M, et al. (2008) Identification of thermo-acidophilic bacteria isolated from the soil of several Japanese fruit orchards. Letters in Applied Microbiology 46: 289–294.

37. Jensen N, Whitfield FB (2003) Role of *Alicyclobacillus acidoterrestris* in the development of a disinfectant taint in shelf-stable fruit juice. Letters in Applied Microbiology 36: 9–14.

38. Sandoval AJ, Barreiro JA, Mendoza S (1992) Thermal Resistance of *Bacillus coagulans* in Double Concentrated Tomato Paste. Journal of Food Science 57: 1369–1370.

39. Clerck ED, Gevers D, Sergeant K, Rodríguez-Díaz M, Herman L, et al. (2004) Genomic and phenotypic comparison of *Bacillus fumarioli* isolates from geothermal Antarctic soil and gelatine. Research in Microbiology 155: 483–490.

40. Ten LN, Im WT, Baek SH, Lee JS, Oh HM, et al. (2006) *Bacillus ginsengihumi* sp nov., a novel species isolated from soil of a ginseng field in Pocheon Province, South Korea. Journal of Microbiology and Biotechnology 16: 1554–1560.

41. Eisele TA, Semon MJ (2005) Best estimated aroma and taste detection threshold for guaiacol in water and apple juice. Journal of Food Science 70: 267–269.

42. Siegmund B, Pöllinger-Zierler B (2006) Odor thresholds of microbially induced off flavor compounds in apple juice. Journal of Agricultural and Food Chemistry 54: 5984–5989.

15

Landscape Factors Facilitating the Invasive Dynamics and Distribution of the Brown Marmorated Stink Bug, *Halyomorpha halys* (Hemiptera: Pentatomidae), after Arrival in the United States

Adam M. Wallner[1]*, **George C. Hamilton**[2], **Anne L. Nielsen**[2], **Noel Hahn**[2], **Edwin J. Green**[3], **Cesar R. Rodriguez-Saona**[2]

1 United States Department of Agriculture – Animal and Plant Health Services – Plant Protection Quarantine, Plant Inspection Station, Miami, Florida, United States of America, **2** Department of Entomology, Rutgers University, New Brunswick, New Jersey, United States of America, **3** Department of Ecology, Evolution and Natural Resources, Rutgers University, New Brunswick, New Jersey, United States of America

Abstract

The brown marmorated stink bug, *Halyomorpha halys*, a native of Asia, has become a serious invasive pest in the USA. *H. halys* was first detected in the USA in the mid 1990s, dispersing to over 41 other states. Since 1998, *H. halys* has spread throughout New Jersey, becoming an important pest of agriculture, and a major nuisance in urban developments. In this study, we used spatial analysis, geostatistics, and Bayesian linear regression to investigate the invasion dynamics and colonization processes of this pest in New Jersey. We present the results of monitoring *H. halys* from 51 to 71 black light traps that were placed on farms throughout New Jersey from 2004 to 2011 and examined relationships between total yearly densities of *H. halys* and square hectares of 48 landscape/land use variables derived from urban, wetland, forest, and agriculture metadata, as well as distances to nearest highways. From these analyses we propose the following hypotheses: (1) *H. halys* density is strongly associated with urban developments and railroads during its initial establishment and dispersal from 2004 to 2006; (2) *H. halys* overwintering in multiple habitats and feeding on a variety of plants may have reduced the Allee effect, thus facilitating movement into the southernmost regions of the state by railroads from 2005 to 2008; (3) density of *H. halys* contracted in 2009 possibly from invading wetlands or sampling artifact; (4) subsequent invasion of *H. halys* from the northwest to the south in 2010 may conform to a stratified-dispersal model marked by rapid long-distance movement, from railroads and wetland rights-of-way; and (5) high densities of *H. halys* may be associated with agriculture in southern New Jersey in 2011. These landscape features associated with the invasion of *H. halys* in New Jersey may predict its potential rate of invasion across the USA and worldwide.

Editor: Joseph Clifton Dickens, United States Department of Agriculture, Beltsville Agricultural Research Center, United States of America

Funding: Funding came from U.S. Hatch Funds and the U.S. Department of Agriculture - National Institute of Food and Agriculture - Specialty Crop Research Initiative Award #2011-51181-30937. The funders had no role in study design, data collection and analysis, decision to publish, or preparation of the manuscript.

Competing Interests: The authors have declared that no competing interests exist.

* E-mail: adam.m.wallner@aphis.usda.gov

Introduction

The brown marmorated stink bug, *Halyomorpha halys* (Stål) (Hemiptera: Pentatomidae), is native to China, Japan, and Korea and is an invasive agricultural pest in the mid-Atlantic United States of America (USA), that was introduced around 1996 into Allentown, Pennsylvania (USA) [17]. Since then it has been found in over 41 states, causing significant reductions in agricultural yields and has become a homeowner nuisance [25]. It is highly polyphagous and able to feed on a wide variety of both agricultural and non-agricultural plants. Some of these plants susceptible to injury include field crops, tree and small fruit, vegetables, and wild and ornamental plants. *H. halys* became a major pest of multiple crops, and its potential for damage is increasing as its range in the USA broadens. Due to high populations in 2010, damage from *H. halys* resulted in over $37 million of loss to mid-Atlantic apples and vegetables [2,24]. Monitoring of this pest and other Pentatomoi-

dea pest species has been conducted using baited pyramid traps, visual sampling, and black lights [25,27,37,38]. Current control relies on frequent insecticide applications, impacting established integrated pest management programs.

H. halys overwinters in forested areas and structures (D–H. Lee, personal comm.). In forested areas, dead trees and standing trees provide overwintering habitat. In urban areas, human-made structures such as residential and commercial buildings also provide overwintering habitat. In some cases, large numbers of overwintering *H. halys* can be found [19]. The ability to overwinter in forests and structures coupled with its wide host range have allowed *H. halys* to spread and establish populations throughout New Jersey and beyond to other states.

Because *H. halys* can overwinter in a variety of habitats, are polyphagous, and have a high rate of dispersal [25,26,35,36], their invasion dynamics may be difficult to examine using conventional statistical models. Thus, spatial analysis (i.e. Geographic Information Systems or GIS) and geostatistics may be more appropriate in examining these dynamics. These analyses have allowed researchers to correlate spatially referenced data with landscape factors through space and time. In fact, the spatial context of individuals and populations has become an important aspect of understanding ecological processes involving insects [28]. This ability is especially useful when characterizing susceptible habitats and analyzing survey data. For example, Shepherd et al. [41] were able to determine how forest type and climate correlated with outbreaks of *Orgyia pseudotsugata* (McDunnough) (Douglas-fir tussock moth) by using historical maps of defoliation, forest type, and climate maps. Similarly, GuoJun et al. [15] investigated the invasion dynamics and dispersal processes of *Lissorhoptrus oryzophilus* Kuschel (rice water weevil) in China using historical maps of pest invasion and distribution of infested rice.

The diverse New Jersey landscape allows for a unique opportunity to investigate invasion dynamics of *H. halys* and landscape features that are facilitating this invasion using spatial analytical tools. New Jersey is comprised of a range of landscapes, from highly urbanized areas, farmland, coastal beaches and marshes to grasslands and wetland forests [33]. Over the past two decades, urban development has rapidly changed New Jersey's landscape and caused extensive fragmentation [16]. This changing mosaic has implications on the spread of a highly polyphagous species like *H. halys*. For example, knowledge of the distribution of *H. halys* and the understanding of its invasion dynamics and colonization processes influenced by landscape parameters would help predict areas susceptible to *H. halys* infestation. In addition, growers in affected areas could target control efforts in areas likely to have elevated populations due to landscape composition and how ecosystem processes such as disturbances caused by pest species are affected by landscape. Examination of stink bug population spatial dynamics has been facilitated by the use of GIS and the compilation of spatial data. These studies have been beneficial in revealing surrounding habitats near Hawaiian macadamia nut orchards that supported high densities of the southern green stink bug *Nezara viridula* (L.) [20]; edge effects of *N. viridula* in peanut-cotton landscapes [47]; population dynamics of *N. viridula* in corn, cotton, and peanut landscapes [46]; and patchy distributions for several pest species of Pentatomidae, such as *N. viridula*, *Euschistus servus* (Say), *Oebalus pugnax* (F.), and *Thyanta custator* (F.) in South Carolina wheat fields [40]. However, few studies on *H. halys*' landscape ecology exist. Kiritani [22] showed that increases in *H. halys* populations in Japan seem to have risen after the planting of conifer plantations following World War II. Zhu et al. [53] developed a model predicting the potential distribution of *H. halys* with respect to climatic factors, but was not able to predict distribution based on landscape features.

Although, some studies [22,53] have examined environmental factors that may facilitate the dispersal of *H. halys* over geographical areas, a state-wide landscape analysis of *H. halys* is lacking. Here, we analyzed our vast dataset of the invasion dynamics of *H. halys* in New Jersey to develop a robust model (i.e. hypothesis) that elucidates the movement of this insect over a large geographical area that is based on natural (e.g., forest, wetland, and grassland habitats) and anthropogenic (e.g., houses, silos, and roads) factors. Nielsen and colleagues [38] identified the yearly population increase of *H. halys* to be 75% based on black light trap catches. The incorporation of landscape features and biological information into a predictive model could be readily tested in other states in the USA and other infested countries. Lastly, these data could offer novel control strategies of this pest insect, for example landscapes that are more susceptible to infestation.

Already, the state of New Jersey has the longest historical data set in the USA of *H. halys* collected from black light traps distributed throughout the state. Although specimens were collected from black light traps beginning in 1999 through 2011, only specimens captured from 2004 to 2011 were used in our analysis because active monitoring of this pest was conducted during these years. In using this dataset, the focus of our study was to 1) examine the invasion dynamics, establishment, and colonization processes of *H. halys* throughout New Jersey from 2004 to 2011 using geostatistical analysis in ArcGIS; 2) determine which environmental factors are facilitating this invasion using Bayesian linear regression analysis; 3) from these results develop a series of hypotheses of *H. halys* dispersal; and 4) discuss landscapes that are susceptible by this pest insect.

Materials and Methods

Insect monitoring with back light traps (110V Gempler's, Madison WI) occurred on 51–71 farms throughout New Jersey and each trap was geo-referenced using a handheld Trimble GeoExplorer II (Trimble Navigation Ltd. Sunnyvale, Calif) GPS unit with 2 to 5m (6.6 to 16.4ft) accuracy after differential correction [18]. Black light traps were placed in open area of the farms in front of sheds or silos. All samples were done on private land as part of the Rutgers Vegetable IPM program. Rutgers has the permission to take these samples. Traps were selected because *H. halys* is attracted to traps in large numbers [38] and provides a standardized sampling effort. Traps were distributed throughout the agricultural areas of the state with fewer traps located in urbanized areas in the northeast, southeastern regions of New Jersey characterized by the sandy soils of the Pinelands (i.e. heavily forested area of coastal plain occupying seven counties in southern New Jersey), and the Atlantic coast (see Figure 1 placement of traps). Monitoring of these insects initiated in 2004 and continues to the present. Because data are still being recorded and analyzed, we focused on *H. halys* collected from 2004 to 2011, an important period of its invasion throughout the state [38].

Insects were collected bi-weekly from May to October through the Rutgers Cooperative Extension Vegetable IPM program for each of these years [35]. Adult stink bug specimens were sorted from these traps and identified to species according to [17,32]. Less than 1% of specimens did not have their gender determined. Voucher specimens are currently housed at the Rutgers Entomology Insect Museum in New Brunswick, New Jersey [38]. Weekly counts of male and female *H. halys* for each year and farm were merged because we were interested in the effects of land use/land cover factors on the total population of *H. halys* throughout New Jersey. These data were compiled into a matrix using Microsoft Excel and imported into ArcGIS 10 [9] and linked to the corresponding trap locations.

Environmental Variables

We selected environmental features (i.e. landscape/land use) that might potentially affect population dynamics of *H. halys* in New Jersey [5,48,53]. These features are classified into five types of landscape/land cover features, which include agriculture, urbanization, wetlands, forests, and distance to nearby roads and highways. Landscape/land use features within the Pinelands and along the Atlantic coast were excluded in the analysis because no traps were placed in these habitats. These five main types of features were obtained by acquiring GIS digital files from the New Jersey Department of Environmental Protection (NJDEP) GIS Land use/Land cover dataset from 2007 [34,9] (see Figure 1). These files were selected because they represent the most current

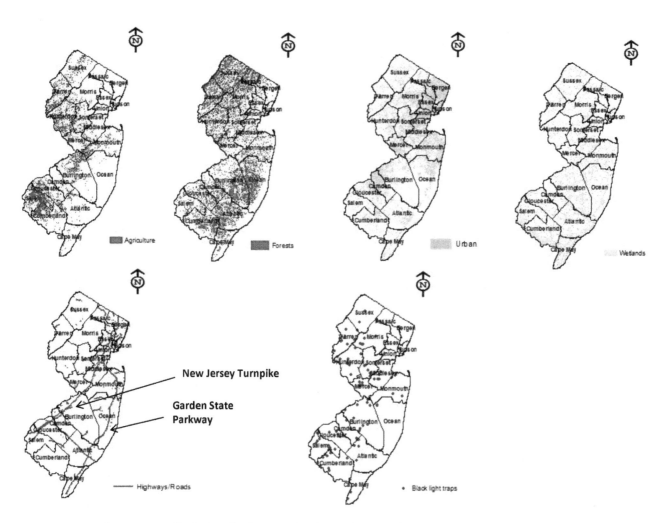

Figure 1. Maps of New Jersey displaying the five main landscape/land use features used in the analysis, including agriculture, forests, urban/residential, wetlands, and major highways and roads; the New Jersey Turnpike and The Garden State Parkway are shown in grey; and distribution of black light traps throughout New Jersey are also provided.

datasets produced by NJDEP. These data were taken from aerial photography captured in the spring of 2007.

We merged datasets from every county within New Jersey to create individual maps representing the five main types of landscape/land use features using the merger function in ArcMap 10. Trap locations were plotted on these maps; buffers of 2 km were placed around each of the traps, representing the total hectares *H. halys* may inhabit [52]; and square acreage of 48 land use/land cover variables within the main landscape/land use features (see Table S1 for list of variables and acreage) was calculated using the calculate geometry function in ArcMap 10. Lastly distance, in meters, from each trap to the nearest highways was calculated in ArcMap 10. Square hectares of all 48 variables, distances to nearest highway, and total population of *H. halys* for each trap were entered into Excel for statistical analysis.

Data Analysis

Kernel Density Estimation

Kernel Density Estimation (KDE) was implemented to examine the overall invasion dynamics and colonization of *H. halys* using the geographical information systems software ArcMap 10. KDE uses probability statistics (i.e. kernel function) to calculate the

estimated density of a population around the actual population data points producing a fitted, smooth, tapered surface that connects all data points used in the analysis [9,42]. We selected KDE because this analysis performs well with small amounts of data, they are robust to autocorrelation, and results are in a utilization distribution (i.e. UD; A UD is a grid where the value for each cell represents the probability of the taxa occurring in that cell) rather than a simple distribution range outline [21].

KDEs were derived from *H. halys* density collected from 2004 through 2011 and bandwidths (i.e. smoothing parameter). In calculating these bandwidths, we randomly measured distances between 20 traps, and then averaged those distances. A bandwidth specifies the maximum distance at which data points (i.e. traps) are distributed over the study area or the search radius of the Kernel Density function [9].

Hot-Spot Analysis

Hot-Spot Analysis was used to detect statistically significant populations of *H. halys* over New Jersey from 2004 to 2011. This was done to provide information on the overall dispersal and distribution of *H. halys* throughout New Jersey and provide insight on the relative rate of spread of *H. halys* populations across the New Jersey landscape. This analysis was used to compliment KDE because it can detect statistically significant (i.e., Z-scores) high (i.e., hot spot) and/or low (cold spot) clustering of *H. halys* populations. To be a statistically significant hot spot, a farm with a high density of *H. halys* has to be surrounded by neighboring farms with high densities of *H. halys*. The local sum of *H. halys* in a farm and its neighbors is then compared to the sum of all farms for each year; when the local sum is very different from the expected local sum, and that difference is too large to be the result of random chance, a statistically significant Z-score will result [9]. Z-scores are standard deviations, associated with p-values found in the tails of the normal distribution.

Bayesian Analysis

Bayesian linear regression was implemented to examine which of the 48 landscape/land use factors are facilitating the invasion, establishment, and dispersal of *H. halys* populations in New Jersey from 2004 to 2011 using the statistical software WinBUGS (Bayesian inference Using Gibbs Sampling; [43]). Bayesian linear regression was selected over univariate [e.g. analysis of variance (ANOVA)] and multivariate [e.g. principal component analysis (PCA)] statistical methods because it is a powerful and robust methodology that allows the observer to combine total *H. halys* population data with additional, independently available information (e.g. landscape/land use variables; the prior) to produce a full probability distribution (posterior distribution) of the relationships between *H. halys* populations and various landscape/land use factors; allows one to make inferences based on small sample sizes; and it appropriately models uncertainty on those relationships [6,8,10,11,23]. Moreover, the Bayesian methodology allows us to ask directly how probable is the hypothesis (e.g. Ho: acreage of different agriculture and urban landscape/land use factors will influence *H. halys* populations over time and space), given the data; whereas, conventional statistical methods lack this advantage. The following Bayesian regression model (Equation 1) was fitted to the hectare and total *H. halys* data, using reversible jump MCMC methods:

Equation 1. Bayesian Regression Model used in the analysis.

$$Y \sim \left(X\beta, \sigma^2 I\right),$$
$$\beta = (\beta_1, \beta_2, \cdots \beta_{48}),$$
$$p(\beta) = \prod_{j=1}^{48} p(\beta_j)$$
$$\beta_j \sim N(0.1000), j = 1, 2, \ldots, 48$$
$$\sigma^2 \sim IG(0.001, 0.001)$$

Reversible jump MCMC (RJMCMC) was selected over other Markov chain algorithms because this algorithm allows for sampling of posterior distributions that are generated from an unequal numbers of traps (i.e., samples) [13,14,29]. RJMCMC computations are generated by computing and converging on the most optimal posterior distributions. These computations are accomplished by fitting complex models (e.g. linear regression model) to our *H. halys* population and landscape/land use data. For relationships between *H. halys* and landscape/land use data to be considered informative, the posterior distribution of that relationship must be 0.05 or greater [i.e., a high posterior probability (i.e., α-level = 0.05), which indicates a statistically significant relationship between total *H. halys* population and a particular environmental variable]. RJMCMC sampled from 1,000,000 iterations and variable selection, which reflected a strong association (i.e., high probability) with the *H. halys* data, was based on the final 500,000 iterations [29–31].

The notations of σ^2 and β in the model above (Equation 1) represent the prior distributions on our parameters, where *IG* (inverse-gamma distribution) represents the marginal posterior distribution for the unknown variance of a normal distribution when an uninformative prior is used. These priors were chosen to be non-informative or vague because we are primarily interested in understanding what relationships are informative between landscape/land use variables and total *H. halys* populations [23]. Other notations in the model include the dependent variables (i.e., $Y_1, Y_2, \ldots, Y_8$). More specifically, Y_n represents the total *H. halys* populations collected from each farm from 2004 through 2011. Lastly, X represents the independent variables or covariates (i.e., 48 landscape/land use variables).

Results

The KDE analysis, generated from data collected in 2004, displayed high densities of *H. halys* in Warren County (northwest region), New Jersey, with no populations occurring in the remainder of the state (Figure 2A). A statistically significant hot-spot was also detected in this county (Figure 3A). In 2005, populations of *H. halys* almost tripled in size to 192 individuals (season total *H. halys* numbers across all black light traps; Table S1). In addition to this increase in population size, the KDE analysis displayed two populations (Figure 2B). The first and largest of these populations was still found in the northwest portion of the state but had subsequently dispersed east and south into most of Warren and Hunterdon Counties. The second population was detected further south in Burlington and Camden Counties. However, the hot-spot analysis only detected one statistically significant hot-spot in Warren and Hunterdon Counties (Figure 3B).

In 2006, the KDE analysis continued to display high densities of *H. halys* in the northwest region of New Jersey (Table S1, Figure 2C). Moreover, probability estimates from the KDE analysis (Figure 2C) showed that these insects spread further east into Middlesex County and south into Mercer, Somerset, and Morris Counties, although populations detected in Burlington and Camden Counties from 2005 were absent. A similar trend of high density and evidence of range expansion of these insects was detected in the hot-spot analysis, but only in Warren and Hunterdon Counties (Figure 3C).

By 2007, *H. halys* populations increased from a total of 473 individuals observed in 2006 to a total of 766 individuals found in 2007, despite a slight decrease in six black light traps between these years (Table S1). Probability estimates from the KDE analysis (Figure 2D) showed that this insect continued to disperse further east into Middlesex and Monmouth Counties, as well as further south into Burlington, Camden, Gloucester, and Salem Counties; and thus yielding a 50% increase in range expansion for *H. halys* from 2006 to 2007 (see Figures 2C–D). This range expansion observed in 2007 is reflected in the hot-spot analysis, which detected an expansion of high clusters of *H. halys* in Warren, Hunterdon, and Morris Counties, as well as detecting statistically low populations of *H. halys* (i.e., contraction of populations) in Salem and Cumberland Counties (Figure 3D).

In 2008 we observed an increase in population size from 766 in 2007 to a total of 1283 in 2008 (Table S1), but a retraction in the distribution of *H. halys* from Camden County (Figure 2E) observed in 2007 (Figure 2D). Although we observed a contraction in range distribution of this insect in 2008, despite an increase in the number of black light traps used, the overall distribution in 2008

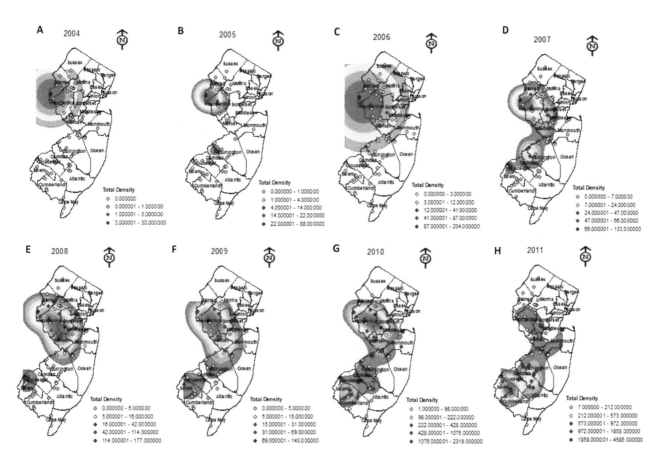

Figure 2. Kernel Density Estimation (KDE) graphs of the density of *Halymorpha halys* **captured from black light traps placed throughout New Jersey from (A) 2004, (B) 2005, (C) 2006, (D) 2007, (E) 2008, (F) 2009, (G) 2010, (H) 2011.** KDE are based on actual and predicted density of *H. halys* where green reflects lowest population density, orange moderate to high population density, and red predicts highest population of *H. halys*. Total density of *H. halys'* for year black lights were monitored is also provided.

was similar, in that *H. halys* density was concentrated in the northwest and spreading south into the center of New Jersey; a small population was present in southern portion of the state (Salem and Gloucester Counties); and the hot-spot analysis detected two high densities of *H. halys* (Figure 3E). One of these hot-spots was still in the northwest and a second hot-spot was observed in the southeast. This result corroborates with the KDE analysis, in that high densities of this insect are spreading further south into the state. Lastly, a cold spot was detected in Salem and Cumberland Counties, suggesting a contraction in *H. halys* density in southern New Jersey (Figure 3E). This trend is similar to the cold spot observed in 2007 (Figure 3D).

By 2009, the KDE analysis (Figure 2F) displayed densities of *H. halys* dispersing into 12 counties in New Jersey from 2008 (Figure 2E); extending their range east into Ocean County; further south in Burlington County; and further extending their range of an isolated population established in 2007 (Figure 3D) in the southernmost part of New Jersey (Salem, Cumberland, and Gloucester Counties; Figure 2F). Also, these data (Figure 2F) indicate that *H. halys* continue to spread further south New Jersey. This trend is reflected in the hot-spot analysis of 2009 (Figure 3F) which displays an increase in hot-spots in the eastern, southern, and northern parts of New Jersey compared to 2008 (Figure 3E). Although, *H. halys* expanded its distribution range the total density of this insect decreased to 677 from 1283 in 2008 (Table S1).

In 2010, we observed a rapid increase in the distribution range and density of *H. halys* from 12 counties in 2009 (Figure 2F) to 14

counties in 2010 (Figure 2G, Figure 3). Other trends observed include a high density of *H. halys* still being maintained in northwest New Jersey but continuing to move south, particularly in Hunterdon and Somerset Counties (Figure 2G) from Warren County observed in previous years (Figures 2A–F); a continued increase in density of *H. halys* in the southernmost portion of New Jersey, particularly in Salem and Cumberland Counties; and additional populations of these insects observed east into Monmouth and Ocean Counties, central in Burlington County, and south into Camden, Gloucester, and Atlantic Counties, as well as in southeast portion of New Jersey (i.e., Cape May County; Figure 2G). This continued dispersal into the southern region of the state is exhibited in the hot-spot analysis (Figure 3G) showed a decreased in the number of hot-spots observed in 2009 (Figure 3F) with no high density of *H. halys* occurring in Warren County as in previous years (Figure 3A–F) and an absence of cold-spots in the southern part of New Jersey (Figure 3G).

Lastly in 2011, we observed an inverse of the trend encountered in 2004 (Figure 2A) in which high densities of *H. halys* occurred in the southernmost part of state, such as Salem, Cumberland, and Gloucester Counties whereas only a few populations of this insect were found in Warren, Hunterdon, and Somerset Counties (Figure 2H). This trend conforms to the hot-spot analysis, which detected significant clusters of *H. halys* in the southernmost part of New Jersey and cold-spots (i.e., contraction) in the northwest part of New Jersey (Figure 3H). Lastly, while the overall distribution

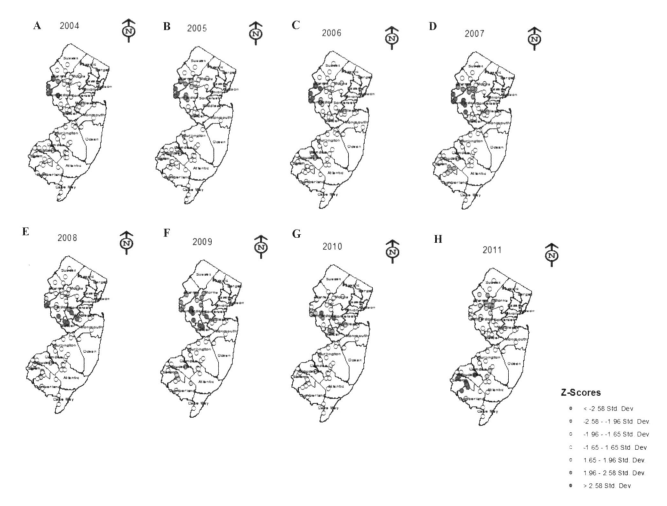

Figure 3. Hot-spot/cold-spot analysis graphs of the densities of *Halyomorpha halys* **captured from black light traps placed throughout New Jersey from (A) 2004, (B) 2005, (C) 2006, (D) 2007, (E) 2008, (F) 2009, (G) 2010, (H) 2011.** Z-scores are standard deviations, associated with p-values found in tails of the normal distribution, with red and orange reflecting high density of *H. halys* that are significantly clustering with one another and light to dark blue reflecting low densities of *H. halys* that are significantly clustering with one another.

range of this insect slightly decreased from 12 counties in 2010 to 11 counties in 2011, the density had doubled (Figure 2H).

Environmental Factors

The Bayesian linear regression identified 41 out of the total 48 landscape/land use variables that displayed a statistically significant relationship with densities of *H. halys* from 2004 through 2011 (Table 1). More specifically, in 2004 we observed intensive land use, for example, communications and utility buildings, suburban communities, and parking lots, displayed a positive relationship with densities of *H. halys*. As the populations dispersed east and south in 2005 (Figure 2B) we observed strong associations with not only intensive land use, but also with commercial land use (e.g. strip malls) and railroads (Table 1, Figure 4). As *H. halys* populations in 2006 increased in density and continued to disperse east and south (Figure 2C), we found this insect displayed strong associations with urban factors, such as railroads, and other landscape/land use features that include deciduous forests and wetlands (e.g. fallow agricultural land and wetland communities either natural or planted that are maintained in residential, commercial, and or industrial areas; Figure 4).

In 2007 as *H. halys* populations continued to move further south they displayed strong positive relationships with commercial structures and wetlands rights-of-way (former wetlands converted to road with some shrub vegetation; Figure 4). By 2008 and 2009, this insect dispersed further south (Figures 2E–F) and showed positive relationships with coniferous forests, and wetland landscapes (e.g., wetland right-of-way, *Phragmites*, fallow agriculture land, and mixed wetlands), and less so with urban factors, for example intensive urban land use and storm water basins (i.e., surface water collection site, associated with new commercial and residential areas), compared to 2005 and 2006 (Table 1, Figure 4). Lastly, from 2010 and 2011, *H. halys* displayed positive significant relationships with the majority (63–92%) of New Jersey landscape features examined. Among these landscape/land use features, agricultural variables, for example vineyards and orchards (Table 1, Figure 4) displayed positive relationships with this insect. Moreover, from 2010 to 2011 we observed a 70% increase in relationships between agriculture variables and *H. halys*. Although, we observed a trend in associations between *H. halys* and various landscape/land use features, relationships between this insect pest and distance to nearest highways and major roads from 2004 through 2011 were absent (Figure 4).

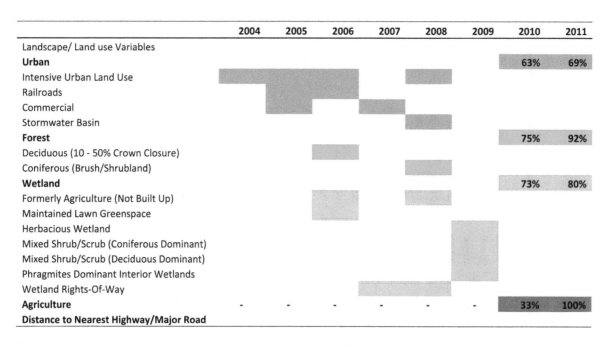

	2004	2005	2006	2007	2008	2009	2010	2011
Landscape/ Land use Variables								
Urban							63%	69%
Intensive Urban Land Use								
Railroads								
Commercial								
Stormwater Basin								
Forest							75%	92%
Deciduous (10 - 50% Crown Closure)								
Coniferous (Brush/Shrubland)								
Wetland							73%	80%
Formerly Agriculture (Not Built Up)								
Maintained Lawn Greenspace								
Herbacious Wetland								
Mixed Shrub/Scrub (Coniferous Dominant)								
Mixed Shrub/Scrub (Deciduous Dominant)								
Phragmites Dominant Interior Wetlands								
Wetland Rights-Of-Way								
Agriculture	-	-	-	-	-	-	33%	100%
Distance to Nearest Highway/Major Road								

Figure 4. Summary table of landscape/land use variables that displayed significant positive relationship with densities of *Halyomorpha halys* **using Bayesian linear regression analysis.** Pink represents urban landscape/land use variables; green represents forest landscape/land use variables; blue represents wetland landscape/land use variables; and red represents agricultural landscape/land use variables. In 2010 and 2011 we calculated the total number of these relationships observed.

Discussion

Understanding landscape factors (i.e., both natural and anthropogenic) that facilitate the spread and establishment of invasive pests is a critical element in invasion biology [50,39]. The invasion into New Jersey by *H. halys* and the spread of this pest throughout the state has created an excellent and unique opportunity to determine factors affecting the invasion dynamics and colonization process of this insect pest, as well as provide insight into developing novel strategies to control this pest. The invasion of *H. halys* can be divided into four phases: 1) initial establishment and dispersal, 2) range expansion, 3) potential lag phase or contraction in population growth, and 4) exponential growth.

First Phase: Initial Establishment and Dispersal

Within the first two years, from 2004 to 2005, the initial dispersal of *H. halys* began in the northwestern portion of New Jersey (Warren and Hunterdon Counties; Figures 2A). These populations were in close proximity to where this insect pest was first detected in Allentown, PA to the east (Somerset and Mercer Counties). A small population in the south (Burlington and Camden Counties) was later detected in 2005 (Figure 2B). Accompanying this initial dispersal we observed an increase in density from 2004 to 2005 (Table S1). Landscape/land use factors displaying positive relationships with this dispersion include: intensive urban land use, commercial developments (e.g., residential houses, strip malls, and supermarkets), and railroads (Figure 4). These factors may have been instrumental during the initial dispersion and establishment because these commercial and residential developments have provided adequate protection and increased the overwintering survival of *H. halys* [22,19].

Another possible explanation for the patterns observed from 2004 to 2005 is the Allee effect. The Allee effect is a positive relationship between the number of individuals in a population and their fitness [1]. Meaning that an individual of a species that is subject to an Allee effect will suffer a decrease in some aspect of its fitness when population density is low [1,44,45,3]. The population of *H. halys* is hypothesized to be severely, genetically bottlenecked and arising from a single introduction with a small propagule size [51] as small as two females. Theoretically, this would cause a significant reduction in fitness caused by inbreeding depression through Allee effect. However, as evidenced by our results, *H. halys* appears to have successfully overcome the Allee effect and spread at a rate of 2.84 farms per year with a 75% annual population increase [38]. The correlation between urban and residential developments during this establishment phase may have reduced the Allee effect of *H. halys*, thus facilitating a higher rate of establishment and dispersion. A similar pattern was observed in French Polynesia for the glassy-winged sharpshooter, *Homalodisca vitrepinnis* (Germar). For example, islands experiencing the early stages of invasion by *H. vitrepinnis* displayed high densities in urbanized areas with substantially lower abundances on remote islands further away from the main population [39].

The small satellite population of *H. halys* observed in Burlington and Camden Counties may also be a result of the Allee effect (Figure 2B) as has been shown in other insects [39]. Brown et al. [4] observed that many species of passerine birds displaying a reduced Allee effect have relatively clustered distribution with many sites of low abundance and few hot-spots of high local abundance.

Human-transportation activity (i.e., the movement of plants and contaminated cargo via ships, cars, boats, and rails) may also explain the increase in *H. halys* density and its establishment and range expansion in northwestern New Jersey. These activities can deliberately move material while simultaneously and unintentionally moving pest species, such as *H. halys* [12,39]. Of these activities, we observed a positive relationship between *H. halys* density and railroads in 2004 and 2005 (Figure 4). This suggests that railroads, also present in northwestern New Jersey (Warren

Table 1. Probability distributions calculated from total yearly densities of *Halyomorpha halys* captured from black light traps placed on farms from 2004 to 2011 using Bayesian linear regression and the reversible jump MCMC algorithm.

Landscape/Land use Variables	2004	2005	2006	2007	2008	2009	2010	2011
CROPLAND AND PASTURELAND	0.00016	0.00029	0.0003	0.0017	0.0015	0.0001	0.0123	**0.9439**
ORCHARDS/VINEYARDS/NURSERIES/HORTICULTURAL AREAS	0.00008	0.00023	0.0006	0.0005	0.0005	0.0003	0.0058	**0.0946**
OTHER AGRICULTURE	0.00030	0.00144	0.0023	0.0036	0.0063	0.0017	**0.9635**	**0.1713**
Distance	0.00000	0.00003	0.0000	0.0000	0.0000	0.0000	0.0003	0.0041
CONIFEROUS BRUSH/SHRUBLAND	0.00079	0.00226	0.0052	0.0049	**0.1062**	0.0065	**0.9998**	**0.1312**
CONIFEROUS FOREST (>50% CROWN CLOSURE)	0.00041	0.00092	0.0033	0.0024	0.0031	0.0018	0.0289	**0.0816**
CONIFEROUS FOREST (10–50% CROWN CLOSURE)	0.00173	0.00672	0.0148	0.0108	0.0146	0.0074	**0.1421**	**0.4825**
DECIDUOUS BRUSH/SHRUBLAND	0.00186	0.01068	0.0195	0.0060	0.0052	0.0046	**0.0544**	**0.1655**
DECIDUOUS FOREST (>50% CROWN CLOSURE)	0.00008	0.00015	0.0003	0.0003	0.0009	0.0002	0.0100	0.0223
DECIDUOUS FOREST (10–50% CROWN CLOSURE)	0.00138	0.01235	**0.0946**	0.0142	0.0051	0.0101	**0.0808**	**0.0927**
MIXED DECIDUOUS/CONIFEROUS BRUSH/SHRUBLAND	0.00038	0.00085	0.0027	0.0051	0.0382	0.0063	**0.3388**	**0.0905**
MIXED FOREST (>50% CONIFEROUS WITH >50% CROWN CLOSURE)	0.00031	0.00122	0.0031	0.0030	0.0053	0.0024	0.0157	**0.0495**
MIXED FOREST (>50% CONIFEROUS WITH 10–50% CROWN CLOSURE)	0.00150	0.00530	0.0147	0.0112	0.0232	0.0112	**0.1337**	0.3840
MIXED FOREST (>50% DECIDUOUS WITH >50% CROWN CLOSURE)	0.00030	0.00132	0.0030	0.0036	0.0081	0.0056	0.0166	**0.0657**
MIXED FOREST (>50% DECIDUOUS WITH 10–50% CROWN CLOSURE)	0.00122	0.00533	0.0133	0.0111	0.0133	0.0154	**0.0854**	**0.3872**
OLD FIELD (<25% BRUSH COVERED)	0.00051	0.00155	0.0043	0.0034	0.0050	0.0026	**0.0517**	**0.1280**
PLANTATION	0.00142	0.00409	0.0207	0.0137	0.0120	0.0053	**0.1484**	**0.2501**
RAILROADS	0.00576	**0.06867**	**0.2849**	0.0495	0.0339	0.0074	**0.1610**	**0.2605**
ATHLETIC FIELDS (SCHOOLS)	0.00099	0.00357	0.0068	0.0131	0.0138	0.0058	**0.1077**	**0.3121**
CEMETERY	0.00194	0.00504	0.0171	0.0127	0.0188	0.0045	**0.1558**	**0.2217**
COMMERCIAL/SERVICES	**0.06679**	**0.06826**	0.0144	**0.7959**	0.0025	0.0048	**0.0708**	**0.0539**
INDUSTRIAL	0.00040	0.00152	0.0029	0.0024	0.0077	0.0009	0.0280	**0.0509**
MAJOR ROADWAY	0.00086	0.00418	0.0045	0.0457	0.0219	0.0025	**0.1000**	**0.1890**
OTHER URBAN OR BUILT-UP LAND	**0.46699**	**0.37560**	**0.1806**	0.0307	**0.0548**	0.0429	**0.3694**	0.0498
RECREATIONAL LAND	0.00219	0.00858	0.0113	0.0040	0.0387	0.0031	0.0274	0.0453
RESIDENTIAL, HIGH DENSITY OR MULTIPLE DWELLING	0.00069	0.00143	0.0036	0.0020	0.0035	0.0011	0.0358	**0.0585**
RESIDENTIAL, RURAL, SINGLE UNIT	0.00016	0.00026	0.0006	0.0008	0.0041	0.0021	0.0156	0.0224
RESIDENTIAL, SINGLE UNIT, LOW DENSITY	0.00015	0.00072	0.0011	0.0016	0.0011	0.0006	0.0203	0.0327
RESIDENTIAL, SINGLE UNIT, MEDIUM DENSITY	0.00011	0.00055	0.0007	0.0024	0.0006	0.0003	0.0270	0.0175
STORMWATER BASIN	0.00263	0.00662	0.0258	0.0148	**0.5166**	0.0067	**0.5515**	**0.2693**
TRANSPORTATION/COMMUNICATION/UTILITIES	0.00415	0.00383	0.0097	0.0135	0.0078	0.0056	**0.0830**	**0.2349**
UPLAND RIGHTS-OF-WAY DEVELOPED	0.00381	0.01138	0.0394	0.0234	**0.7640**	0.0389	**0.5384**	**0.4097**
UPLAND RIGHTS-OF-WAY UNDEVELOPED	0.00168	0.00583	0.0242	0.0106	0.0078	0.0036	**0.2672**	**0.3562**
AGRICULTURAL WETLANDS (MODIFIED)	0.00010	0.00045	0.0010	0.0011	0.0011	0.0011	0.0221	**0.0768**

Table 1. Cont.

Landscape/Land use Variables	2004	2005	2006	2007	2008	2009	2010	2011
CONIFEROUS WOODED WETLANDS	0.00047	0.00170	0.0043	0.0044	0.0078	0.0265	0.0510	0.1391
DECIDUOUS SCRUB/SHRUB WETLANDS	0.00137	0.00396	0.0171	0.0083	0.0102	0.0072	0.0905	0.2712
DECIDUOUS WOODED WETLANDS	0.00063	0.00317	0.0039	0.0100	0.0072	0.0004	0.5888	0.0238
DISTURBED WETLANDS (MODIFIED)	0.00064	0.00250	0.0059	0.0089	0.0069	0.0077	0.0902	0.1200
FORMER AGRICULTURAL WETLAND (SOME SHRUBS, NOT BUILT-UP)	0.00785	0.02948	0.0574	0.0409	0.0518	0.0227	0.4830	0.4956
HERBACEOUS WETLANDS	0.00113	0.00339	0.0079	0.0060	0.0079	0.2885	0.0889	0.2401
MANAGED WETLAND IN BUILT-UP MAINTAINED REC AREA	0.00282	0.01193	0.0265	0.0238	0.0246	0.0143	0.2547	0.3276
MANAGED WETLAND IN MAINTAINED LAWN GREENSPACE	0.02417	0.03335	0.0536	0.0285	0.0457	0.0231	0.3317	0.5552
MIXED SCRUB/SHRUB WETLANDS (CONIFEROUS DOM.)	0.00282	0.00974	0.0185	0.0167	0.0431	0.5924	0.2557	0.3274
MIXED SCRUB/SHRUB WETLANDS (DECIDUOUS DOM.)	0.00083	0.00299	0.0061	0.0046	0.0077	0.0509	0.1157	0.1749
MIXED WOODED WETLANDS (CONIFEROUS DOM.)	0.00019	0.00077	0.0016	0.0019	0.0025	0.0085	0.0188	0.0470
MIXED WOODED WETLANDS (DECIDUOUS DOM.)	0.00031	0.00108	0.0026	0.0045	0.0061	0.0337	0.0291	0.0646
PHRAGMITES DOMINATE INTERIOR WETLANDS	0.00096	0.00276	0.0067	0.0088	0.0095	0.0520	0.0779	0.1623
WETLAND RIGHTS-OF-WAY	0.00352	0.00963	0.0208	0.0502	0.0836	0.0253	0.4247	0.6943

For relationships between *H. halys* and landscape/land use data to be considered informative, the posterior distribution of that relationship must be 0.05 or greater [(i.e. α-level = 0.05), indicating a statistically significant relationship].

and Hunterdon Counties), may have eased barriers to establishment resulting in the movement of this insect into new habitats. The association with railroads has been demonstrated in the establishment and dispersion of other insect pests, such as the glassy-winged sharpshooter in French Polynesia [39] and the rice water weevil in China [15]; although additional monitoring and analysis is required to adequately elucidate the relationship between *H. halys* and railroads.

In addition, these railroads and urban/residential developments may be surrounded by high host-plant diversity for *H. halys*. Two such hosts are *Paulownia tomentosa* (Thunb.) Sieb. and Zucc. ex Steud. (i.e., princess tree) and *Ailanthus altissima* (Mill.) Swengel (i.e., tree of heaven), natives to China [7,35,25]. These tree species are ornamentals, found in parks, gardens, and can be found in highly disturbed environments (e.g. intensive urban development and potentially near railroads). Because of the potential close proximity of these trees to railroads and urban/residential developments the following effect may have been produced: *P. tomentosa* and *A. altissima* are acting as a sink for *H. halys*, which could accelerate their invasion into urban/residential developments thereby increasing its abundance and reducing the Allee effect; and *H. halys* could readily move from *P. tomentosa* and *A. altissima* to nearby rail cars.

Lastly, orchard trees may be facilitating the movement of this pest insect to rail cars. These orchard trees are another potential source of establishment and dispersion because they are known hosts of *H. halys* [26]. Moreover, these orchards occupy approximately 434 hectares in Warren and Hunterdon Counties [49], which were the first counties to harbor this pest insect. Despite this evidence, additional investigations are needed to examine the associations between orchards, railroads, and *H. halys* in early establishment.

Second Phase: Population Growth and Range Expansion

After the initial establishment and dispersal of *H. halys* in 2004 and 2005, we observed rapid population growth and range expansion from 2006 to 2008 (Figure 2D–F). This growth and range expansion was associated with an increase in the insect's relationship with urban, forest, and wetland landscape/land use variables than the previous years (Table 1, Figure 4). The agriculture/urban interface, which is common in New Jersey, may have facilitated population growth of *H. halys* by offering agricultural and cultivated host plants for development and natural and human-made overwintering structures. The association with human habitat may be specifically important and has been suggested to increase overwintering survival relative to pentatomid species that overwinter in natural habitats [22]. As a consequence, we hypothesize that populations of *H. halys* may have rapidly increased, overcoming a critical minimum density threshold, followed by the establishment of new populations observed in eastern and southern New Jersey (Figures 2C–E, 3C–E), and thus reducing a possible Allee effect.

In addition to overcoming the Allee effect, the long distance movement from northwestern New Jersey to the southernmost portion of the state from 2006 to 2008 (Figures 2C–E) may be a consequence of stratified diffusion. Stratified diffusion refers to the movement of an invasive species by rapid long-distance transport and local diffusion around a point of infestation [39]. Long-distance movement may have been facilitated by rail cars to central New Jersey in 2006 (Figure 4). Moreover, these railroads are located in areas where high densities of *H. halys* have been observed in 2006 (Figure 2C). In 2007 and 2008 long distance-movement from central New Jersey to the southernmost part of the state may be attributed to close proximity of commercial

developments (e.g., strip malls, motels, and gas stations) acting as stepping-stones for invasion and fruit tree orchards or other adjacent suitable habitats near wetland rights-of-way (Figure 4). It should be noted that no relationships were detected between *H. halys* and highways/roads because black lights may have been too far away from this pathway.

The second component of stratified dispersion is local diffusion which occurs when a nascent colony becomes established and begins to expand naturally. This diffusion of *H. halys* after the long distance movement, appearing in 2007 (3 years after initial establishment to the southernmost region of New Jersey), may have been facilitated by encountering new habitats for overwintering, such as deciduous forests [i.e., dead, standing trees (D–H. Lee, unpublished data)]. Other factors facilitating this diffusion may include suitable host plants near fallow agriculture that has undergone succession into wetlands, and the increased hectares of orchard trees located in the southernmost region of New Jersey, such as Gloucester and Burlington Counties [49].

Phase 3: Potential Spurious Lag Phase

Following this population growth and expansion phase observed from 2006 to 2008 (Figures 2C–E), was a subsequent reduction in population size of *H. halys* in 2009 (Figure 2F, Table S1). This reduction coincides with a decrease in the number of positive significant relationships observed between *H. halys* and landscape/land use variables (Table 1, Figure 4). More specifically, only wetland variables, such as herbaceous wetlands, mixed shrub/scrub deciduous and coniferous wetlands, and Phragmites wetlands (Table 1, Figure 4) displayed significant positive relationships with this insect. These results suggest these wetland habitats could be barriers for mate finding of *H. halys* [45], although additional data are needed to examine the impact of these wetlands on the reproduction of this insect. Another more probable explanation in the decrease of *H. halys* observed in 2009 may be a sampling artifact, caused by 51 black light traps used in 2009 compared to 71 black light traps used in 2008. This potential spurious result is further supported in Nielsen et. al [38] study, which showed a continuing increase in populations of *H. halys* from 2004–2011 in New Jersey when black light traps are kept constant.

Phase 4: Exponential Growth

An exponential growth phase was observed from 2010 to 2011 (Table S1), as well as *H. halys* possibly dispersing and/or populations increasing from the north to the south (Figures 1G–H, 2G–H). This pattern was associated with all the main landscape/land use categories, with the exception of major highways and roads (Table 1, Figure 4). Positive density increases of *H. halys* and their subsequent expansion into the southernmost part of the state may be attributed to higher acreage of suitable habitats, in particular orchards. This was supported by a positive relationship observed between orchards and *H. halys* (Table 1). The high acreage of orchard trees in southern New Jersey [49], which are important hosts of this insect [25,26,36], may have supported high densities of *H. halys*. Although, continued monitoring in and near these orchards is needed to increase confidence in this trend.

Although there are many positive aspects of our dataset, our conclusions are dependent upon the location of the black light traps themselves. In this study, the black light traps are always placed in a vegetable farm and thus we have minimal data in non-agricultural areas such as densely populated urban areas in northeastern New Jersey, the Pinelands, and coastal regions. Additional black light traps in these habitats are needed to provide a more complete picture of the invasion and colonization dynamics of *H. halys* in New Jersey and other regions with diverse landscape/land use features.

Conclusions

We identified four phases of *H. halys* invasion in New Jersey from this study: (1) initial establishment (2004) and dispersal (1 year after invasion was being monitored in New Jersey) of the invasive population is confined to urban/residential developments and railroads, populations densities are low (92–189 individuals), and there is absence of this pest on the rest of the state, with the exception of a satellite population in the south. (2) An invasion of intermediate duration (2–3 years after becoming established in northwestern New Jersey) is characterized by an expansion of its range further east and south, and exhibiting relationships with urban, forest, and wetland habitats. Pest density in these habitats increased from 473 to 1283. (3) A potential contraction in population growth in 2009 after approximately 5 years from the initial establishment of this pest, although this pattern may be attributed to reduction in black light traps. Associations with wetland habitats in the northwestern region of New Jersey were present. (4) Finally, after 8 years of monitoring the invasive *H. halys* population reached high densities of 17690 in 2010 and 34241 in 2011 with high abundance in urban, wetland, forest, and agricultural habitats. Human-mediated long-distance dispersal events (e.g., railroads, wetland rights-of-way, and potentially trucks) may have facilitated the movement of this pest. In addition, high adaptability to novel overwintering sites and alternative host plants, as well as areas with large hectares of agriculture in the northwest and particularly in the south of New Jersey may have contributed to the high density, establishment, and dispersal of this pest. Stricter regulations preventing plant movement combined with monitoring and implementation of integrated pest management strategies near railroads, wetland rights-of-way, urban/residential developments, and orchards with known high density of *H. halys* may be needed to greatly reduce the spread of this pest throughout New Jersey and other states with documented invasions of *H. halys*.

Acknowledgments

This study would not have been possible without the black light data provided by Kristian Holmstrom. We would also like to thank Joseph Ingerson-Mahar, Amy Willmott, Andrea Wagner, Thomas Pike, John Cambridge, and Thomas Konar for assisting in the identification and recording of *H. halys*.

Author Contributions

Conceived and designed the experiments: GCH ALN CR-S. Performed the experiments: AMW GCH ALN CR-S. Analyzed the data: AMW NH EJG. Contributed reagents/materials/analysis tools: AMW NH EJG. Wrote the paper: AMW GCH ALN NH EJG CR-S. Provided invaluable comments and criticisms to the manuscript: CR-S GCH ALN NH EJG.

References

1. Allee WC (1931) Animal Aggregations, a Study in General Sociology. Chicago: The University of Chicago Press. 431 p.
2. American/Western Fruit Grower (2011) Brown marmorated stink bug causes $37 million in losses to Mid-Atlantic apple growers. April 14, 2011. Available: http://www.growingproduce.com/article/21057/brown-marmorated-stink-bug-causes-37-million in-losses-to-mid-atlantic-apple-growers. Accessed 15 October 2012.
3. Arim M, Abades R, Neil PE, Lima M, Marquet PA (2006) Spread dynamics of invasive species. PNAS 103: 374–378.
4. Brown JH, Mehlman DW, Stevens GC (1995) Spatial variation in abundance. Ecology 76: 2028–2043.
5. Capinha C, Anasácio P (2010) Assessing the environmental requirements of invaders using ensembles of distribution models. Divers Distrib 17: 13–24.
6. Carlin BP, Louis TA (2000) Bayes and empirical bayes methods for data analysis. Second Edition. New York: Chapman and Hall. 440 p.
7. Chung BK, Kang SW, Kwon JH (1995) Damages, occurrences, and control of hemipterous insects in non-astringent persimmon orchards. J Agr Sci 37: 376–382.
8. Congdon P (2001) Bayesian statistical modeling. Chichester: John Wiley and Sons.
9. ESRI (2011) ArcGIS Desktop: Release 10. Redlands CA: Environmental Systems Research Institute.
10. Fordyce JA, Gompert Z, Forister ML, Nice CC (2011) A hierarchical Bayesian approach to ecological count data: a flexible tool for ecologist. PLos One 6: 1–7.
11. Gelman A, Carlin JB, Stern HS, Rubin DB (1995) Bayesian data analysis. Boca Raton: Chapman and Hall/CRC. 552 p.
12. Gilbert M, Fielding N, Evans HF, Grégoire JC, Freise JF, et al. (2004) Long-distance dispersal and human population density allow the prediction of invasive patterns in the horse chestnut leafminer *Cameria ohridella*. J Anim Ecol 73: 459–468.
13. Green PJ (1995) Reversible jump Markov chain Monte Carlo computationand Bayesian model determination. Biometrika 82: 711–732.
14. Green PJ, Hastie D (2009) Reversible jump MCMC. Available: http://www.maths.bris.ac.uk/mapjg/Papers.html. Accessed 11 May 2013.
15. GuoJun Q, Yan G, DeChao H, LüLiHua (2012) Historical invasion, expansion process and the potential geographic distributions for the rice water weevil, *Lissorhoptrus oryzophilus* in China based on MAXENT. Acta Phytophylacica Sinica 39: 129–136.
16. Hasse J, Lathrop R (2010) Urban Growth and Open Space Loss in NJ 1986 thru 2007. Rowan University, Glassboro, NJ, and Rutgers University, Grant F. Walton Center for Remote Sensing and Spatial Analysis. New Brunswick, NJ.
17. Hoebeke ER, Carter ME (2003) *Halyomorpha halys* (Stål) (Heteroptera: Pentatomidae): a polyphagous plant pest from Asia newly detected in North America. Proc Entom Soc Wash 105: 225–237.
18. Holmstrom KE, Hughes MG, Walker SD, Kline WL, Ingerson-Mahar J (2001) Spatial mapping of adult corn earworm and European corn borer populations in New Jersey. HortTechnology 11: 103–109.
19. Inkley DB (2012) Characteristics of home invasion by the brown marmorated stink bug (Hemiptera: Pentatomidae). J Entom Sci 47: 125–130.
20. Jones VP, Westcott DM, Finson NN, Nishimoto RK (2001) Relationship between community structure and southern green stink bug (Heteroptera: Pentatomidae) damage in macadamia nuts. Environ Entomol 30: 1028–1035.
21. Kernohan BJ, Gitzen RA, Millspaugh JJ (2001) Analysis of animal space use and movements. In: Millspaugh JJ, Marzluff JM editors. Radio Tracking and Animal. San Diego: Populations Academic Press. 125–166.
22. Kirtani K (2006) Predicting impacts of global warming on population dynamics and distribution of arthropods in Japan. Popul Ecol 48: 5–12.
23. Latimer AM, Shanshan W, Gelfand AE, Silander Jr JA (2006) Building statistical models to analyze species distribution. Ecol Appl 16: 33–50.
24. Leskey T, Hamilton GC (2010) Brown Marmorated Stink Bug Working Group Meeting. Available: http://projects.ipmcenters.org/Northeastern/FundedProjects/ReportFiles/Pship2010/Pship2010-Leskey-ProgressReport-237195-Meeting-2010_11_17.pdf. Accessed 10 May 2013.
25. Leskey TC, Hamilton GC, Nielsen AL, Polk DF, Rodriguez-Saona C, et al. (2012) Pest status of the brown marmorated stink bug, *Halyomorpha halys* in the USA. Outlooks on Pest Management DOI: 10.1564/23aug00.
26. Leskey TC, Short BD, Butler BR, Wright SE (2012) Impact of the invasive brown marmorated stink bug, *Halyomorpha halys* (Stål), in mid-Atlantic tree fruit orchards in the United States: Case studies of commercial management. Psyche 2012: 1–14.
27. Leskey TC, Hogmire HW (2005) Monitoring stink bugs (Hemiptera: Pentatomidae) in mid-Atlantic apple and peach orchards. J Econ Entom 98: 143–153.
28. Levin SA (1992) The problem of pattern and scale in ecology. Ecology 73: 1943–1966.

29. Lunn DJ, Best N, Whittaker JC (2008) Generic reversible jump MCMC using graphical models. Statistics and Computing. DOI: 10: 1007/S11222–008–9100–0.

30. Lunn DJ, Thomas A, Best N, Spiegelhalter D (2000) WinBugs – A Bayesian modeling framework: concepts, structure, and extensibility. Stat Comput 10: 325–337.

31. Lunn DJ, Whittaker JC, Best N (2006) A Bayesian toolkit for genetic association studies. Genetic Epidemiology 30: 213–247.

32. McPherson JE, McPherson JE (2000) Stink bugs of economic importance in America North of Mexico. Boca Raton: McPherson CRC Press. 256 p.

33. New Jersey Division of Fish and Wildlife (2012) New Jersey Landscape Project, Version 3.1. New Jersey Department of Environmental Protection, Division of Fish and Wildlife, Endangered and Nongame Species Program. 33 p.

34. New Jersey Department of Environmental Protection (2010) Geographic Information Systems, Land use/Land cover. Available: http://www.state.nj.us/dep/gis/lulc07shp.html. Accessed 15 October 2012.

35. Nielsen AL, Hamilton GC (2009a) Life history of the invasive species *Halyomorpha halys* (Hemiptera: Pentatomidae) in northeastern United States. Ann Entom Soc Am 102: 608–616.

36. Nielsen AL, Hamilton GC (2009b) Seasonal occurrence and impact of *Halyomorpha halys* (Hemiptera: Pentatomidae) in tree fruit. J Econ Entomol 102: 113–1140.

37. Nielsen AL, Hamilton GC, Shearer PW (2011) Seasonal phenology and monitoring of the non-native *Halyomorpha halys* (Hemiptera: Pentatomidae) in soybean. Environ Entomol 40: 231–238.

38. Nielsen AL, Holmstrom K, Hamilton GC, Cambridge J, Ingerson-Mahar J (2013) Use of black light traps to monitor the abundance, spread, and flight behavior of *Halyomorpha halys* (Hemiptera: Pentatomidae). J Econ Entomol 106: 1495–1502.

39. Petit NJ, Hoddle MS, Grandgirard J, Roderick GK, Davies N (2008) Invasion dynamics of the glassy-winged sharpshooter *Homalodisca vitripennis* (Germar) (Hemiptera: Cicadellidae) in French Polynesia. Biol Invasions 10: 955–967.

40. Reay-Jones FPF (2010) Spatial and temporal patterns of stink bugs (Hemiptera: Pentatomidae) in wheat. Environ Entomol 39: 944–955.

41. Shepherd RF, Bennett DD, Dale JW, Tunnock S, Dolph RE, et al. (1988) Evidence of synchronized cycles in outbreak patterns of Douglas-fir tussock moth, *Orygia pseudotsugata* (McDunnough) (Lepidoptera: Lymantriidae). Mem Ent Soc Can 106: 107–121.

42. Silverman BW (1986) Density estimation for statistics and data analysis. London: Chapman and Hall. 176 p.

43. Spieglehalter DJ, Thomas A, Best N, Lunn D (2003) WinBugs 1.4 Manual. Available: http://www.mrc-bsu.cam.ac.uk/bugs/winbugs/contents.shtml. Accessed 16 October 2012.

44. Stephens PA, Sutherland WJ, Freckleton RP (1999) What is the Allee effect? Oikos 87: 185–190.

45. Taylor MCZ, Hastings A (2005) Allee effects in biological invasions. Ecol Lett 8: 895–908.

46. Tillman PG (2011) Influence of corn on stink bugs (Heteroptera: Pentatomidae) in subsequent crops. Environ Entomol 40: 1159–1176.

47. Tillman PG, Northfield TD, Mizell RF, Riddle TC (2009) Spatiotemporal patterns and dispersal of stink bugs (Heteroptera: Pentatomidae) in peanut-cotton farmscapes. Environ Entomol 38: 1038–1052.

48. Toyama M, Ihara F, Yaginuma K (2011) Photo-response of the brown marmorated stink bug, *Halyomorpha halys* (Stål) (Heteroptera: Pentatomidae), and its role in the hiding behavior. Appl Entomol and Zool 46: 37–40.

49. USDA (2007) Census of Agriculture – County Data. Available: http://www.agcensus.usda.gov/Publications/2007/Full_Report/Volume_1,_Chapter_2_County_Level/New_Jersey/st34_2_001_001.pdf. Accessed 10 May 2013.

50. Van Driesche RG, Bellows Jr TS (1996) Biological Control. New York: Chapman and Hall. 560 p.

51. Xu J, Fonseca DM, Hamilton GC, Hoelmer KA, Nielsen AL (2014) Tracing the origin of US brown marmorated sting bugs, *Halyomorpha halys*. Biol Invasions 16(1): 153–166 DOI 10.1007/s10530–013–0510–3.

52. Zhang C, Daluan L, Haifeng S, Guoliang X (1993) A study of the biological characteristics of *Halyomorpha picus* and *Erthisna fullo*. Forest Research 6: 271–275.

53. Zhu G, Bu W, Gao Y, Liu G (2012) Potential geographic distribution of brown marmorated stink bug invasion (*Halyomorpha halys*). PLos One 7: 1–10.

16

Testing the Enemies Hypothesis in Peach Orchards in Two Different Geographic Areas in Eastern China: The Role of Ground Cover Vegetation

Nian-Feng Wan, Xiang-Yun Ji, Jie-Xian Jiang*

Eco-environment Protection Research Institute, Shanghai Academy of Agricultural Sciences, Shanghai Key Laboratory of Protected Horticultural Technology, Shanghai, China

Abstract

Many studies have supported the enemies hypothesis, which suggests that natural enemies are more efficient at controlling arthropod pests in polyculture than in monoculture agro-ecosystems. However, we do not yet have evidence as to whether this hypothesis holds true in peach orchards over several geographic locations. In the two different geographic areas in eastern China (Xinchang a town in the Shanghai municipality, and Hudai, a town in Jiangsu Province) during a continuous three-year (2010–2012) investigation, we sampled arthropod pests and predators in *Trifolium repens* L. and in tree canopies of peach orchards with and without the ground cover plant *T. repens*. No significant differences were found in the abundances of the main groups of arthropod pests and predators in *T. repens* between Hudai and Xinchang. The abundance, richness, Simpson's index, Shannon-Wiener index, and Pielou evenness index of canopy predators in ground cover areas increased by 85.5, 27.5, 3.5, 16.7, and 7.9% in Xinchang, and by 87.0, 27.6, 3.5, 17.0 and 8.0% in Hudai compared to those in the controls, respectively. The average abundance of Lepidoptera, Coleoptera, Homoptera, true bugs and Acarina canopy pests in ground cover areas decreased by 9.2, 10.2, 17.2, 19.5 and 14.1% in Xinchang, and decreased by 9.5, 8.2, 16.8, 20.1 and 16.6% in Hudai compared to that in control areas, respectively. Our study also found a higher density of arthropod species resources in *T. repens*, as some omnivorous pests and predators residing in *T. repens* could move between the ground cover and the orchard canopy. In conclusion, ground cover in peach orchards supported the enemies hypothesis, as indicated by the fact that ground cover *T. repens* promoted the abundance and diversity of predators and reduced the number of arthropod pests in tree canopies in both geographical areas.

Editor: Youjun Zhang, Institute of Vegetables and Flowers, Chinese Academy of Agricultural Science, China

Funding: This study was funded by the grants from Shanghai Agriculture Committee for young-talent growth (No. 2014.1-22), and from Shanghai Municipal Science and Technology Commission (12ZR1449100, 08dz1900401, 123919N0400). The funders had no role in study design, data collection and analysis, decision to publish, or preparation of the manuscript.

Competing Interests: The authors have declared that no competing interests exist.

* Email: jiangjiexian@163.com

Introduction

The enemies hypothesis argues that natural enemies are more effective at controlling arthropod pests in diverse ecosystems than in simple ones [1–3]. This hypothesis has received support from studies of diversified agro-ecosystems [4–6], but evidence from diversified orchard ecosystems is scarce, even though there have been reports on pest management in orchards [7–8].

Ground cover vegetation, as an important diversified plantation model [9–11], has played a key role in enhancing crop yield, productivity, and stability in ecosystems [12–14]. Although the use of ground cover vegetation in peach orchards for pest management has been investigated [15–16], few such studies have evaluated the enemies hypothesis.

China has the largest peach plantings and the largest production of peaches in the world. Shanghai municipality and Jiangsu province in eastern China are the two most important peach-growing regions: both Shanghai, with an annual peach cultivation area of 1.5×10^4 ha and peach yield of 1.8×10^8 kg, and Jiangsu, with an annual peach cultivation area of 3.2×10^4 ha and a peach

yield of 3.5×10^8 kg, have acquired national recognition as peach growing areas.

The use of ground cover vegetation in peach orchards began in China in the 1990s and was gradually applied nationwide. Recently, the use of a clover ground cover (*Trifolium repens* L.) was selected as a useful approach to enhance soil fertility in orchards [17–18], and the use of this clover has been widely applied in eastern China. However, whether its use also enhances natural enemies in pest control (the enemies hypothesis) has not been tested.

In our former study in Shanghai municipality and Jiangsu province in eastern China, we observed that the arthropod pests in *T. repens* mainly included Homoptera (in this paper Homoptera is a suborder of the Hemiptera insect, including Psyllidae, whitefly, cicadas, aphids, scale insects, i.e.), Lepidoptera, true bugs, Acarina and Coleoptera, while the predators in *T. repens* mainly included Araneida, Coleoptera, Neuroptera, Diptera and Hemiptera. In additional, we found that these species provided an excellent pool for canopy arthropods in peach orchards [19]. The outbreaks of canopy arthropod pests, mainly including Homoptera, true bugs, Coleoptera, Lepidoptera and Acarina pests, were severely

Table 1. Variance analysis of different sites and years against the abundance of the main groups of arthropod pests in *Trifolium repens* L.

Treatment	Lepidoptera		Homoptera		True bug		Acarina		Coleoptera	
	F	Sig.	F	Sig.	F	Sig.	F	Sig.	F	Sig.
Site	0.764	0.384	1.983	0.162	0.853	0.358	0.510	0.476	0972	0326
Year	0.426	0.654	1.103	0.336	1.483	0.231	1.302	0.276	0522	0595
Site × year	0.037	0.964	0.295	0.745	0.372	0.690	0.091	0.913	2.029	0136

Notes: here Homoptera is a suborder of the Hemiptera insect, including Psyllidae, whitefly, cicadas, aphids, scale insects, i.e.; in *Trifolium repens* L. Lepidoptera pests mainly included Noctuidae, Pyralidae, Psychidae, Pieridae, Papilionidae and Lycaenidae; Homoptera pests mainly included Aphididae, Aleyrodidae and Cicadellidae; true bugs mainly included Miridae, Coreidae, Pentatomidae and Tingidae pests; Coleoptera pests mainly included Melolonthidae, Rutelidae, Cetoniidae, Chrysomelidae, Curculionidae, Meloidae and Elateridae; Acarina mainly included Tetranychidae. The same below.

Table 2. Variance analysis of different sites and years against the abundance of the main groups of arthropod predators in *T. repens*.

Treatment	Araneida		Coleoptera		Neuroptera		Diptera		Hemiptera	
	F	Sig.	F	Sig.	F	Sig.	F	Sig.	F	Sig.
Site	1.482	0.226	1.604	0.208	0.242	0.624	3.676	0.058	2.036	0.156
Year	0.291	0.748	1.241	0.293	0.007	0.994	1.747	0.179	2.205	0.115
Site × year	0.053	0.948	0.028	0.972	0.868	0.423	0.883	0.416	1.295	0.278

Notes: in *T. repens* Araneida predators mainly included Thomisidae, Erigonidae, Salticidae, Araneidae, Lycosidae, Thomisidae, Tetragnathidae, Dictynidae, Theridiidae and Oxyopidae; Coleoptera predators mainly included Coccinellidae and Carabidae; Neuroptera predators mainly included Chrysopidae; Diptera predators mainly included Syrphidae and Tachinidae; Hemiptera predators mainly included Anthocoridae. The same below.

Table 3. Comparison of the abundance of the main groups of arthropod pests in *T. repens*.

Site	Lepidoptera	Homoptera	True bug	Acarina	Coleoptera
Xinchang, Shanghai	14.93±1.51a	14.73±0.88a	6.74±0.36a	23.74±2.76a	3.58±0.09a
Hudai, Jiangsu	16.95±1.71a	16.44±0.83a	7.23±0.39a	26.71±3.08a	3.46±0.08a

Notes: here Homoptera is a suborder of the Hemiptera insect, including Psyllidae, whitefly, cicadas, aphids, scale insects, i.e.; the same letters in the same column indicate that the means are not significantly different at $P<0.05$ (independent samples t-test) between Xinchang and Hudai during the three years (57 sample dates were considered as 57 replicates). Means of five groups of arthropod pests were calculated from a sample size of one square meter of *T. repens*.

suppressed by Araneida, Coleoptera, Neuroptera, Diptera and Hemiptera predators in canopy peach trees [19–20]. However, whether the abundances of the arthropod pests and predators in *T. repens* and in the peach canopy differing between the two different geographic locations in eastern China have not been examined, and whether *T. repens* ground cover enhances the level of control exerted by predators on the abundance of key pests in the orchard canopies has not been tested.

This study, therefore, sought to test the enemies hypothesis in peach orchards in the two different sites in eastern China. In particular, we sought to determine: (1) what effect the pool of predators and pests in *T. repens* played, and whether the abundance of the main groups of arthropod pests and predators in *T. repens* were different between the two geographical areas studied; (2) whether the abundance and diversity of predators in canopy orchards were enhanced by the use of clover ground cover and whether this difference varied between the two locations; and (3) whether the abundance of the main groups of arthropod pests in canopy orchards was inhibited by manipulation of *T. repens* ground cover and whether this differed between the two locations. If this hypothesis proved true, the use of ground cover to favor indigenous predators would be a more sustainable approach to controlling arthropod pests than the input of insecticides or exotic control agents in eastern China, or indeed in other agricultural areas.

Methods

Ethical statement

The Guoyuan village of Xinchang town in Shanghai municipality and Longyan village of Hudai town in Jiangsu province issued permission for the experiments in each location. No vertebrates were studied, as our study involved only arthropods (insects, mites and spiders).

Study sites

This experiment was conducted in two locations: one in Xinchang town, Pudong district, Shanghai municipality of China (31.03°N, 121.41°E, elevation 4.3 m), and the other in Hudai town, Wuxi district, Jiangsu Province of China (31.34°N,

121.18°E, elevation 3.5 m). The two sites were both located in the Yangtze River delta alluvial plain, the Hudai site near Taihu Lake (the second largest freshwater lake in China), and the Xinchang site near the Yangtze River, 20 kilometers away from the first site. The sites share a similar climate, both belonging to the prevailing zone of the East Asian monsoon in the southern rim of North Asian tropics. Peach varieties were "Hujing" honey peach in Hudai and "Xinfeng" honey peach in Xinchang, both mid-season maturing varieties with 8–10 year-old trees that were 2–2.5 m high, and arranged in a 4×4 m grid.

Treatment and management

A randomized block design was used in this experiment, and each treatment and the control were replicated three times in both Xinchang and Hudai. Thirty peach trees were sampled for each replicate to calculate the abundance and diversity of predators and the abundances of pest arthropods. Each plot was 100 m wide and 52 m long (5200 m²), and separated from adjacent plots by a 100 m long isolation belts. Treatment areas were seeded with a ground cover of the biennial clover *T. repens*, which was mowed twice a year to a height of about 10 cm. Control areas in the orchards were bare ground without weeds (which were pulled by hand) under the trees. Farm management and operation methods, including pest management, were identical between treatment and control plots in each orchard. Pest management in peach orchards mainly relies on physical and manual means, such as trimming of diseased and pest-infested peach branches and shoots in winter, fruit bagging in June, trapping insect pests in pheromone traps, and spraying with 4–5 Baume lime sulfur during the dormant period.

Sampling methods

Sampling methods in peach orchards to measure densities of the predatory and pest arthropods in the plots followed those of Bi et al. [21]. Within each replicate plot (treatment and control areas), thirty adjacent peach trees with checkerboard type distribution and similar to each other in height and vigor, were selected as permanent sampling points to monitor the abundance of predators and pests. At each sampling date, each tree was sampled in each of

Table 4. Comparison on the abundance of the main groups of predators in *T. repens*.

Site	Araneida	Coleoptera	Neuroptera	Diptera	Hemiptera
Xinchang, Shanghai	21.77±1.60a	17.65±1.35a	6.81±0.38a	3.22±0.11a	4.22±0.18a
Hudai, Jiangsu	19.18±1.36a	15.33±1.22a	7.08±0.38a	3.51±0.11a	3.90±0.15a

Notes: the same letters in the same column indicate that the means are not significantly different at $P<0.05$ (independent samples t-test) between Xinchang and Hudai during the three years (57 sample dates were considered as 57 replicates). Means of five groups of predators were calculated from a sample size of one square meter of *T. repens*.

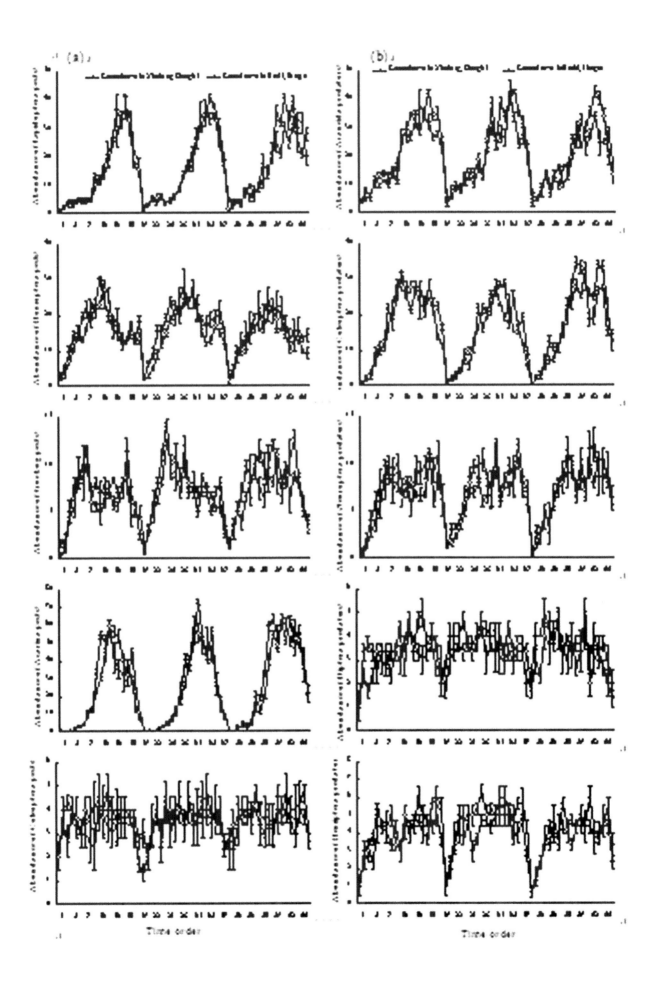

Figure 1. Dynamics of the abundance of the main groups of arthropod pests (a) and arthropod predators (b) in plots with *Trifolium repens* **L. ground cover from a sample size of 1 m² in peach orchards.** Vertical bars denote SE. The numbers on the X-axis indicate the sampling times, i.e., the first 19 times (1 to 19) were conducted from late-March to early-October 2010, the second 19 times (20 to 38) from late-March to early-October 2011, and the third 19 times (39 to 57) from late-March to early-October 2012.

the four cardinal directions (east, south, west and north) at three levels each (upper, middle and lower), such that each tree canopy was split into 12 zones [22]. The branch beating method was used to collect predators and pests from the canopy peach according to Simon et al. [23]. Branches in the tree canopy were struck with a rubber hose over a 25 cm dia collection funnel. Predators and pests falling from the branches were identified and counted immediately, but mobile arthropods on the exterior of the tree were first located and collected before beating branches.

Sampling to measure the densities of the predators and pests in the *T. repens* clover ground cover followed a "Z"-shaped sampling plan, with seven sampling sites (plots) in the area beneath each treatment in each peach orchard. Sample locations were spaced equidistant along the Z transect, at 12 m intervals. Each sampling location was a 1 m² area defined by a wire frame. The arthropods in the *T. repens* ground cover were beaten by hand into a white porcelain tray (0.3 m×0.4 m) and were then immediately classified to species and the numbers of individuals within orders was counted.

Approximately 3–5 minutes were needed to collect the canopy arthropods and count the individuals of each species at each canopy level, choosing one representative twig (20–30 cm long) from each zone in each tree. Each twig from the base to the tip was intensively examined to count all the arthropods present. In addition, 3–5 minutes were taken to collect the arthropods in the *T. repens* ground cover at each of the seven sample points per treatment replicate. Any unidentified species in the orchards and in the *T. repens* were collected in vials of 80% alcohol, counted and labeled for later identification in the laboratory. Sampling was done ca every 10 days, from late March to early October in 2010, 2011, and 2012 (with some delays because of rain or other contingencies), for a total of nineteen sample dates per year.

Data analysis

In most cases, we classified the arthropod pests collected in the *T. repens* ground cover into five main groups (Homoptera mainly including Aphididae, Aleyrodidae and Cicadellidae pests, true bugs mainly including Miridae, Coreidae, Pentatomidae and Tingidae pests, Lepidoptera mainly including Noctuidae, Pyralidae, Psychidae, Pieridae, Papilionidae and Lycaenidae pests, Coleoptera mainly including Melolonthidae, Rutelidae, Cetoniidae, Chrysomelidae, Curculionidae, Meloidae and Elateridae pests, and Acarina mainly including Tetranychidae pests) and similarly for the predators from the *T. repens* ground cover as Araneida, Coleoptera, Neuroptera, Diptera, or Hemiptera. The species composition of arthropod pests and predators collected from the *T. repens* ground cover can be found in Table S1.

Data from each canopy sample in each plot were collated and entered into a relational database with basic information on the predators (biological characteristics, feeding guild, development mode, host range, and taxonomic status) and on the samples (date, location in peach orchard, direction in tree canopy, and sampling area). Results from the thirty sampled peach trees for each replicate in each of treatment and control area over three years (2010–2012) in Xinchang and Hudai were used to calculate the diversity indices of predators in the canopy of peach orchards with or without the ground cover plant *T. repens*. Abundance, species richness, Simpson's index, Shannon-Wiener index, and the Pielou

evenness index were used to measure the diversity of predators. We also classified the arthropod pests collected in tree canopies into five main groups as Lepidoptera, mainly including Ortricidae, Lyonetiidae, Pyralididae, Sphingidae, Carposinidae, Noctuidae, Cossidae, Limacodidae, Psychidae, Lymantridae, Saturniidae, Pieridae and Carposinidae: Coleoptera, mainly including Curculionidae, Cetoniidae, Rutelidae, Melolonthidae and Cerambycidae: Homoptera mainly including Aphididae, Cicadellidae, Cicadidae, Fulgoridae, Coccidae, Margarodidae, Diaspididae and Aleyrodidae: true bugs, mainly including Pentatomidae, Miridae, Tingidae and Coreidae and Acarina, mainly including Tetranychidae. The species composition of arthropod pests and predators in tree canopies can be found in Table S2.

Statistical analyses were performed using the Statistical Analysis Systems software (SigmaStat Statistical Software, SPSS Science, Chicago, IL, USA). Normal distribution and homocedasticity of all data were checked by the Kolmogorow-Smirnov test and Levene test, respectively. Three-factor analysis of variance with General Linear Model was adopted to analyze the diversity indices of canopy predators and the number of the main groups of canopy arthropod pests, so as to compare the interactive effects of different sites, years, and orchard types on the diversity of canopy predators and the abundance of the main arthropod pests. Two-factor (site and year) analysis of variance with General Linear Model was used to analyze the number of the main groups of arthropod pests and predators in *T. repens* ground cover. The sites were Xinchang and Hudai; the years were 2010, 2011, and 2012; the orchard types were ground cover or bare ground, and the interactive effects involved were site × year, site × orchard type, year × orchard type, and site × year × orchard type. If the values of the diversity indices of canopy predators and the abundance of canopy pests were not significantly affected by the years with the above three-factor analysis of variance, we took the three years of data as a whole to compare the differences among means related to treatment and control in Xinchang and in Hudai, respectively, with Tukey's Honestly Significant Difference (HSD) test at the 0.05 level. If the values of the abundance of the main groups of arthropod pests and predators in *T. repens* ground cover plots were not significantly affected by the years with the above two-factor analysis of variance, we again took the three years of data as a whole to compare the differences among means between Xinchang and Hudai, respectively, with independent samples t-test at the 0.05 level.

Results

Effect of ground cover on the abundance of the main groups of arthropod pests and predators in the *Trifolium repens* clover ground cover

The factors of site and year had no significant effects on the abundance of arthropod pests of Lepidoptera, Homoptera, true bugs, Acarina or Coleoptera, and the same effects were indicated by the interactions of the two factors (Table 1). Neither the single factors (site and year) nor the interaction of two factors (site × year) significantly affected the abundance of predators of Araneida, Coleoptera, Neuroptera, Diptera or true bugs (Table 2).

The abundances of Lepidoptera, Homoptera, true bugs, and Acarina arthropod pests in plots with *T. repens* ground cover in

Table 5. Variance analysis of different sites, years and orchard types against the diversity indices of canopy predators in peach orchards.

Treatment	Abundance		Richness		Simpson's index		Shannon-Wiener index		Pielou evenness index	
	F	Sig.	F	Sig.	F	Sig.	F	Sig.	F	Sig.
Site	0.145	0.704	0.145	0.703	0.060	0.807	0.250	0.618	0.012	0.912
Year	0.132	0.876	0.707	0.494	2.854	0.060	2.450	0.089	2.757	0.066
Orchard type	92.282	<0.001	435.572	<0.001	728.062	<0.001	903.413	<0.001	2.052×10^3	<0.001
Site × year	0.103	0.902	0.063	0.939	0.285	0.752	0.038	0.963	1.211	0.300
Site × orchard type	0.003	0.958	0.034	0.853	0.048	0.828	0.025	0.875	0.037	0.848
Year × orchard type	0.583	0.559	0.114	0.892	1.221	0.297	0.165	0.848	0.683	0.506
Site × year × orchard type	0.058	0.943	0.023	0.977	0.250	0.779	0.181	0.834	0.869	0.421

Xinchang were all lower, while that of Coleoptera was slightly higher, than that in Hudai, but these differences were not significant (t-test: Homoptera, $t=1.416$, df$=112$, $P=0.603$; Lepidoptera, $t=0.886$, df$=112$, $P=0.377$; true bugs, $t=0.925$, df$=112$, $P=0.986$; Acarina, $t=0.718$, df$=112$, $P=0.157$; Coleoptera, $t=0.981$, df$=112$, $P=0.759$) (Table 3).

While the numbers of Araneida, Coleopteran and Hemipteran predators were all higher, and the numbers of Neuropteran and Dipteran predators in plots with *T. repens* ground cover were both lower in Xinchang than in Hudai, these differences were not significant (t-test: Araneida, $t=1.236$, df$=112$, $P=0.072$; Coleoptera, $t=1.275$, df$=112$, $P=0.319$; Neuroptera, df$=112$, $t=0.497$, $P=0.918$; Diptera, $t=1.907$, df$=112$, $P=0.616$; Hemiptera $t=1.408$, df$=112$, $P=0.160$) (Table 4).

The dynamics of the abundance of the main groups of arthropod pests and predators in plots with *T. repens* ground cover were similar for Xinchang and Hudai. In both sites, the abundances of pest Homoptera, Lepidoptera, true bugs, Acarina, and Coleoptera, as well as that of the predacious Araneida, Coleoptera, Neuroptera, Diptera, and Hemiptera increased during spring but stabilized thereafter (Figure 1).

Effect of ground cover on the diversity of canopy predators

The single factors of site and year had no significant effect on the diversity indices of canopy predator communities, while the factor of orchard type did have a significant effect. The interactive effects of two factors (site × year, site × orchard type, and year × orchard type) and of all three factors (site × year × orchard type) on the diversity indices were also not significant, respectively (Table 5).

The abundance, richness, Simpson's index, Shannon-Wiener index, and Pielou evenness index of predator communities in ground cover areas were all significantly greater than those in the control areas in both Xinchang and Hudai (abundance: $F_{3, 224}=31.694$, $P<0.001$; richness: $F_{3, 224}=149.374$, $P<0.001$; Simpson's index: $F_{3, 224}=241.408$, $P<0.001$; Shannon-Wiener index: $F_{3, 224}=304.400$, $P<0.001$; Evenness: $F_{3, 224}=674.945$, $P<0.001$) (Table 6).

The five diversity indices of canopy predator communities were consistently higher in ground cover areas than control areas. Irrespective of ground cover, all diversity indices increased during spring but stabilized thereafter in both sites (Figure 2).

Effect of ground cover on the abundance of the main groups of canopy arthropod pests

Ground covers of *T. repens* in peach orchards had significant effects on the numbers of pest Coleoptera, Homoptera, and true bugs, but no significant effects on the numbers of pest Lepidoptera or Acarina. The single factor of site had a significant effect on the number of Coleoptera pests, but no significant effect on the number of the other four groups of pests, while the single factor of year had no significant effect on the numbers of pests in any of the five groups. Similarly, none of the interactive factors (site × year, site × orchard type, year × orchard type, and site × year × orchard type) had any significant effects on the numbers of pests in any of the five groups (Table 7).

In neither Xinchang nor Hudai did ground cover have any significant effect on the number of pests in any of the groups except Coleoptera (one-way ANOVA: Lepidoptera, $F_{3, 224}=0.496$, $P=0.685$; Coleoptera, $F_{3, 224}=23.533$, $P<0.001$; Homoptera, $F_{3, 224}=2.220$, $P=0.087$; true bugs, $F_{3, 224}=2.542$, $P=0.057$; Acarina, $F_{3, 224}=0.963$, $P=0.411$) (Table 8). However,

Table 6. Comparison of diversity indices of canopy predators in peach orchards with and without ground cover by *T. repens* (means ± SE).

Orchard type (site)	Abundance	Richness	Simpson's index	Shannon-Wiener index	Pielou evenness index
Ground cover (Xinchang, Shanghai)	315.0±16.8a	26.0±0.2a	0.9645±0.0003a	4.6716±0.0094a	0.9932±0.0005a
Bare ground (Xinchang, Shanghai)	169.8±10.2b	20.4±0.3b	0.9319±0.0016b	4.0025±0.0245b	0.9201±0.0021b
Ground cover (Hudai, Jiangsu)	308.4±19.6a	25.9±0.2a	0.9650±0.0004a	4.6640±0.0117a	0.9933±0.0005a
Bare ground (Hudai, Jiangsu)	164.9±10.3b	20.3±0.4b	0.9320±0.0018b	3.9878±0.0340b	0.9196±0.0024b

Notes: Different letters in the same column indicate that the means are significantly different at $P<0.05$ (*HSD* test) within groups of ground cover (treatment) and bare ground (control) in Xinchang and Hudai during the three years (57 sample dates were considered as 57 replicates). Means of five diversity indices were calculated from a total of 30 trees per treatment or control.

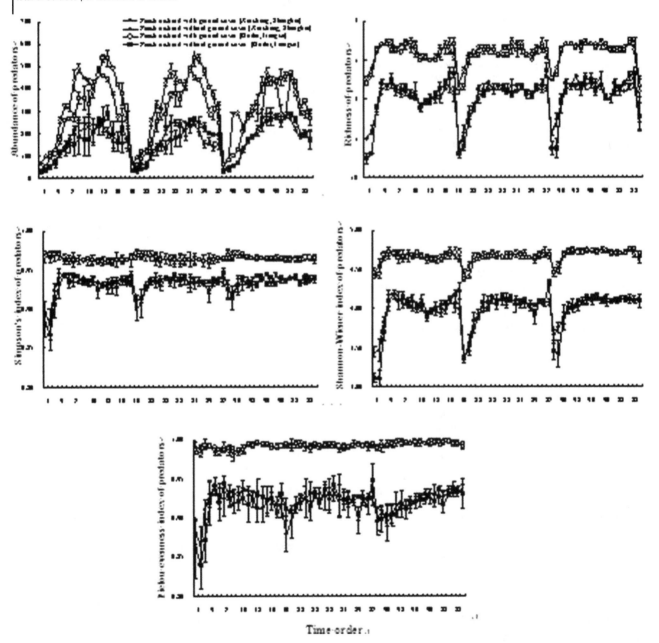

Figure 2. Dynamics of the diversity indices of predatory canopy arthropod in peach orchards with and without ground cover of *T. repens*. Vertical bars denote SE. The numbers on the *X*-axis indicate the sampling times, i.e., the first 19 times (1 to 19) were conducted from late-March to early-October 2010, the second 19 times (20 to 38) from late-March to early-October 2011, and the third 19 times (39 to 57) from late-March to early-October 2012.

Table 7. Variance analysis of different sites, years and orchard types against the abundance of the main groups of canopy arthropod pests in peach orchards.

Treatment	Lepidoptera		Coleoptera		Homoptera		True bug		Acarina	
	F	Sig.	F	Sig.	F	Sig.	F	Sig.	F	Sig.
Site	0.233	0.630	12.955	<0.001	0.350	0.555	0.158	0.691	0.288	0.592
Year	0.311	0.733	0.228	0.797	0.015	0.986	0.419	0.659	0.349	0.706
Orchard type	1.218	0.271	55.040	<0.001	6.090	0.014	7.227	0.008	2.477	0.117
Site × year	0.678	0.509	0.636	0.530	0.323	0.725	0.011	0.989	0.043	0.958
Site × orchard type	<0.001	0.998	1.036	0.310	0.007	0.932	0.007	0.935	0.035	0.851
Year × orchard type	0.040	0.961	0.512	0.600	0.028	0.972	0.136	0.873	0.178	0.837
Site × year × orchard type	0.084	0.919	0.133	0.875	0.049	0.952	0.004	0.996	0.031	0.969

Notes: here Homoptera is a suborder of the Hemiptera insect, including Psyllidae, whitefly, cicadas, aphids, scale insects, i.e.; in tree canopies Lepidoptera pests mainly included Ortricidae, Lyonetiidae, Pyralididae, Sphingidae, Carposinidae, Noctuidae, Cossidae, Limacodidae, Psychidae, Lymantridae, Saturniidae, Pieridae and Carposinidae, Coleoptera pests mainly included Curculionidae, Cetoniidae, Rutelidae, Melolonthidae and Cerambycidae, Homoptera pests mainly included Aphididae, Cicadellidae, Cicadidae, Fulgoridae, Coccidae, Margarodidae, Diaspididae and Aleyrodidae, True bug pests mainly included Pentatomidae, Miridae, Tingidae and Coreidae, and Acarina mainly included Tetranychidae.

compared to those in control areas, the numbers of pest Lepidoptera, Coleoptera, Homoptera, true bugs, and Acarina in ground cover areas decreased by 9.2, 10.2, 17.2, 19.5 and 14.1%, respectively, in Xinchang and by 9.5, 8.2, 16.8, 20.1, and 16.6%, respectively, in Hudai.

Meanwhile, there were no significant differences in the number of Lepidoptera, Coleoptera, Homoptera, true bug, or Acarina pests between Xinchang and Hudai in ground cover areas (HSD: Xinchang vs. Hudai: Lepidoptera, $P = 0.986$; Coleoptera, $P = 0.255$; Homoptera, $P = 0.984$; true bug, $P = 0.996$; Acarina, $P = 0.994$). However, in control areas, while the difference in the number of Lepidoptera, Homoptera, true bugs, or Acarina pests between Xinchang and Hudai was not significant, the difference in the number of Coleoptera was significant (HSD: Xinchang vs. Hudai: Lepidoptera, $P = 0.986$; Coleoptera, $P = 0.006$; Homoptera, $P = 0.962$; true bug, $P = 0.986$; Acarina, $P = 0.954$) (Table 8).

The abundances of pests of all five groups of arthropods were consistently lower in ground cover than control areas. Irrespective of ground cover, the abundances of Coleoptera and true bugs increased during spring but stabilized thereafter, while the abundance of Acarina pests was relatively higher in the 7th–9th month with higher temperature, the abundance of Homoptera pests was highest in June, and the abundance of Lepidoptera pests was highest in August at both sites (Figure 3).

Discussion

Root [24] proposed the enemies hypothesis, which states that predators are more abundant and effective in diverse systems than in simple ones. Andow [4] maintained that diversified vegetational ecosystems would support the enemies hypothesis, as both the richness and abundance of predators would increase, as was found in ecosystems with ground covers [13,15,25]. On the other hand, some experiments have indicated that predators are not significantly influenced by cropping pattern [26–27], and some results do not support the enemies hypothesis [28–30].

Our results indicated that when the ground beneath peach orchards was covered with *T. repens*, the richness and abundance of canopy predators increased by 27.5 and 85.5% in Xinchang and by 27.6 and 87.0% in Hudai, respectively. Similar studies found that the richness and abundance of predators increased by ground cover in pecan orchards [31], apple orchards [32], and lemon trees [33], as well as by grass cover (*Elymus trachycaulus*) in maize fields [34].

Ground cover vegetation can create a suitable ecological structure within the agricultural landscape to provide food resources for predators to feed and reproduce [35–36]. Our study found the average abundance of the main predators in *T. repens* ground cover in Xinchang and in Hudai to be 53.7 and 49.0

Table 8. Comparison on the abundance of the main groups of canopy arthropod pests in peach orchards with ground cover and without ground cover.

Orchard type (site)	Lepidoptera	Coleoptera	Homoptera	True bug	Acarina
Ground cover (Xinchang, Shanghai)	161.6±15.1a	41.27±0.50c	115.9±8.6a	86.3±6.8a	110.1±11.9a
Bare ground (Xinchang, Shanghai)	177.9±16.2a	45.98±0.61a	140.0±11.0a	107.2±8.1a	128.1±13.3a
Ground cover(Hudai, Jiangsu)	154.5±12.8a	39.84±0.54c	111.1±8.3a	88.8±7.4a	114.7±11.8a
Bare ground (Hudai, Jiangsu)	170.7±13.7a	43.40±0.55b	133.6±8.9a	111.1±9.0a	137.5±13.9a

Notes: here Homoptera is a suborder of the Hemiptera insect, including Psyllidae, whitefly, cicadas, aphids, scale insects, i.e.; different letters in the same column indicate that the means are significantly different at $P<0.05$ (HSD test) within groups of ground cover (treatment) and bare ground (control) in Xinchang and Hudai during the three years (57 sample dates were considered as 57 replicates). Means of the five orders of arthropod pests were calculated from a total 30 trees per treatment or control.

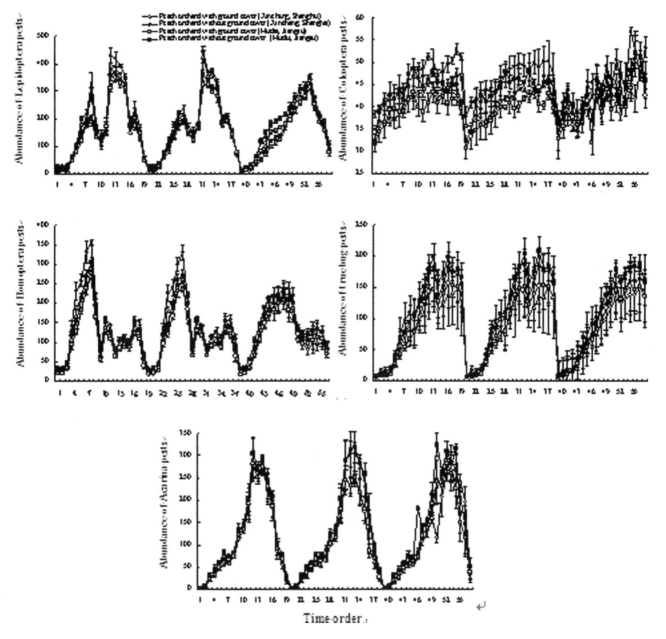

Figure 3. Dynamics of the five-group arthropod pests in peach orchards with and without ground cover of *T. repens*. Vertical bars denote SE. The numbers on the *X*-axis indicate the sampling times, i.e., the first 19 times (1 to 19) were conducted from late-March to early-October 2010, the second 19 times (20 to 38) from late-March to early-October 2011, and the third 19 times (39 to 57) from late-March to early-October 2012.

individuals per m², respectively, which provided a strong pool of predators potentially able to forage in tree canopies in peach orchards. The principal value of ground cover in orchards might be the movement of predators from the orchard floor to the trees, thus augmenting the richness and abundance of predators in trees. We observed, for instance, that predators such as spiders, ladybirds, lacewings, carabid beetles, and hoverflies could readily disperse from *T. repens* ground cover to the peach tree canopy, which might explain the higher diversity of predators in the orchard canopy.

In addition, ground cover vegetation could also offer refuge for predators avoiding adverse influences from alien disturbance to ecosystems [37]. Similar research found that uncut strips of lucerne (*Medicago sativa*) provided refuge to a range of coccinellid

and hemipteran predators [38]. In practice, we observed that predators always actively transferred into the *T. repens* ground cover to avoid man-made disturbance when pathogen-infested peach branches or shoots were trimmed, and that adverse meteorological conditions (such as an abrupt drops of temperatures or severe rain) likewise caused predators to take shelter in the ground cover stratum.

Some omnivorous arthropod pests resident in *T. repens* ground cover, such as harmful mites, noctuids, and whitefly *Bemisia tabaci* (Gennadius) (the average abundance of the three harmful groups was 23.7, 5.1 and 5.8 per m² in Xinchang, and 26.7, 5.2 and 4.9 per m² in Hudai, respectively), could transfer from *T. repens* to peach trees for feeding. Similarly, Gruys [39] reports that tarnished plant bug could disperse from cover crops to damage field, orchard, and row crops, and Tedders et al. [40] notes that *Tetranychus urticae* Koch moved from clover into pecan trees.

T. repens ground cover enhanced the ecological functions of predators to control pests in peach orchards [41]. In peach orchards, hoverflies mainly prey on aphids, scale insects and leafhoppers, while spiders, ladybirds and lacewings mainly prey on aphids, leafhoppers and bugs. Our study indicated that ground cover promoted the biological control service function as the abundance of canopy spiders, ladybirds, lacewings and hoverflies increased by 38.1%, 81.5%, 93.0% and 80.2% in Xinchang and by 43.8%, 91.2%, 126.7% and 117.1% in Hudai, respectively.

Among the main groups of arthropod pests, the aphids which make peach leaves curl and the oriental fruit moth (*Grapholitha molesta* (Busck)) which makes peach shoots wither are the most serious insect pests in eastern China [41]. Compared to bare ground plots, the abundances of aphids and *G. molesta* decreased respectively by more than 30% both in Shanghai and Jiangsu. Meanwhile, the rate of pests bored into the peach sarcocarp (a typical indicator of fruit quality) was decreased by more than 8% and the peach yield was increased by more than 5% in both sites.

While geographical distribution might have a certain influence on species diversity and abundance [42], our study found that the richness of arthropod pests and predators in *T. repens* in the two sites was very similar, with 44 species of arthropod pests belonging to 26 families (8 orders) and 27 species of predators belonging to 16 families (5 orders) in Xinchang, and 45 species of arthropod pests belonging to 27 families (8 orders) and 29 species of predators belonging to 16 families (5 orders) in Hudai. The abundance of the main groups of arthropod pests and predators in *T. repens* was likewise not significantly affected by the two geographical

locations. A more comprehensive, long-term and large-scale geographical distribution investigation of the potential ground cover vegetation in peach orchards is the logical next area of study.

Acknowledgments

We would like to thank Prof. Roy Van Driesche at the University of Massachusetts for improving the English of this manuscript. This study was funded by the grants from Shanghai Agriculture Committee for young-talent growth (No. 2014.1-22), and from Shanghai Municipal Science and Technology Commission (12ZR1449100, 08dz1900401, 123919N0400).

Author Contributions

Conceived and designed the experiments: NFW JXJ. Performed the experiments: NFW XYJ JXJ. Analyzed the data: NFW. Contributed reagents/materials/analysis tools: NFW. Wrote the paper: NFW.

References

1. Castagneyrol B, Jactel H, Vacher C, Brockerhoff EG, Koricheva J (2014) Effects of plant phylogenetic diversity on herbivory depend on herbivore specialization. J Appl Ecol 51: 134–141.

2. Letourneau DK, Armbrecht I, Rivera BS, Lerma JM, Carmona EJ, et al. (2011) Does plant diversity benefit agroecosystems? A synthetic review. Ecol Appl 21: 9–21.

3. Zou Y, Sang W, Bai F, Axmacher JC (2013) Relationships between plant diversity and the abundance and α-diversity of predatory ground beetles (Coleoptera: Carabidae) in a mature Asian temperate forest ecosystem. PLoS ONE 8: e82792.

4. Andow DA (1991) Vegetational diversity and arthropod population response. Annu Rev Entomol 36: 561–568.

5. Giffard B, Corcket E, Barbaro L, Jactel H (2012) Bird predation enhances tree seedling resistance to insect herbivores in contrasting forest habitats. Oecologia 168: 415–424.

6. Riihimäki J, Koricheva J, Vehviläinen H (2005) Testing the enemies hypothesis in forest stands: the important role of tree species composition. Oecologia 142: 90–97.

7. Bone NJ, Thomson LJ, Ridland PM, Cole P, Hoffmann AA (2009) Cover crops in Victorian apple orchards: effects on production, natural enemies and pests across a season. Crop Prot 28: 675–683.

8. Mailloux J, Bellec FL, Kreiter S, Tixier MS, Dubois P (2010) Influence of ground cover management on diversity and density of phytoseiid mites (Acari: Phytoseiidae) in Guadeloupean citrus orchards. Exp Appl Acarol 52: 275–290.

9. Podgaiski LR, Joner F, Lavorel S, Moretti M, Ibanez S, et al. (2013) Spider Trait Assembly Patterns and Resilience under Fire-Induced Vegetation Change in South Brazilian Grasslands. PLoS ONE 8: e60207.

10. Schut AGT, Wardell-Johnson GW, Yates CJ, Keppel G, Baran I, et al. (2014) Rapid characterisation of vegetation structure to predict refugia and climate change impacts across a global biodiversity hotspot. PLoS ONE 9: e82778.

11. Wasser L, Day R, Chasmer L, Taylor A (2013) Influence of vegetation structure on lidar-derived canopy height and fractional cover in forested Riparian buffers during leaf-off and leaf-on conditions. PLoS ONE 8: e54776.

12. Carvalheiro LG, Veldtman R, Shenkute AG, Tesfay GB, Pirk CWW, et al. (2011) Natural and within-farmland biodiversity enhances crop productivity. Ecol Lett 14: 251–259.

13. Costello MJ, Daane KM (1998) Influence of ground cover on spider populations in a table grape vineyard. Ecol Entomol 23: 33–40.

14. Gray DM, Swanson J, Dighton J (2012) The influence of contrasting groundcover vegetation on soil properties in the NJ pine barrens. Appl Soil Ecol 60: 41–48.

15. Dong J, Wu XY, Xu CX, Zhang QW, Jin XH, et al. (2005) Evaluation of the Lucerne cover crop for improving biological control of Lyonetia clerkella (Lepidoptera: Lyonetiidae) by means of augmenting its predators in peach orchards. Great Lakes Entomol 28: 186–199.

16. Penvern S, Bellon S, Fauriel J, Sauphanor B (2010) Peach orchard protection strategies and aphid communities: towards an integrated agroecosystem approach. Crop Prot 29: 1148–1156.

17. Uliarte EM, Schultz HR, Frings C, Pfister M, Parerad CA, et al. (2013) Seasonal dynamics of CO_2 balance and water consumption of C_3 and C_4-type cover crops compared to bare soil in a suitability study for their use in vineyards in Germany and Argentina. Agr Forest Meteorol 181: 1–16.

18. Wilson AR, Nzokou P, Cregg B (2010) Ground covers in fraser fir (*Abies fraseri* [Pursh] Poir.) production systems: effects on soil fertility, tree morphology and foliar nutrient status. Eur J Hortic Sci 75: 269–277.

19. Wan NF, Ji XY, Jiang JX, Dan JG (2011) Effects of ground cover on the niches of main insect pests and their natural enemies in peach orchard. Chin J Ecol 30: 30–39.

20. Jiang JX, Wan NF, Ji XY, Dan JG (2011) Diversity and stability of arthropod community in peach orchard under effects of ground cover vegetation. Chin J Ecol 22: 2303–2308.

21. Bi SD, Zhou XZ, Li L, Ding CC, Gao CQ, et al. (2003) Seasonal dynamics of the relative abundance of arthropod communities in peach orchards. Chin J Ecol 22: 113–116.

22. Song BZ, Wu HY, Kong Y, Zhang J, Du YL, et al. (2010) Effects of intercropping with aromatic plants on the diversity and structure of an arthropod community in a pear orchard. BioControl 55: 741–751.

23. Simon S, Defrance H, Sauphanor B (2007) Effect of codling moth management on orchard arthropods. Agric Ecosyst Environ 122: 340–348.

24. Root RB (1973) Organization of plain arthropod association in simple and diverse habitats: the fauna of collards (*B. oleracea*). Ecol Monogr 43: 95–124.

25. Frechette B, Cormier D, Chouinard G, Vanoosthuyse F, Lucas É (2008) Apple aphid, *Aphis spp.* (Hemiptera: Aphididae), and predator populations in an apple orchard at the non-bearing stage: the impact of ground cover and cultivar. Eur J Entomol 105: 521–529.

26. Björkman M, Hambäck P, Hopkins R, Rämert B (2010) Evaluating the enemies hypothesis in a clover-cabbage intercrop: effects of generalist and specialist natural enemies on the turnip root fly (*Delia floralis*). Agr Forest Entomol 12: 123–132.

27. Letourneau DK (1987) The enemies hypothesis: tritrophic interactions and vegetational diversity in tropical agroecosystems. Ecology 68: 1616–1622.

28. Chen LL, You MS, Chen SB (2011) Effects of cover crops on spider communities in tea plantations. Biol Control 59: 326–335.

29. Risch SJ (1981) Insect herbivore abundance in tropical monocultures and polycultures: an experimental test of two hypotheses. Ecology 62: 1325–1340.

30. Schuldt A, Both S, Bruelheide H, Härdtle W, Schmid B, et al. (2011) Predator diversity and abundance provide little support for the enemies hypothesis in forests of high tree diversity. PLoS ONE 6: e22905.

31. Smith MW, Arnold DC, Eikenbary RD, Rice NR, Shiferaw A, et al. (1996) Influence of ground cover on beneficial arthropods in pecan. Biol Control 6: 164–176.

32. Wyss E (1996) The effects of artificial weed strips on diversity and abundance of the arthropod fauna in a Swiss experimental apple orchard. Agric Ecosyst Environ 60: 47–59.

33. Silva EB, Franco JC, Vasconcelos T, Branco M (2010) Effect of ground cover vegetation on the abundance and diversity of beneficial arthropods in citrus orchards. B Entomol Res 100: 489–499.

34. Lundgren JG, Fergen JK (2010) The effects of a winter cover crop on *Diabrotica virgifera* (Coleoptera: Chrysomelidae) populations and beneficial arthropod communities in no-till maize. Environ Entomol 39: 1816–1828.

35. Landis DA, Wratten SD, Gurr GM (2000) Habitat management to conserve natural enemies of arthropod pests in agriculture. Annu Rev Entomol 45: 175–201.

36. Norris RF, Kogan M (2005) Ecology of interactions between weeds and arthropods. Ann Rev Entomol 50: 479–503.

37. Hughes AR (2012) A neighboring plant species creates associational refuge for consumer and host. Ecology 93: 1411–1420.

38. Hossain Z, Gurr GM, Wratten SD (2000) The potential to manipulate the numbers of insects in lucerne by strip cutting. Aust J Entomol 39: 39–41.

39. Gruys P (1982) Hits and misses. The ecological approach to pest control in orchards. Entomol Exp Appl 31: 70–87.

40. Tedders WL, Payne JA, Inman J (1984) A migration of *Tetranychus urticae* from clover into pecan trees. J Geor Entomol Soc 19: 498–502.

41. Wan NF, Ji XY, Gu XJ, Jiang JX, Wu JH, et al. (2014) Ecological engineering of ground cover vegetation promotes biocontrol services in peach orchards. Ecol Eng 64: 62–65.

42. Downey RV, Griffiths HJ, Linse K, Janussen D (2012) Diversity and distribution patterns in high southern latitude sponges. PLoS ONE 7: e41672.

Permissions

List of Contributors

Yunbin Qin, Zhongbao Xin, Xinxiao Yu and Yuling Xiao
Institute of Soil and Water Conservation, Beijing Forestry University, Beijing, China

Wei Li and Yue Li
National Engineering Laboratory for Forest Tree Breeding, Key Laboratory for Genetics and Breeding of Forest Trees and Ornamental Plants of Ministry of Education, Beijing Forestry University, Beijing People's Republic of China

Xiaoru Wang
State Key Laboratory of Systematic and Evolutionary Botany, Institute of Botany, Chinese Academy of Sciences, Beijing, People's Republic of China

Ying-Hong He, Sayaka Isono, Makoto Shibuya, Masaharu Tsuji, Charith-Raj Adkar Purushothama, Kazuaki Tanaka and Teruo Sano
Faculty of Agriculture and Life Science, Hirosaki University, Hirosaki, Japan

Thomas Guillemaud and Auré lie Blin
Equipe "Biologie des Populations en Interaction", UMR 1301 I.B.S.V. INRA-UNSA-CNRS, Sophia Antipolis, France

Sylvaine Simon and Karine Morel
UE695 Recherche Intégreé, INRA, Domaine de Gotheron, Saint-Marcel-lès-Valence, France

Pierre Franck
UR1115 Plantes et Systèmes de Culture Horticoles, INRA, Avignon, France

David W. Crowder, Emily Martin, Wade Peterson and Jeb P. Owen
Department of Entomology, Washington State University, Pullman, Washington, United States of America

Elizabeth A. Dykstra, Jo Marie Brauner, Anne Duffy and Caitlin Reed
Washington State Department of Health, Olympia, Washington, United States of America

Yves Carrière
Department of Entomology, University of Arizona, Tucson, Arizona, United States of America

Pierre Dutilleul
Department of Plant Science, McGill University, Macdonald Campus, Ste-Anne-de-Bellevue, Quebec, Canada

Jared G. Ali, Lukasz L. Stelinski and Larry W. Duncan
Entomology and Nematology Department, Citrus Research and Education Center, University of Florida, Lake Alfred, Florida, United States of America

Hans T. Alborn and Fatma Kaplan
Center for Medical, Agricultural, and Veterinary Entomology, Agricultural Research Service, U.S. Department of Agriculture, Gainesville, Florida, United States of America

Raquel Campos-Herrera
Entomology and Nematology Department, Citrus Research and Education Center, University of Florida, Lake Alfred, Florida, United States of America
Departamento de Contaminación Ambiental, Instituto de Ciencias Agrarias, Centro de Ciencias Medioambientales, Madrid, Spain

Cesar Rodriguez-Saona and Albrecht M. Koppenhöfer
Department of Entomology, Rutgers University, New Brunswick, New Jersey, United States of America

Federico Martinelli
Plant Sciences Department, University of California Davis, Davis, California, United States of America

Sandra L. Uratsu, Russell L. Reagan, My L. Phu and Abhaya M. Dandekar
Plant Sciences Department, University of California Davis, Davis, California, United States of America

Ute Albrecht and Kim D. Bowman
U.S. Horticultural Research Laboratory, United States Department of Agriculture, Agricultural Research Service, Fort Pierce, Florida, United States of America

Monica Britton, Vincent Buffalo, Joseph Fass and Dawei Lin
Bioinformatics Core, Genome Center, University of California Davis, Davis, California, United States of America

Elizabeth Leicht and Raissa D'Souza
Mechanical and Aerospace Engineering Department, University of California Davis, Davis, California, United States of America

Center for Computational Science and Engineering, University of California Davis, Davis, California, United States of America

Weixiang Zhao and Cristina E. Davis
Mechanical and Aerospace Engineering Department, University of California Davis, Davis, California, United States of America

Erika Yashiro and Patricia S. McManus
Department of Plant Pathology, University of Wisconsin-Madison, Madison, Wisconsin, United States of America

Luc Leblanc, Daniel Rubinoff and Mark G. Wright
Department of Plant and Environmental Protection Sciences, University of Hawai'i, Honolulu, Hawai'i, United States of America

Ting Wu, Yi Wang, Changjiang Yu, Rawee Chiarawipa, Xinzhong Zhang and Zhenhai Han
College of Agronomy and Biotechnology, China Agricultural University, Beijing, China

Lianhai Wu
Rothamsted Research, North Wyke, Okehampton, United Kingdom

Blas Lavandero and Angela Mendez
Laboratorio de Interacciones Insecto-Planta, Universidad de Talca, Talca, Chile

Christian C. Figueroa
Facultad de Ciencias, Instituto de Ecología y Evolución, Universidad Austral de Chile, Valdivia, Chile

Pierre Franck
Plantes et Systèmes de culture Horticoles, INRA, Avignon, France

Jian Sun, Qiang Zhang, Jia Zhou and Qinping Wei
Institute of Forestry and Pomology, Beijing Academy of Agriculture and Forestry Sciences, Beijing, China

Luis Garcia-Torres, Juan J. Caballero-Novella, David Gómez-Candón and Ana Isabel De-Castro
Institute for Sustainable Agriculture (IAS), Spanish Council for Scientific Research (CSIC), Cordoba, Spain

Jiangbo Zhang, Tianli Yue and Yahong Yuan
College of Food Science and Engineering, Northwest A&F University, Yangling, PR China

Adam M. Wallner
United States Department of Agriculture – Animal and Plant Health Services – Plant Protection Quarantine, Plant Inspection Station, Miami, Florida, United States of America

George C. Hamilton, Anne L. Nielsen, Noel Hahn and Cesar R. Rodriguez-Saona
Department of Entomology, Rutgers University, New Brunswick, New Jersey, United States of America

Edwin J. Green
Department of Ecology, Evolution and Natural Resources, Rutgers University, New Brunswick, New Jersey, United States of America

Nian-Feng Wan, Xiang-Yun Ji and Jie-Xian Jiang
Eco-environment Protection Research Institute, Shanghai Academy of Agricultural Sciences, Shanghai Key Laboratory of Protected Horticultural Technology, Shanghai, China

Index

Printed in the USA
CPSIA information can be obtained
at www.ICGtesting.com
JSHW051445221024
72173JS00006B/1584